DATE DUE
Unless Recalled Earlier

DEMCO, INC. 38-2931

Experimental Electrochemistry for Chemists

EXPERIMENTAL ELECTROCHEMISTRY FOR CHEMISTS

DONALD T. SAWYER

Department of Chemistry
University of California
Riverside, California

JULIAN L. ROBERTS, JR.

Department of Chemistry
University of Redlands
Redlands, California

A WILEY-INTERSCIENCE PUBLICATION

JOHN WILEY & SONS

New York · London · Sydney · Toronto

Copyright © 1974, by John Wiley & Sons, Inc.

Library of Congress Cataloging in Publication Data:

Sawyer, Donald T
   Experimental electrochemistry for chemists.

   "A Wiley-Interscience publication."
   Includes bibliographical references.
   1. Electrochemistry. I. Roberts, Julian L.,
joint author. II. Title.

QD553.S32   541'.37   74-3235
ISBN 0-471-75560-5

Printed in the United States of America

10 9 8 7 6 5 4 3 2

PREFACE

The goal of this book is to outline the basic principles and modern methodology of electrochemistry, in such a way that the uninitiated may gain sufficient background to use electrochemical methods for the study of chemical systems. Thus chemical problems that are amenable to an electrochemical approach are introduced as representative examples. We hope by this approach to give the reader the necessary background to assess intelligently the most appropriate electrochemical technique for the study of a specific chemical problem.

This monograph has been developed for the non-electrochemist who is interested in the advantages and use of electrochemical methods for research and laboratory measurements. The approach is pragmatic, with the hope that chemists in general will gain a better appreciation of the usefulness of electrochemical methodology for chemical characterization. We are convinced that modern electrochemistry should take its place along with infrared spectroscopy, NMR spectroscopy, and UV-visible spectroscopy as a standard tool for chemical characterization. With such a pragmatic goal, our purpose has not been to develop the theory in detail. Therefore, we have limited our presentation to those relationships that allow the effective treatment of experimental data. Sufficient references are given for those who wish to pursue the theory.

The writing and preparation of a book not only is a demanding task, but also requires the support and encouragement of colleagues and family. Special thanks is extended to Mrs. Marian Mann for her patience and care in typing the manuscript, and to Mr. Karoly Fogassy for the preparation of the figures. The editorial staff of Wiley-Interscience have been helpful and understanding in their assistance to the authors. Finally, a sincere and personal acknowledgment to our wives, Shirley and Jane, for their indulgence, patience, and support to what appeared to be a never-ending project.

DONALD T. SAWYER

JULIAN L. ROBERTS, JR.

Riverside, California
Redlands, California
April, 1974

CONTENTS

Experimental Electrochemistry for Chemists

One

INTRODUCTION

I. Present State of Electrochemistry

Electrochemistry is a well-developed specialty area of chemistry with a complete set of theories and quantitative relationships. In many respects, it is one of the oldest specialties of classical physical chemistry and traces its origins to the midnineteenth century. Relationships that describe the techniques of potentiometry and polarography derive directly from solution thermodynamics. In the case of polarography, there is a further dependence on the diffusion of ionic species in solution. The latter is the basis of conductivity measurements, another area that traces its origin to the nineteenth century. These quantitative relationships make it possible to apply electrochemistry to the detailed characterization of chemical species and processes in the solution phase.

During the past two decades the dynamics and mechanisms of electron-transfer processes have been studied by numerous groups throughout the scientific world. This has been made possible by applying transition state theory to electrochemical kinetic processes. As a result, both the kinetics of the electron-transfer process (from solid electrode to the solution species), as well as of pre- and postchemical homegeneous processes, can be characterized quantitatively. This has brought about a much better understanding of heterogeneous electron-transfer mechanisms.

By the use of various transient methods, electrochemistry has found extensive new applications for the study of chemical reactions and adsorption phenomena. Thus a combination of thermodynamic and kinetic measuremements has made it possible for electrochemists to characterize the chemistry of heterogeneous electron-transfer reactions. Furthermore, heterogeneous adsorption processes (liquid-solid) have been the subject of intense investigations. The mechanisms of metal-ion complexation reactions also have been ascertained through the use of various electrochemical impulse techniques.

The so-called Renaissance of electrochemistry has come about through a combination of modern electronic instrumentation and the development of a more pragmatic theory. Many challenging applications of modern electrochemistry have been undertaken during the past 20 years as a result of this Renaissance. Within the area of physical chemistry there have been numerous thermodynamic studies of unstable reaction intermediates. In addition, there have been extensive studies of the kinetics of electron-transfer processes both in aqueous and in nonaqueous media. The electrochemical characterization of adsorption phenomena has been of immense benefit to the understanding of catalytic processes.

Some of the most exciting applications of electrochemistry have occurred in the areas of organic and inorganic chemistry, as well as very recently in the area of biochemistry. The applications have ranged from mechanistic studies to the synthesis of unstable or difficultly obtainable species. The control of an oxidation or reduction process through electrochemistry is much more precise than is possible with chemical reactants. Within the area of inorganic chemistry, electrochemistry has been especially useful for the determination of formulas of coordination complexes and the electron-transfer stoichiometry of new organometallic compounds. Electrochemical systhesis is increasingly important to the field of organometallic chemistry.

During the past 10 years numerous exciting extensions of electrochemistry to the field of analytical chemistry have occured. A series of selective-ion potentiometric electrodes have been developed, such that most of the common ionic species can be quantitatively monitored in aqueous solution. A highly effective electrolytic moisture analyzer has become the method for the quantitative moisture analysis of gases. Another practical development has been the polarographic oxygen membrane electrode, which responds linearly to the partial pressure of oxygen either in the gas or in the solution phase. A final example of the extension of electrochemistry to practical analysis is the pesticide analyzer which uses a coulometric electrochemical system as a gas chromatographic detector.

II. Nomenclature and Classes of Electrochemical Methodology

Modern electrochemistry has evolved to the extent that it has a diverse set of specialized terms and symbols. The latter are defined in Table 1-1 as used in most contemporary electrochemical literature and in this book. Because of the rapid expansion in specialized electrochemical methodology and its application to chemical problems, a nomenclature has evolved for their categorization. This is outlined in Table 1-2 and provides an overview of the complete realm of electrochemical methodology. Within this table, key references to the major monographs for each specialized type of electrochemistry are included. These references provide the theory and details of application to complement the introductory and practical presentation of this book.

III. Sign and Graphical Conventions

Nothing has caused more difficulty than the sign conventions of electrochemistry. The problem probably began with Benjamin Franklin's experiments and has evolved through many historical and regional stages. The result is that the electrochemical literature requires exceptional expertise or a "guide book" to avoid confusion. Although the approach followed in this book is outlined in subsequent chapters, the basic rationale can be stated here.

All cells are considered as the combination of two half-cells, with each of the latter represented by a half-reaction written as a reduction

$$Ox + ne^- \rightarrow Red \tag{1-1}$$

By use of an electromotive series (E^0 values) for standard half-reactions written as reductions (see Chapter 6) the potential of each half-cell can be calculated by means of the Nernst equation

$$E = E^0 + \frac{RT}{nF} \ln \frac{[Ox]}{[Red]} \tag{1-2}$$

On this basis the half-cell with the more positive value will be the positive electrode (and the other the negative electrode); the algebraic difference of the two half-cell potentials equals the cell potential.

Another area of confusion in electrochemical literature is the way "electrochemograms" are oriented. Because the polarograms of polarography dominate books and manuals concerned with electrochemistry, we have adopted a convention of graphical presentation that is consistent with historical precedent. Figure 1-1 illustrates this with increasingly negative potentials to the right of the origin, and with cathodic (reducing) currents above the origin.

Table 1-1 Definitions of Symbols Used in Electrochemistry

A	amperes
A	area
a	activity
a	nFv/RT
A-C	alternating current
C	concentration
C	capacitance
D	diffusion coefficient
D-C	direct current
DME	dropping mercury electrode
d	density
E	potential
E^0	standard potential
E'	formal potential
$E_{1/2}$	half-wave potential (polarography)
E_p	peak potential
e	electron
erf	error function
F	faraday
F	farad
f	frequency
G	Gibbs free energy
H	enthalpy
h	height
i	current
i_d	diffusion current
i_{lim}	limiting current
I	diffusion current constant (polarography)
K	equilibrium constant
k_f	forward homogeneous rate constant
k_b	backward homogeneous rate constant
$k_{f,h}{}^0$	forward heterogeneous electrochemical rate constant at 0.0 V versus NHE
$k_{b,h}{}^0$	backward heterogeneous electrochemical rate constant at 0.0 V versus NHE
k_s	heterogeneous electrochemical rate constant at E$'$
m	mass flow rate of DME

Table 1-1 *(Continued)*

NHE	normal hydrogen electrode
n	number of electrons
n_a	number of electrons in the rate determining step
Ox	oxidized species
pH	negative logarithm of hydrogen ion activity
Q	coulombs of charge
R	gas constant
R	resistance
r	radius
Red	reduced species
S	entropy
SCE	saturated calomel electrode
T	temperature (^{0}K)
t	time
t_d	drop time (polarography)
$\left.\begin{array}{l} t_+ \\ t_- \end{array}\right\}$	transference numbers
$\left.\begin{array}{l} u_+ \\ u_- \end{array}\right\}$	ion mobilities
V	voltage (volts)
V	volume
v	velocity
z	charge of an ion
α	transfer coefficient (symmetry parameter)
γ	activity coefficient
$\gamma_\pm$	mean activity coefficient
δ	mobility of an ion
Δ	thickness of a diffusion layer
Δ	differential
ϵ	dielectric constant
Λ	equivalent conductance
λ	conductance of an ion
η	overpotential
ν	scan rate
ν	kinematic viscosity
τ	transition time (chronopotentiometry)
ω	angular velocity
Ω	ohms

Table 1-2 **Outline and Nomenclature for the Methodology of Electrochemistry**[a]

	Methods	Controlled Variable	Measured Variable	Refs.
I.	Potentiometry	$i = 0$	E	1, 2
II.	Controlled potential			
	A. Voltammetry	E	i	3
	1. Polarography (DME)	E	i	4–6
	2. Single-sweep	$E(t)$	i	3, 7
	3. Cyclic-sweep	$E(t)$	i	3, 7–9
	4. Rotating disc and ring disc	E, ω	i	10, 11
	5. Pulse	$\Delta E(t)$	Δi	12, 13
	6. A-C	$E + \Delta E(A\text{-}C)$	$\Delta i(A\text{-}C)$	14
	B. Chronoamperometry	E-step	$i(t)$	15
	1. Chronocoulometry	E-step	$\int i\,dt$	16
	C. Controlled potential	E	$\int i\,dt$	17, 18
	D. Electrolysis	V	Weight	18, 19
III.	Controlled current			
	A. Chronopotentiometry	i	$E(t)$	20, 21
	B. Galvanostatic	i	$E(t=0)$	22–24
	C. Coulometric titrations	i	$(it)/F$	25
IV.	Conductivity	$V(A\text{-}C)$	$i(A\text{-}C)$	26, 27

[a] Literature citations refer to key monographs or reviews which treat the method and its applications in detail.

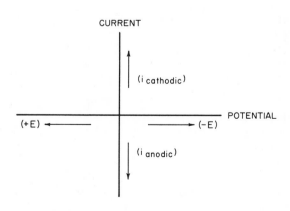

CONVENTION FOR ELECTROCHEMOGRAMS

Figure 1. Graphical system for electrochemical data presentation in this monograph.

IV. Utilization of Electrochemistry for Chemical Characterization

An increasing number of chemists use electrochemistry as a characterization technique in a fashion analogous to their use of infrared, UV-visible, NMR, and ESR spectroscopy. One of the main purposes of this book is to encourage this trend, and to provide practical insights to electrochemical methodology. At this point a brief outline of the approach for solving chemical questions by means of electrochemical measurements may be helpful and illustrative.

Some of the chemical questions that are amenable to treatment by electrochemistry include (a) the standard potentials (E^0) of the compound's oxidation-reduction reactions, (b) evaluation of the solution thermodynamics of the compound, (c) determination of the electron stoichiometry of the compound's oxidation-reduction reactions, (d) evaluation of the heterogeneous electron-transfer kinetics and mechanisms of the compound, (e) determination of the effect of solvent and electrode material and its preconditioning upon the electron-transfer kinetics, (f) study of reaction and product adsorption processes in relation to heterogeneous catalysis, (g) study of pre- and postchemical reactions (thermodynamics and kinetics) that are associated with the electron-transfer reaction of a compound, (h) preparation and study of unstable intermediates, (i) evaluation of the formal charge of new compounds and of their metal component, (j) determination of the formulas and stability constants of metal complexes, and (k) studies of the effects of solvent, supporting electrolyte, and solution acidity upon oxidation-reduction reactions. To answer these questions, the investigator must select the proper electrochemical method and sample conditions. Frequently the solubility and stability of the sample dictate the solvent system that must be used. After selection of an electrochemically compatible solvent, a noninterfering soluble supporting electrolyte must be added (Chapter 3 discusses solvents and supporting electrolytes that are commonly utilized in electrochemistry).

As the later chapters indicate, a given question concerning a chemical system usually can be answered by any one of several electrochemical techniques. However, experience has demonstrated that there is a most convenient or reliable method for a specific kind of data. For example, classical polarography remains the most reliable electrochemical method for the quantitative determination of trace metal ion concentrations. This is true for two reasons: (a) the reproducibility of the dropping mercury electrode is unsurpassed; and (b) the reference literature for analysis by polarography surpasses that for any other electrochemical method by at least an order of magnitude.

Often the first step in the electrochemical characterization of a compound is to ascertain its oxidation-reduction reversibility. In our opinion, cyclic voltammetry is the most convenient and reliable technique for this and related qualitative characterizations of a new system. The discussion in Chapter 7 outlines the specific procedures and relationships. The next step in the characterization usually is the determination of the electron stoichiometry of the oxidation-reduction steps of the compound. Controlled potential coulometry (discussed in Chapter 7) provides a rigorously quantitative means to such evaluations.

With respect to chemical steps prior to the electron-transfer step, chronopotentiometry offers a convenient technique. The methods of measurement and the quantitative relationships are outlined in Chapter 8. Postchemical reactions to the electron-transfer step are most conveniently characterized by cyclic voltammetry (see Chapter 7). Although the techniques of cyclic voltammetry and chronopotentiometry both provide a means to detect qualitatively adsorption processes at an electrode, the coulostatic method and chronocoulometry are the methods of choice for quantitative measurements of adsorption.

Because electrochemistry provides a unique controlled means of adding or of subtracting electrons to or from a compound, it can be used to produce transiently stable species for study by other physical methods such as optical and ESR spectroscopy and mass spectrometry. Conversely, electrochemistry is an especially sensitive means for the detection of reaction products from photolysis and pyrolysis reactions.

Many other applications of electrochemical methods for chemical characterization are presented in the following chapters. As is true with all physical methods of characterization, new techniques and examples are outlined in each month's journal. The state of utilization is such that for many research groups in the fields of organic chemistry and inorganic chemistry, electrochemistry has become a characterization tool as essential as infrared and NMR spectroscopy. This is quickly becoming true for several areas of biochemistry, especially enzymology.

References

1. I. M. Kolthoff and N. H. Furman, *Potentiometric Titrations*, 2nd ed., John Wiley and Sons, Inc., New York, 1931.

2. N. H. Furman, "Potentiometry," in *Treatise on Analytical Chemistry*, Vol. 4, I. M. Kolthoff and P. J. Elving, eds., Chap. 45, Interscience Publishers, Inc., New York, 1963.

3. P. Delahay, *New Instrumental Methods in Electrochemistry*, Interscience Publishers, Inc., New York, 1954.

4. I. M. Kolthoff and J. J. Lingane, *Polarography*, 2nd ed., Interscience Publishers, Inc., New York, 1952.

5. J. Heyrovsky and J. Kůta, *Principles of Polarography*, Academic Press, New York, 1966.

6. L. Meites, *Polarographic Techniques*, 2nd ed., Interscience Publishers, Inc., New York, 1965.

7. R. N. Adams, *Electrochemistry at Solid Electrodes*, Marcel Dekker, Inc., New York, 1969.

8. R. S. Nicholson and I. Shain, *Anal. Chem.*, **36**, 706 (1964).

9. R. S. Nicholson, *Anal. Chem.*, **37**, 1351 (1965).

10. V. G. Levich, *Physiochemical Hydrodynamics*, Prentice-Hall, Englewood Cliffs, N.J., 1963.

11. W. J. Albery, S. Bruckenstein, and D. T. Napp, *Trans. Faraday Soc.*, **62**, 1932 (1966).

12. D. E. Burge, *J. Chem. Ed.*, **47**, A81 (1970); E. P. Parry and R. A. Osteryoung, *Anal. Chem.*, **37**, 1634 (1965)

13. W. O. Dean and R. A. Osteryoung, *Anal. Chem.*, **43**, 1879 (1971).

14. D. E. Smith, *CRC Crit. Rev. Anal. Chem.*, **2**, 247 (1971).

15. P. Delahay, "The Study of Fast Electrode Processes by Relaxation Methods," in *Advances in Electrochemistry and Electrochemical Engineering*, Vol. 1, P. Delahay, ed., Interscience Publishers, Inc., New York, 1961, pp. 247–254.

16. J. Osteryoung and R. A. Osteryoung, *Electrochim. Acta*, **16**, 525 (1971).

17. J. J. Lingane, *Electroanalytical Chemistry*, 2nd ed., Interscience Publishers, Inc., New York, 1958, pp. 351–391, 450–483.

18. D. G. Davis, *Anal. Chem.*, **44**, 79R (1972).

19. J. J. Lingane, *Electroanalytical Chemistry*, 2nd ed., Interscience Publishers, Inc., New York, 1958, pp. 196–233, 392–415.

20. J. J. Lingane, *Electroanalytical Chemistry*, 2nd ed., Interscience Publishers, Inc., New York, 1958, pp. 617–638.

21. P. Delahay, "Chronoamperometry and Chronopotentiometry," in *Treatise on Analytical Chemistry*, Vol. 4, I. M. Kolthoff and P. J. Elving, eds., Interscience Publishers, Inc., New York, 1963, Chap. 44.

22. P. Delahay, *New Instrumental Methods in Electrochemistry*, Interscience Publishers, New York, 1954, pp. 186–189.

23. P. Delahay and T. Berzins, *J. Am. Chem. Soc.*, **75**, 2486 (1953).

24. P. Delahay and C. C. Mattax, *J. Am. Chem. Soc.*, **76**, 874 (1954).

25. J. J. Lingane, *Electroanalytical Chemistry*, 2nd ed., Interscience Publishers, Inc., New York, 1958, pp. 484–616.

26. J. J. Lingane, *Electroanalytical Chemistry*, 2nd ed., Interscience Publishers, Inc., New York, 1958, pp. 168–195.

27. T. Shedlovsky and L. Shedlovsky, "Conductometry," in *Physical Methods of Chemistry*, Vol. 1, Part IIA, A. Weissberger and B. W. Rossiter, eds., Interscience Publishers, Inc., New York, 1971, Chap. III.

Two

INDICATOR ELECTRODES

I. Introduction

A. MEASUREMENT OF ELECTRODE POTENTIALS

The potential of a single electrode or half-cell cannot be directly measured in a simple way. Any potential measuring device such as a voltmeter or electrometer has two metallic terminals that must be connected across the two points between which the potential difference is measured. One terminal can be connected to the metallic electrode, but the other terminal must make connection to the solution through a wire as illustrated in Figure 2-1. The immersion of the connecting wire in the solution creates a second metal/solution interface whose potential difference will be included in the measurement. A fuller discussion of this point can be found in Ref. 1.

Because the second interface created by immersing a connecting wire into the solution would have an erratic and ill-defined potential, the potential of a working electrode must be measured with respect to some reference electrode whose potential is stable and reproducible. This measurement will therefore include at least two single-electrode or half-cell potentials. Although this measurement is simple to make experimentally, its interpretation can be complicated by the presence of junction potentials between solutions of different composition, by resistive (iR) drops in cells in which a net current is flowing, and by internal polarization of the electrodes caused by net chemical changes produced by the passage of a current.

Figure 2-2 represents the two most common configurations used to make potential measurements. The two-electrode configuration (Figure 2-2a) is a

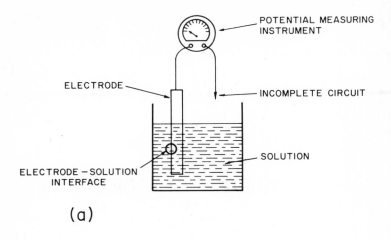

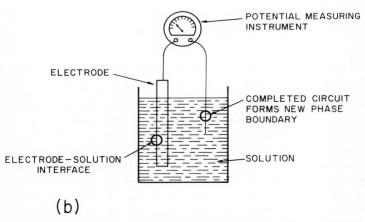

Figure 2-1. Cell and circuit elements for the measurement of electrode potentials. The upper system illustrates the dilemma of attempts to measure single electrode potentials.

system in which the cell current passes through both the working electrode and the reference electrode. In potentiometric measurements with glass electrodes or other specific ion electrodes of high internal impedance, a potential measuring device is necessary with a correspondingly high input impedance. Under these conditions, the net current flowing in the cell is very small (10^{-10}–10^{-15} A), and the internal resistive drop in the cell is negligible (see Chapter 5 for more details).

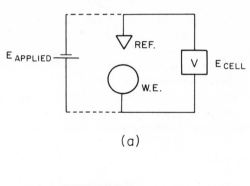

(a)

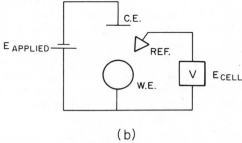

(b)

Figure 2-2. Circuits for the measurement of cell potentials. (*a*) Two-electrode system with current passing through both working and reference electrodes. (*b*) Three-electrode system with current passing through the working and counter electrodes but not the reference electrode.

A two-electrode configuration also can be used in a voltammetric or polarographic cell in which the current is measured as a function of the applied potential. In this case the working electrode potential will be less than the applied potential because of the iR drop in the cell. In addition, the current passing through the reference electrode may cause its potential to deviate from its equilibrium (zero-current) value due to changes in concentration of the electroactive species at the metal/solution interface. Both of these effects act to reduce the potential of the working electrode

$$E_{\text{working electrode}} = E_{\text{cell}} - iR_{\text{cell}} - E_{\text{polarization}} \tag{2-1}$$

To avoid serious errors, the cell current and internal cell resistance must be kept as small as possible, and the reference electrode must be designed to have low internal resistance and a metal/solution interface of sufficient area to minimize internal polarization. Under ordinary polarographic conditions ($10\,\mu\text{A}$ current and 1000-Ω internal cell resistance) the error amounts to 10 mV.

To minimize errors in voltammetric work, the three-electrode configuration (Figure 2-2b) is commonly used. The cell current flows between the working electrode and the counter or auxiliary electrode, while the potential of the working electrode is measured with respect to the reference electrode using a high impedance measuring device. This avoids internal polarization of the reference electrode and compensates for the major portion of the iR drop in the cell. This point is discussed more fully in Chapter 7, and the use of positive feedback to effect compensation is discussed in Chapter 5.

B.   JUNCTION POTENTIALS

The reference electrode is often isolated from the working electrode compartment by a salt bridge or by a Luggin capillary so that one or more junctions exist in the cell. A potential difference will arise between two different ionic solutions in contact because of the differential mobility of the ions across the junction. This is illustrated in Figure 2-3, where HCl of concentration C_{HCl} and KCl of concentration C_{KCl} form a junction. The HCl and KCl will diffuse in opposite directions across the junction from the region of their greater activity (concentration) to the one of lower activity. Because the H^+ ion has a greater mobility than either K^+ or Cl^- ion, it diffuses more rapidly across the junction. This separation of charge (the liquid-junction potential) decelerates the H^+ ions and accelerates the K^+ ions until their velocities are equal. If the junction is constructed so that a stable diffusion geometry is established, a steady-state junction potential will be established that is approximately constant with time.

C.   THE PROBLEM OF A SOLVENT INDEPENDENT EMF SCALE

A junction potential also exists across the interface between electrolyte solutions of the same concentration in two different solvents. Because of this, potential measurements in two different solvents cannot be related even though the same reference electrode is used for both solvents. The dilemma can only be resolved by making extrathermodynamic assumptions because of the impossibility of making a measurement of a single-electrode potential. A number of systems have been proposed as the basis for a suitable reference electrode scale to relate directly potential measurements in one solvent to those made in another solvent. These include $Rb^+/Rb(2)$, ferrocinium$^+$/ferrocene (3), and $(o$-phenanthroline$)_3Fe(III)/(o$-phenanthroline$)_3Fe(II)$. At present there is no generally accepted solution; it is one of several interrelated problems that have been succinctly stated and critically reviewed by Popovych (5,6). These interrelated problems include (a) the establishment of a single solvent-independent scale for pH and for other ion activities; (b)

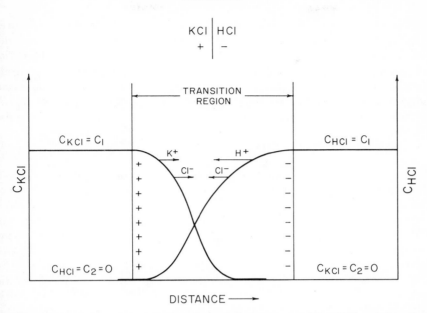

Figure 2-3. Junction between two electrolyte solutions, one containing KCl and the other HCl. The concentration profiles indicate the higher mobility of H^+ ions relative to K^+ ions.

the formulation of a solvent-independent standard potential series having a single reference point $E^0_{H^+/H_2} = 0$ V in water only; and (c) the evaluation of liquid-junction potentials at aqueous-nonaqueous interfaces. Popovych (5) summarizes the problem:

By convention, the activity of a solute is referred to infinite dilution in the given solvent as the standard state. Similarly, the EMF series in any solvent has as its arbitrary zero the standard hydrogen electrode (SHE) in the same medium. As a result, there can be as many independent activity scales and EMF series as there are solvents. The key to the solution of the above interrelated problems lies in the knowledge of medium effects for single ions. The medium effect, $_m\gamma_i$, usually expressed in logarithmic form, is a measure of the difference between the standard free energy of a solute i in water $(_wG^0_i)$ and in the given nonaqueous solvent $(_sG^0_i)$

$$_sG^0_i - _wG^0_i = RT\ln {_m\gamma_i} \qquad \text{(molal scale of concentrations)} \qquad (2\text{-}2)$$

Subscripts s and w denote nonaqueous and aqueous standard states, respectively. For example, using a known medium effect, $_m\gamma_H$, for the proton in a given nonaqueous solvent, we could convert a conventional pa_H^* value, referred to infinite dilution in that solvent, to its counterpart on a single pa_H scale, referred to the aqueous standard state

$$p a_H = p a_H{}^* - \log {}_m\gamma_H \qquad (2\text{-}3)$$

Because the difference between the solvation energies of an ion in two solvents is directly related to the difference between the corresponding standard electrode potentials, the potential of the SHE in a nonaqueous medium SH on the single aqueous scale, ${}_wE^0{}_H(\text{SH})$, can be calculated from

$$_wE_H{}^0(\text{SH}) = \frac{RT}{F}\ln {}_m\gamma_H + {}_wE_H{}^0(\text{H}_2\text{O}) \qquad (2\text{-}4)$$

where ${}_wE_H{}^0(\text{H}_2\text{O})$ is the potential of the SHE in water, assumed to be 0 V. Finally, from the potential E of the following galvanic cell

$$H\text{-electrode}/a_H = 1/a_H^* = 1/H\text{-electrode}$$

$$\text{aqueous}/\text{nonaqueous} \qquad (2\text{-}5)$$

$$E_j$$

and a known $\ln {}_m\gamma_H$, it would be possible to determine the liquid-junction potential E_j at the aqueous-nonaqueous interface

$$E = \frac{RT}{F}\ln {}_m\gamma_H + E_j \qquad (2\text{-}6)$$

Medium effects for complete electrolytes can be calculated from their solubility products or from the E^0's of cells reversible to their ions. For single ions, medium effects can be estimated only using extrathermodynamic assumptions.

The medium effect, expressed as ${}_m\gamma_i$, should be distinguished from salt effects, expressed as ${}_s\gamma_i$. As a rule, medium effects are significantly greater than salt effects, often by several orders of magnitude. Thus an activity coefficient in a nonaqueous solvent when referred to the aqueous standard state is really a product of the salt and medium effects

$$_w\gamma_i = {}_s\gamma_i {}_m\gamma_i \qquad (2\text{-}7)$$

As a corollary, the medium effect for a given solute can be viewed as a conversion factor from the nonaqueous to the aqueous activity scale

$$a_i = a_i^* {}_m\gamma_i \qquad (2\text{-}8)$$

For aqueous solutions, ${}_m\gamma_i$ will have a value of unity.

In the last two decades some 20 methods have been proposed in the literature for the estimation of single-ion medium effects, and the present activity in this area suggests that a suitable solution to these problems may be forthcoming in the next few years. One of the most promising methods is based on the assumption that a reference anion and a reference cation have equal medium effects in a given solvent. Ideally, the electrolyte should be composed of large symmetrical ions of equal size and solvation properties, whose medium effects can be apportioned equally between the anion and the cation. The central atom and the charge of such ions should be shielded by large organic residues to minimize both the surface charge density and any specific interactions with solvent. Several electrolytes that approach these criteria have been proposed, including tetraphenylphosphonium tetraphenylborate (7), triisoamyl-*n*-butylammonium tetraphenylborate (8), and tetraphenylarsonium tetraphenylborate (9). Kolthoff and Chantooni (10) have critically assessed the last assumption for five different solvents and have gathered solubility product data for more than 20 salts in these solvents.

D. CALCULATION AND INDIRECT MEASUREMENT OF JUNCTION POTENTIALS

The direct potentiometric measurement of a junction potential is not possible because of the impossibility of directly measuring a single-electrode potential. Therefore the fraction of the total cell potential due to the junction potential cannot be unambiguously assigned. However, it is possible to estimate junction potentials indirectly or to make calculations based on assumptions about the geometry and distribution of ions in the region of the junction. For a junction between two dilute solutions of the same univalent electrolyte, the calculated values appear to be quite exact and independent of the way in which the junction is formed.

The basic equation relating the junction potential, E_j, to the transport number, charge, and activity of the ions forming the junction is (11)

$$E_j = \frac{-RT}{F} \int_1^2 \sum_n \frac{t_i}{z_i} d\ln a_i \qquad (2\text{-}9)$$

where t_i is the transport number (related to the mobility of the ion) of the ith ion, z_i the algebraic value of the charge on the ion, and a_i the activity of the ion, and n implies that the summation is carried out over the n kinds of ions present. Equation 2-9 can be derived from both a thermodynamic and a diffusional approach (12); implicit in the equation is the fact that solvent transport by the solvated ions has been neglected. This equation is not rigorously thermodynamic because single ion activities cannot be measured directly.

To carry out the integration specified by Equation 2-9 requires that one know: (a) how the concentration of each ion varies in the transition region; (b) how the activity coefficients vary with the concentrations, c_i (where $a_i = \gamma_i c_i$); and (c) how the transport number varies with the concentrations of the ions, c_i. The general case is too difficult to solve analytically, but a solution can be obtained readily when simplifying assumptions are made.

If the activity coefficients are taken to be unity, and the concentration of each ion assumed to vary linearly from c_1 to c_2 (dropping the subscript i), then the equation first derived by Henderson can be obtained. In the region of the junction the concentration of each ion is assumed to be given by mixing a fraction x of electrode solution 2 with a fraction $(1 - x)$ of electrode solution 1 to give

$$c = xc_2 + (1 - x)c_1 \tag{2-10}$$

$$= c_1 + (c_2 - c_1)x \tag{2-11}$$

Hence

$$d\ln c = \frac{dc}{c} = \frac{(c_2 - c_1)dx}{c} \tag{2-12}$$

In the transition region the transport number of each ion is defined by

$$t = \frac{cuz}{\left[x\Sigma c_2 uz + (1 - x)\Sigma c_1 uz \right]} \tag{2-13}$$

where u is the ionic mobility [$cm^2/(sec)(V)$] and z is the valence of the ion. Both u and z are taken as positive for cations and negative for anions. Multiplying Equation 2-12 and 2-13 together gives

$$\frac{t}{z}d\ln c = \frac{u(c_2 - c_1)dx}{\Sigma c_1 uz + x\Sigma uz(c_2 - c_1)} \tag{2-14}$$

Substitution in Equation 2-9 then leads, on integration, to

$$E_j = \frac{-RT}{F} \int_{x=0}^{x=1} \frac{\Sigma u(c_2 - c_1)dx}{\Sigma c_1 uz + x\Sigma uz(c_2 - c_1)} \tag{2-15}$$

$$E_j = \frac{-RT}{F} \frac{\Sigma u(c_2 - c_1)}{\Sigma uz(c_2 - c_1)} \ln \frac{\Sigma uzc_2}{\Sigma uzc_1} \tag{2-16}$$

which is the Henderson equation (13) for liquid-junction potential. Two cases are considered for junctions formed by dilute solutions of simple electrolytes.

1. Liquid Junctions between Two Different Concentrations of the Same $z:z$ Valent Electrolyte

If the charges on the anion and cation of the electrolyte are equal, then Equation 2-16 reduces to

$$E_j = \frac{-RT}{zF} \frac{(u_+ - u_-)}{(u_+ + u_-)} \ln \frac{c_2}{c_1} \tag{2-17}$$

where z and the ionic mobilities of the cation and anion (u_+ and u_-, respectively) are now taken as positive. This equation predicts that as the mobilities of the cation and anion approach the same value, the junction potential will approach zero. The reasonableness of junction potentials calculated from Equation 2-17 can be examined by considering the following cell with liquid junction

$$\underset{E_1}{\text{Ag/AgCl/MCl}(c_1)} \; : \; \underset{E_2}{\text{MCl}(c_2)/\text{AgCl/Ag}} \tag{2-18}$$

Defining E_{cell} as the potential of the right-hand electrode, E_2, measured with respect to the left-hand electrode, E_1, we may write E_{cell} as

$$E_{cell} = E_2 - E_1 + E_j \tag{2-19}$$

The reversible half-cell potentials E_1 and E_2 may be calculated from the Nernst equation for the half reaction

$$\text{AgCl} + e^- \rightarrow \text{Ag} + \text{Cl}^-$$

$$E = E_{\text{AgCl/Ag}}^0 - \frac{RT}{F} \ln a_{\text{Cl}^-} \tag{2-20}$$

so that E_{cell} will be given by

$$E_{cell} = \frac{-RT}{F} \ln \frac{(a_{\text{Cl}^-})_2}{(a_{\text{Cl}^-})_1} + E_j \tag{2-21}$$

In dilute solutions the activity of the chloride ion may be calculated as

$$a_{\text{Cl}^-} = \gamma_\pm c_{\text{Cl}^-} \tag{2-22}$$

where $\gamma_\pm$ is the mean ionic activity coefficient in the electrolyte and can be measured in a thermodynamically rigorous fashion in cells without liquid junction.

To calculate the junction potential, values of the mobilities (or of the transport numbers) are necessary. Because the junction involves the same

Table 2-1 Computed Potentials of the Liquid Junction in Cells of the Type Ag/AgCl/MCl(c_1):
MCl(c_2)/AgCl/Ag

Concentrations (M)		Electrolyte	$\gamma_\pm$ at c_1	$\gamma_\pm$ at c_2	t_+	E_j Calculated Eq. 2-24 (mV)	E_{cell} Calculated Eq. 2-21 (mV)	E_{cell} Observed (mV)
c_1	c_2							
0.005	0.01	NaCl	0.928	0.904	0.392	3.84	−13.3	−13.4
		KCl	0.927	0.902	0.490	0.36	−16.7	−16.8
		HCl	0.928	0.906	0.825	−11.57	−28.8	−28.3
0.005	0.04	NaCl	0.835	0.928	0.391	11.64	−39.1	−39.6
		KCl	0.832	0.927	0.490	1.07	−49.6	−49.6
		HCl	0.844	0.928	0.826	−34.83	−85.8	−84.2
0.04	0.06	NaCl	0.812	0.835	0.388	2.33	−7.4	−7.6
		KCl	0.807	0.832	0.490	0.21	−9.4	−9.6
		HCl	0.824	0.844	0.829	−6.85	−16.7	−16.3

From Ref. 11, pp. 162, 164, 226.

electrolyte on both sides of the junction the expressions

$$t_+ = \frac{u_+}{u_+ + u_-} \quad \text{and} \quad t_- = \frac{u_-}{u_+ + u_-} \tag{2-23}$$

are valid and $t_+ + t_- = 1$, where t_+ is the fraction of current carried by the cation and t_- is the fraction carried by the anion. Substituting the transport numbers for the mobilities in Equation 2-17, the junction potential can be expressed as

$$E_j = \frac{-RT}{zF}(2t_+ - 1)\ln\frac{c_2}{c_1}. \tag{2-24}$$

The transport numbers measured by MacInnes and Longsworth (14) can be used to calculate the junction potentials from Equation 2-24. Table 2-1 gives the experimental E_{cell} values and the values of E_{cell} calculated from Equations 2-21, 2-22, and 2-24. The reasonable agreement indicates that junction potentials for dilute solutions of $1:1$ electrolytes calculated from Equation 2-17 or 2-24 are credible.

2. Liquid Junctions between Two Different Univalent Electrolytes Both at the Same Concentration, with One Ion in Common.

The experimental evidence cited by Rock (15) indicates that only when certain restrictive conditions are met is it possible to make a reliable (± 1 mV) estimate of the liquid-junction potential between solutions containing different electrolytes. These conditions are: (a) the junction must be formed from the same charge type salts having an anion or cation in common (e.g., KCl(aq)/KBr(aq), or Na_2SO_4(aq)/K_2SO_4(aq)); (b) the solvent must be the same, and the salt concentrations in the two solutions must be equal and less than 0.1 M; and (c) the transport numbers of the ions which are not common to both solutions should not be very different in the two solutions.

If the junction is formed between univalent electrolytes of the same concentration as in the cell

$$\text{Ag/AgCl/M}_1\text{Cl}(c):\text{M}_2\text{Cl}(c)/\text{AgCl/Ag} \tag{2-25}$$

where M_1 and M_2 are different univalent cations, Equation 2-17 reduces to the Lewis and Sargent formula (16)

$$E_j = -\frac{RT}{F}\ln\frac{u_{M_2} + u_{Cl^-}}{u_{M_1} + u_{Cl^-}} = -\frac{RT}{F}\ln\frac{\Lambda_{M_2Cl}}{\Lambda_{M_1Cl}} \tag{2-26}$$

where Λ_{M_2Cl} and Λ_{M_1Cl} are the equivalent conductances of the solutions concerned. The potential of the cell specified by Equation 2-25 will be given by Equation 2-21, and if the chloride ion activities are the same in two different chlorides at the same concentration, E_{cell} will equal E_j. MacInnes and Yeh (17) have tested this predicted result by comparing experimental values of E_{cell} with E_j values calculated from the Lewis and Sargent equation. The results are tabulated in Table 2-2; the agreement is far from perfect but generally better than 1 to 2 mV.

**Table 2-2   Liquid Junction Potentials, E_j, Calculated by the Lewis
and Sargent Equation** (Eq. 2-26) **for the Cell**
$$Ag/AgCl/M_1Cl(c): M_2Cl(c)AgCl/Ag$$

Concentration (M)	Junction	E_j Calculated Eq. 2-26 (mV)	E_j Observed (mV)
0.1	HCl : KCl	+28.52	+26.78
	HCl : NaCl	33.38	33.09
	HCl : LiCl	36.14	34.86
	HCl : NH$_4$Cl	28.57	28.40
	KCl : LiCl	7.62	8.79
	KCl : NaCl	4.86	6.42
	KCl : NH$_4$Cl	0.046	2.16
	NaCl : LiCl	2.76	2.62
	NaCl : NH$_4$Cl	−4.81	−4.21
	LiCl : NH$_4$Cl	−7.57	−6.93

3. More Exact Treatments of Liquid-Junction Potentials

In deriving the Henderson equation, the activity coefficients were assumed to be unity, and the transport numbers were assumed to be constant. In addition the junction region was assumed to have a composition given by a linear combination of the solutions on either side of the boundary. Calculations for a number of other models have been made by Smyrl and Newman (18) and the results compared with the continuous mixture model (including calculation of activity coefficients). The agreement in most cases is within 1 to 2 mV.

Covington recently has noted the renewal of interest in the problems associated with liquid-junction potentials in his review of reference electrodes (19).

E. CELLS WITH LIQUID JUNCTIONS AND ELIMINATION OF JUNCTION POTENTIALS

When electrochemical cells are employed to obtain thermodynamic data, high accuracy ($\pm 0.05\,\text{mV}$) requires the use of cells that are free from liquid junction (in the sense that the construction of the cell does not involve bringing into contact two or more distinctly different electrolyte solutions). Otherwise, the previously discussed uncertainties in the calculation of liquid-junction potentials will limit the accuracy of the data.

The difference between cells with and without liquid junctions is one of degree rather than of kind. The cell used to establish the mean ionic activity coefficients of HCl

$$(\text{Pt})/\text{H}_2,\ \text{H}^+\text{Cl}^-/\text{AgCl}/\text{Ag} \qquad (2\text{-}27)$$

really contains two or more solutions. Because silver ion is reduced by hydrogen, in practice the half reactions are separated and should, strictly, be represented as follows

$$(\text{Pt})/\text{H}_2,\ \text{H}^+\text{Cl}^-(\text{satd. with H}_2):\text{H}^+\text{Cl}^-(\text{satd. with AgCl})/\text{AgCl}/\text{Ag}$$

$$(2\text{-}28)$$

In addition a solution of pure hydrochloric acid might be present between the two saturated solutions. In this particular case, the considerations outlined in previous sections would lead us to believe that the liquid-junction potential in such a cell is negligible. If, however, hydrogen and silver chloride were more soluble, the resulting liquid-junction potential might not be negligible.

Several types of cells of necessity involve liquid junctions. Examples (20) include: (*a*) half-cells of the second kind

$$\text{M}(s)/\text{MX}(s)/\text{X}^-(\text{soln}) \qquad (2\text{-}29)$$

where the anion, X^-, is protonated in acidic solution, and the salt, MX, reacts in basic solution; for example,

$$\text{Hg(liq)}/\text{Hg}_2\text{CrO}_4(s)/\text{CrO}_4^{2-} \qquad (2\text{-}30)$$

Such electrodes cannot be investigated in cells of the type

$$(\text{Pt})/\text{H}_2(g),\ \text{H}^+\text{X}^-/\text{MX}(s)/\text{M} \qquad (2\text{-}31)$$

or

$$\text{Hg}/\text{HgO}(s)/\text{K}^+\text{OH}^-,\ \text{K}^+\text{X}^-/\text{MX}(s)/\text{M} \qquad (2\text{-}32)$$

because X^- is protonated in cell 2-31 and $MX(s)$ reacts with hydroxide ion in cell 2-32; (b) electrodes for which both the oxidized and reduced forms of the half-cell reaction are soluble. For example, the cell

$$(Au)/K_4Fe(CN)_6, K_3Fe(CN)_6, KCl/AgCl(s)/Ag \qquad (2\text{-}33)$$

is not usable because the reaction

$$Fe(CN)_6^{4-} + AgCl \rightarrow Ag + Cl^- + Fe(CN)_6^{3-} \qquad (2\text{-}34)$$

occurs spontaneously in the solution.

Various attempts have been made to circumvent these problems and to eliminate junction potentials, including: (a) extrapolation procedures designed to eliminate the difference between the compositions of the two solutions in the appropriate limit; (b) separation of the two solutions by means of a double-junction salt bridge; (c) the use of double cells with dilute alkali metal amalgam connectors; and (d) the use of glass or other types of ion specific electrodes as "bridging" reference electrodes.

The extrapolation procedures used for cells with liquid junction are time consuming, and the method is not entirely free of theoretical pitfalls (20).

Because salt bridges usually involve double junctions, an important distinction needs to be made between the behavior of single-junction and double-junction salt bridges (21). In a single junction the residual junction potential can be reduced, in principle, to a low value by using a large concentration of an equitransferent electrolyte on one side of the junction. However, the use of equitransferent solutions (containing mixtures of KCl and KNO_3) in a double-junction bridge actually increases the residual junction potential between 0.1 F KCl and 0.1 F HCl. Thus although the ions of the concentrated solution may contribute the major portion of the two junction potentials at either end of the bridge, the residual full-bridge potential is mainly determined by the dilute solutions at either end of the bridge (21). A properly designed double-junction salt bridge can reduce but not completely eliminate the junction potential. This conclusion results from the work of Guggenheim (22) and Finkelstein and Verdier (21) on the cell

$$Hg/Hg_2Cl_2/HCl(0.1\ F) : bridge: KCl(0.1\ F)/Hg_2Cl_2/Hg \quad (2\text{-}35)$$

In the absence of a bridge the cell potential (mainly attributable to the junction potential) is about 27 mV. When a KCl bridge is inserted, the cell potential decreases with increasing concentration of the KCl bridge electrolyte as shown in Figure 2-4. For a concentration of 3.5 to 4.0 F KCl in the bridge, the residual junction potential has decreased to about 1 to 2 mV.

The use of dilute amalgam electrodes goes back to G. N. Lewis and

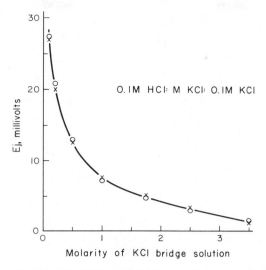

Figure 2-4. Junction potential between a 0.1 M HCl solution and a 0.1M KCl solution as a function of the concentration of KCl in the connecting salt bridge.

coworkers who used them to determine the standard electrode potentials of the alkali metal electrodes. Such electrodes have been employed as connectors in double cells to eliminate junction potentials (20). The use of glass electrodes and other ion specific electrodes as "bridge" electrodes for elimination of liquid-junction potentials also has been discussed (19,20,23). A single glass pH electrode has been transferred between the following cells

$$(Pt)/H_2/H^+Cl^-/glass\ electrode \qquad\qquad (2\text{-}36)$$

$$Ag/AgCl/H^+Cl^-/glass\ electrode \qquad\qquad (2\text{-}37)$$

and gave excellent agreement with measurements obtained from a cell of the type shown in Equation 2-28 for HCl concentrations less than 0.5 molal. Glass electrodes tend to drift with time, but when a glass electrode is transferred from cell 2-36 to cell 2-37 and then back again to cell 2-36, the extrapolated potentials at the moment of transfer (as obtained from the two drift rates) agree within 0.03 to 0.05 mV.

A solid-state iodide-sensitive electrode has been used to estimate changes in the liquid-junction potential in a cell which employed a double-junction salt bridge (24). Because double-junction salt bridges are widely employed in specific ion electrode measurements, this technique promises to be valuable.

1. Design of Junction Devices

Considerable effort has been applied to the design of junction devices for reference electrodes and salt bridges. For highest stability and reproducibility, the junction should be formed within a cylindrical tube by one of the methods described in Ref. 21 or 22. Such junctions can be categorized as the "constrained diffusion" type with cylindrical symmetry. However, these rather elaborate methods of forming junctions have not become popular for routine use because of their inconvenience.

Most of the commercially manufactured electrodes have junctions that can be classified as "continuous impeded flow" types; a number of different junction devices are illustrated in Figure 2-5. Their relative merits are discussed in the following sections.

2. Flow Rate of Junctions

The flow rate of a junction is of concern to the user for two reasons: (*a*) a high flow rate may tend to contaminate the sample with the filling electrolyte and (*b*) too low a flow rate may lead to erratic junction potentials or a tendency of the junction to clog if the sample contains colloidal material. Some typical flow rates and recommended applications for the different junction styles are summarized in Table 2-3.

The electrolyte of the impeded-flow type junctions should stream into the test solution at a constant rate by a single leakage path, preferably with cylindrical symmetry. The presence of several separated openings through which the bridge solution may flow at varying rates is not conducive to the establishment of a reproducible steady state. A constant and substantial flow rate is probably the most critical factor for obtaining reproducible junction potentials. For pH measurements of ordinary precision (± 0.05 pH unit) a reproducibility of about ± 3 mV is tolerable, and almost any of the junctions described will work satisfactorily provided the junction is not clogged.

When specific ion electrodes are employed, an error of less than ± 1 to 2% of the activity of the ion is desirable. This requires that the potential measurement have an error of less than 0.1 to 0.2 mV; the reproducibility of the junction potential becomes a critical factor in the measurement. On the basis of extensive experience, Orion Research, Inc., recommends (25) that a sleeve type junction be employed for specific ion electrode measurements. This junction has a relatively high flow rate that tends to produce a flow velocity sufficient to overcome back diffusion of sample ions into the region of the junction. In addition, the ability to flush the junction makes restoration convenient should it become contaminated.

Some of the problems of clogging of fiber or ceramic frit junctions, or erratic junction potentials from too low a flow rate, may be remedied by

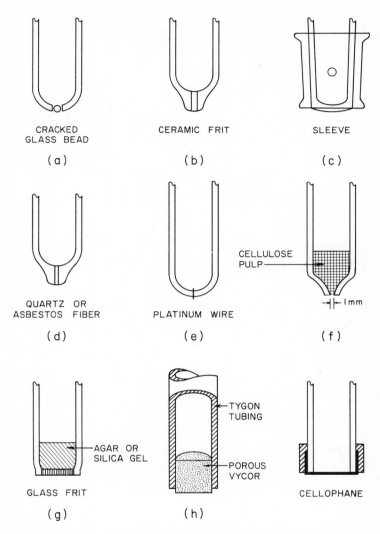

Figure 2-5. Junction devices for reference electrodes and salt bridges.

Table 2-3 Typical Liquid Junction Characteristics

Junction	Approximate Resistance (Ω) 3.5 F aq. KCl	Electrolyte Flow Rate ($\mu l/hr$)	Comments
Soft glass in Pyrex	1000	3–30	General use; avoid precipitates and colloids
Quartz in Pyrex	1000	3–30	General use; avoid precipitates and colloids
Platinum wire	1000	3–30	General use; avoid strong oxidizing and reducing agents, precipitates and colloids
Ceramic frit	1000	3–30	Can be used in strong bases; not recommended for precipitates and colloids
Asbestos fiber	1000–3000	10–200	General use; avoid precipitates and colloids
Carborundum frit	100–500	100–500	General use; not recommended for low ionic strength solutions
Ground sleeve	100	100–500	General use; recommended for slurries and colloids; not recommended for strong acid solutions

increasing the flow rate. This can be accomplished by pressurizing the electrolyte through a side arm connected to an elevated reservoir of electrolyte (20-40-cm head). A dramatic improvement was obtained by pressurizing with air a reference electrode with a ceramic frit junction to about 5 psi (equivalent to a 300-cm head of electrolyte) (26). This increased the flow rate through the ceramic junction to about 80 μl/hr and reduced erratic 5-mV potential excursions to a negligible level.

Another form of a reference electrode (27) employs a conducting organic polymer as the junction device. It has a low electrolyte flow rate, resists clogging, and appears to be useful for routine measurements in industrial applications.

3. Choice of Electrolyte for Salt Bridges and Reference Electrodes

Many of the difficulties encountered in potentiometric measurements can be attributed to erratic or drifting junction potentials caused by clogged junctions. Certain elementary rules should be observed in choosing the filling solution for a salt bridge or reference electrode, particularly when they will be used in organic solvents or solutions that are only partially aqueous.

1. The electrolyte should not react with any species in the sample. For example, a reference electrode solution which contains silver chloride or mercurous chloride should not be used in contact with a solution containing sulfide ion because metallic sulfides may precipitate in the junction and clog it. The formation of soluble reaction products (such as a complex ion) may alter the mobility of a species in the bridge electrolyte, upsetting a desirable equitransferent condition. Sodium perchlorate is often used in aqueous solutions to maintain a high ionic strength. If it is in contact with a low flow-rate junction that contains a high concentration of potassium chloride, potassium perchlorate will precipitate in the junction and clog it.

2. In an analytical measurement that employs a specific ion electrode, the filling electrolyte should not contain the ion being measured in the sample, or an ion which produces a significant interference at the sensing electrode. This rule is particularly important in trace determinations on small volume samples where contamination of the sample by the filling solution may be significant.

3. The ionic strength of the filling solution should be at least tenfold that of the sample so that the bridge solution will largely determine the junction potential and "swamp out" the effect of the sample ions.

4. The filling solution should be equitransferent. If the rates at which positive and negative charge diffuse into the sample are nearly equal then a minimum junction potential results. The best choice is usually a 1:1 electrolyte with the ionic mobilities (refer to Tables 4-1 and 4-2) of the

cation and anion nearly the same. An aqueous KCl solution fulfills this condition and is commonly used as a bridge electrolyte, when K^+ and Cl^- ions do not pose problems. A mixture of salts may be used to more nearly approach the equitransferent condition (25).

Solutions that contain high levels of strong acids or bases are particularly troublesome because the high mobilities of H^+ and OH^- ions make it impossible to "swamp out" their effect on the junction potential. If the level of H^+ or OH^- ions is approximately constant in the test solution, then the junction potential can be minimized by use of a filling solution with about the same level of acid or base.

5.   In a junction formed between different solvents, the electrolyte must be soluble in both solvents. Potassium chloride is insoluble in many commonly used organic solvents such as acetonitrile and dimethylsulfoxide. Therefore, aqueous reference electrodes that employ 3 to 4 F KCl cannot be used without an intervening salt bridge and an electrolyte soluble in both solvents. Quaternary ammonium perchlorate or halide salts (see Chapter 4) commonly are used as bridge electrolytes in organic solvents. Cox (28) has suggested the use of tetraethylammonium picrate or tetra-n-butylammonium tetraphenylborate ($NBu_4^+BPh_4^-$) as bridge electrolytes because they are nearly equitransferent in many organic solvents.

4. Unusual Junction-Potential Problems

There is the possibility of serious errors in potentiometric pH measurements when a platinum or palladium junction device (Figure 2-5) is used in the presence of strong oxidizing agents (permanganate ion) or reducing agents (stannous chloride) (29). In the junction the shorted cell

$$\overline{(Pt)/KCl(aq):test\ solution/(Pt)}$$

is set up. Any redox reaction between the two solutions that is mediated by the metal-junction device gives rise to a continuous current along the relatively high resistance path of the salt-bridge solution; even a small current may generate sufficient ohmic drop to cause an error of 0.1 to 2 pH units.

Significant errors also can occur in pH measurements of tris(hydroxymethyl)aminomethane ("Tris") buffers when linen fiber-junction reference electrodes are used (30); improved procedures for such pH measurements have been developed (31).

The special problems of liquid-junction potentials in plasma and other protein containing solutions have been reviewed (32,33) as have the problems encountered in using electrolyte-filled microelectrodes with orifices smaller than 1 μ (34–37).

F. SOME PRACTICAL CONSIDERATIONS IN THE USE OF SALT BRIDGES

Salt bridges are most commonly used to diminish or stabilize the junction potential between solutions of different composition and to minimize cross contamination between solutions. For example, in working with nonaqueous solvents an aqueous reference electrode often is used that is isolated from the test solution by a salt bridge containing the organic solvent. However, this practice cannot be recommended, except on the grounds of convenience, because there is no way at present to relate thermodynamically potentials in different solvents to the same aqueous reference electrode potential; furthermore, there is a risk of contamination of the nonaqueous solvent by water.

Salt bridges (particularly those that employ saturated KCl) also have been widely used in the conversion of polarographic and other potentials from one reference electrode scale to another. This conversion is commonly made by direct comparison of the two reference electrodes in the cell

$$\text{ref electrode 1 : KCl salt bridge : ref electrode 2} \qquad (2\text{-}38)$$

However, there is convincing experimental evidence that this can introduce substantial errors because the residual junction potential errors are not negligible (38). It is recommended that the reference electrodes be connected to one another through the sample solution in the cell

$$\text{ref electrode 1 : sample solution : ref electrode 2} \qquad (2\text{-}39)$$

The potential difference of cell 2-39 will more closely represent the potential difference of the two reference electrode scales.

With voltammetric techniques like polarography or coulometry a salt bridge often is used to isolate the working and counter electrode compartments. Because the cell current flows through the salt bridge, the resistance of the salt bridge must be kept as low as possible to minimize iR drop. Many of the junction devices shown in Figure 2-5 are more suited for potentiometric work where current flow is negligible. For voltammetric work, a larger junction area is needed to obtain low resistance. This is accomplished by the use of glass-frit or porous Vycor ("thirsty glass") separators. Several designs are illustrated in Figure 2-6, and salt bridges that are incorporated into cell designs are described in Chapter 3.

Gels are sometimes used in conjunction with fritted tubes because they lower flow rates while maintaining a low resistance. Aqueous solutions that are not strongly acidic or basic are readily gelled with a 4% solution of agar. The agar is first dissolved in hot water, then the salt is added, and the hot

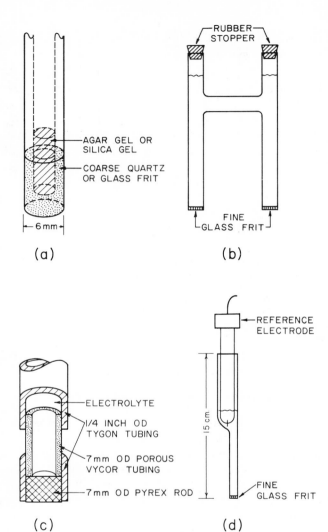

Figure 2-6. Junction devices for voltammetric measurements.

fluid gel is drawn or pipetted into the salt bridge where it is allowed to harden. A convenient device for filling the cross tube of an H-type cell is shown in Figure 2-7. Once prepared, an agar bridge should not be allowed to dry out and should be stored in contact with the bridge electrolyte when not in use. Agar will not gel properly in many organic solvents, and agar bridges are not recommended for connecting aqueous and organic solvent solutions. Five grams of methyl cellulose (Dow Methacel, 15 cP viscosity) in 100 ml of 0.5 F LiClO$_4$ in pyridine gives a colorless, transparent, conducting gel (39); this material may be suitable for gelling other organic solvents.

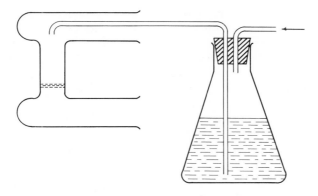

Figure 2-7. System for filling H-cell salt bridge with agar electrolyte.

The use of a 7% fumed silica gel with silica frit bridges of the style shown in Figure 2-6a has been recommended (40). Fumed silica forms stable gels with water and a number of organic solvents, but the aqueous gel is not stable above pH 8.

Salt bridges employing Corning Code 7930 Porous Glass (an unfused form of Code 7900 Vycor glass) have been described (41,42). This material gives a low leak rate with moderate resistance, but it cannot be worked in a flame without shattering and will shatter if allowed to dry with a salt solution inside. It strongly adsorbs dyes and proteins, and exhibits a "chemical memory" which necessitates a long wash-out operation when the filling solution is changed.

Simple low-resistance salt bridges, suitable for student laboratory use, can be made by closing the end of a length of glass tubing with cellophane (dialysis tubing) that is held in place with rings of rubber or polyvinyl tubing or with a plug of cellulose tissue.

A salt bridge prepared from fluorocarbon tubing and fittings has been described that is suitable for use in concentrated HF solutions (where glass or silica cannot be used) (43).

II. Reference Electrodes

A. PROPERTIES OF THE IDEAL REFERENCE ELECTRODE

An ideal reference electrode should show the following properties: (a) it should be reversible and obey the Nernst equation with respect to some species in the electrolyte; (b) its potential should be stable with time; (c) its potential should return to its initial value after small currents are passed through the electrode (no hysteresis); (d) if it is an electrode of the second kind (e.g., Ag/AgCl), the solid phase must not be appreciably soluble in the electrolyte; and (e) it should show low hysteresis with temperature cycling.

Because the flow of electric current always involves the transport of matter in solution and chemical transformations at the solution/electrode interface, ideal behavior can only be approached. It can be approximated, however, by a reference electrode whose potential is controlled by a well-defined electron transfer process in which the essential solid phases are present in an adequate amount and the solution constituents are present at sufficiently high concentrations. The electron transfer is a dynamic process, occurring even when no net current flows, and the larger the anodic and cathodic components of this *exchange* current, the more nearly reversible and non-polarizable the reference electrode will be. A large exchange current increases the slope of the current-potential curve so that the potential of the electrode is more nearly independent of the current. The current-potential curves (polarization curves) are frequently used to characterize the reversibility of reference electrodes.

Reference electrodes can be classified into several types: (a) electrodes of the first kind—a metallic or soluble phase in equilibrium with its ion

$$H^+/H_2(Pt); \text{ ferrocinium/ferrocene; } Ag^+/Ag;$$

$$\text{amalgam electrodes of the type } M^+/M(Hg)$$

(b) electrodes of the second kind—a metallic phase in equilibrium with its sparingly soluble metal salt

$$AgCl/Ag; \ Hg_2Cl_2/Hg; \ Hg_2SO_4/Hg; \ HgO/Hg$$

(c) miscellaneous—glass electrodes, ion specific electrodes, electrodes of the third kind.

B. REFERENCE ELECTRODES FOR USE IN AQUEOUS SOLUTION

The extensive studies of reference electrodes have been critically surveyed in the comprehensive monograph edited by Ives and Janz (37). Bates (44) and Covington (19) have reviewed recent studies; these three sources provide a wealth of detailed information on the preparation and use of reference electrodes.

1. The Hydrogen Electrode

The hydrogen electrode is discussed first because it is the primary reference electrode used to define an internationally accepted scale of standard potentials in aqueous solution. By convention, the potential of an electrode half reaction measured with respect to the standard hydrogen electrode (SHE, where H^+ and H_2 are at unit activity) is defined as the electrode potential of the half reaction. This convention amounts to arbitrarily assigning the standard potential of the hydrogen electrode as zero for all temperatures. Thus there is in effect a separate scale of electrode potentials at each temperature.

The hydrogen electrode also is one of the most reproducible electrodes which is available; properly prepared electrodes will show bias potentials of less than 10 μV when compared with one another.

The hydrogen electrode commonly consists of a platinum foil, the surface of which is able to catalyze the reaction

$$H^+ + e^- \rightarrow \tfrac{1}{2}H_2 \qquad (2\text{-}40)$$

Platinum black is the most useful catalyst and will function either dispersed in the solution (45) or deposited on the surface of a bright platinum electrode. Platinum black usually is deposited from a 1 to 3% solution of chloroplatinic acid (H_2PtCl_6) containing a small amount of lead acetate. The lead acetate is not essential, but it is desirable, because electrodes prepared with lead acetate have a longer life and are less susceptible to poisoning. The electrode is poisoned by traces of sulfide and cyanide, but when this occurs the electrode potential changes so markedly that malfunction is obvious. Poisons apparently act by displacing adsorbed hydrogen (46). Drifting potentials are observed when species such as benzoic acid and nitrophenols are present that are readily hydrogenated.

Two forms of the hydrogen electrode are shown in Figure 2-8 (47). In the cell of Figure 2-8a the electrodes are supported in a stopper and a flexible polyvinyl chloride tube is used to make the connection to the second half-cell. The cell of Figure 2-8b is equipped with a presaturator and a hydrogen bypass and is designed to be immersed completely in a thermo-

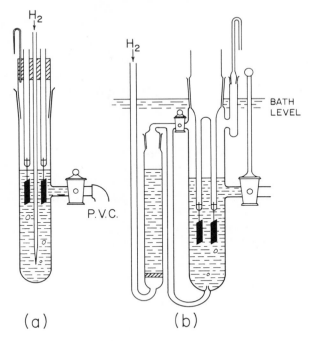

(a)   (b)

Figure 2-8. Hydrogen electrodes.

statted water or oil bath. Although the half-cell of Figure 2-8*b* shows the platinum-glass seal immersed in the solution, a 2 to 3-cm "stalk" of platinum wire is recommended by some (47) so that the glass-metal seal can be kept out of solution; a strained glass-metal junction immersed in solution can give rise to spurious potentials. To avoid this problem, Bates (48) recommends covering with glass the whole region of the spot weld between the platinum foil and wire as shown in Figure 2-9. The connection in the contact tube should be made by spot welding or silver soldering copper wire to the platinum or by use of Wood's metal (a low melting alloy). The use of mercury is considered objectionable by many authorities because of the possibility that it might work its way through the seal. The techniques for making platinum electrodes and sealing platinum in glass are described in Chapter 3.

Before platinization, electrodes should be cleaned by brief immersion in a solution prepared by combining three volumes of 12 F hydrochloric acid with one volume of 16 F nitric acid and 4 volumes of water (50% aqua regia). This mixture also can be employed to strip the platinum black from

Figure 2-9. Platinum electrode sealed in "soft" glass with the wire-foil spot weld covered by glass.

used electrodes. The electrodes are then washed in 16 F nitric acid, rinsed in water, cathodized in 0.01 F sulfuric acid, and given a final thorough rinse in distilled water. The cathodization step is designed to remove surface oxides and should be carried out immediately before platinization.

Feltham and Spiro (49) have surveyed platinizing procedures and recommend a solution of 0.072 F (3.5%) chloroplatinic acid plus 1.3×10^{-4} F (0.005%) lead acetate, at a current density of 30 mA/cm^2 for up to 10 min. A deposition time of 5 min should be adequate for hydrogen and for conductance electrodes; smaller deposits speed equilibration and reduce adsorption from the electrolyte. Good stirring is essential, and no gas should be evolved at the platinum cathode. The chlorine evolved at the anode can easily be prevented from interacting with the cathode by employing a salt bridge between two beakers or an H-type cell with a glass-frit separator.

The surface coat of platinum black should appear to be a uniform, jet black compact deposit. After platinization the electrode should be washed and stored in water. If dry electrodes are exposed to air for a prolonged period, their catalytic activity is lost and they must be replatinized before use.

The active surface of the platinized foil electrode should be covered completely by the solution when in use. It should be supplied with pure hydrogen gas at the rate of 1 to 2 bubbles/sec from a jet about 1 mm in diameter. To avoid changes in concentration, the hydrogen can be passed through a saturator that contains water or cell solution before it enters the cell.

Electrolytic grade hydrogen (available commercially in cylinders) is a convenient source of gas. It typically contains 0.15 to 0.20% of oxygen and 0.03 to 0.05% nitrogen. The potential of a hydrogen electrode will be biased by the presence of dissolved oxygen in the solution because the oxygen is readily reduced at the catalytically active surface (50). Therefore, oxygen should be excluded from the cell during measurements and should be removed from the hydrogen gas supply by one of the methods described in Chapter 3. The palladium-membrane electrolytic generators provide a convenient source of dry, oxygen-free hydrogen and may be less expensive than cylinders in the long run.

The partial pressure of hydrogen in most experiments will not be exactly 1 atm. In aqueous solutions containing no volatile solutes such as ammonia or carbon dioxide, the partial pressure of hydrogen gas at the electrode surface is obtained by subtracting the vapor pressure of water and adding the *depth effect* (51) to the barometric pressure. The solubility of hydrogen is determined by the depth of immersion of the jet through which the gas enters; this excess pressure slightly supersaturates the solution. An empirical expression for the partial pressure of hydrogen in the cell is given by

$$P_{H_2} = P_{barometric} - P_{H_2O} + 0.030\,h \qquad (2\text{-}41)$$

where all pressures are expressed in mm Hg and h is the depth of immersion of the jet in mm.

2. The Palladium-Hydrogen Electrode

Substrates other than platinum have been used for the preparation of hydrogen electrodes. Although a palladium electrode is normally a poor substitute for platinum because the catalytic activity of palladium is lower and it decays faster, the capacity of palladium to dissolve up to 30 volumes of hydrogen allows the preparation of a unique kind of hydrogen electrode that can be used in solutions containing no dissolved hydrogen. A palladium-hydrogen microreference electrode suited for use in fast potentiostatic current measurements has been described (52) that uses a palladium wire welded to platinum and charged with hydrogen electrolytically. The exposed area of the palladium wire can be as small as 10^{-4} cm^2, and it can be placed within 0.002 cm of the working electrode. The normal Luggin capillary cannot be placed this close to the electrode surface without "screening" the electrode, which results in a nonuniform current density at the electrode surface. The palladium microreference electrode gives a steady potential of $+50$ mV with respect to a hydrogen electrode in the same solution and is stable for about 24 hr. The electrode is easily renewed by recharging it electrolytically with hydrogen. The potentials of the palladium-hydride reference electrode with respect to the platinum-hydrogen electrode between 25 and 195°C have been determined (53).

3. The Silver-Silver Chloride Reference Electrode

Next to the hydrogen electrode, the silver-silver chloride electrode is probably the most reproducible and reliable reference electrode, and it is certainly one of the most convenient electrodes to construct and use. The chemical processing industry has by and large standardized on the use of silver chloride reference electrodes (54), and they have begun to displace the ubiquitous calomel reference electrodes in pH and other specific ion electrodes.

The preparation and properties of silver chloride reference electrodes have been reviewed recently (55,56); the electrode reaction is

$$AgCl + e^- \rightarrow Ag + Cl^- \qquad (2\text{-}42)$$

The solubility of silver chloride in water is about 10^{-5} F at 25°C, which sets a lower limit on the use of the electrode as an ion specific electrode for chloride ion. The solubility in saturated KCl solution increases to about 6×10^{-3} F due to the formation of soluble complexes of the type $AgCl_2^-$. For this reason the saturated KCl electrolyte must be presaturated with silver chloride; otherwise the electrode becomes stripped of its AgCl coating.

Because of this tendency to form anionic complexes, from the point of view of establishing satisfactory reference electrodes of the second kind, the most significant constant is not necessarily the solubility product constant, but the equilibrium constant for the reaction

$$AgCl \text{ (solid)} + Cl^- = AgCl_2^- \qquad (2\text{-}43)$$

In the half-cell of Equation 2-42, the concentration of $AgCl_2^-$ must be small compared to that of Cl^-, or a liquid junction potential will result because the mobilities of $AgCl_2^-$ and Cl^- are not the same. Thus for a reference electrode of the second kind to be effective in cells without appreciable junction potentials, the equilibrium constant for reaction 2-43 must be smaller than unity (preferably less than 0.1). In water, methanol, formamide, and N-methylformamide this criterion is met, but in most organic solvents the equilibrium constant for reaction 2-43 ranges from 30 to 100. The silver chloride electrode is not recommended for general use in organic solvents (57).

The potential of the silver-silver chloride electrode is sensitive to traces of bromide in the solution used to deposit AgCl. The presence of 0.01 mole percent of bromide in a KCl electrolyte is sufficient to alter the potential of electrodes immersed in the solution by 0.1 to 0.2 mV (58). The potentials are not greatly affected by traces of iodide or cyanide. Light of ordinary intensities does not have a marked effect on the potential of the electrodes, but exposure to direct sunlight should be avoided.

Of the several methods for preparation of silver-silver chloride electrodes, three are widely used:

1. A silver wire is made the anode in a 0.1 F HCl or KCl solution and coated with AgCl. Electrodes prepared in this way are suitable for potentiometric titrations or routine voltammetric use. Mechanical strains in the silver wire introduced in the drawing process, and impurities in the silver or in the chloride electrolyte generally cause the bias potentials of such electrodes to be rather high (± 5 mV), but they are reasonably stable with time.

To coat the silver electrode with AgCl it is first cleaned with 3 F nitric acid, washed thoroughly with water, and put in an H-type cell with a glass frit separator or the cathode is separated from the silver anode by a salt bridge. The silver is coated with silver chloride at a current density of about 0.4 mA/cm^2 for about 30 min, using 0.1 F HCl as the electrolyte. The electrodes are thoroughly washed and allowed to age for 1 to 2 days. The color of the fresh electrodes is variable, ranging from a sepia to a pale tan or brown. After washing, the color will usually change to a pink, grayish-pink, or plum color.

2. The electrolytic type of AgCl electrode is formed by the electrodeposition of silver on a platinum wire, foil, or gauze. The surface of the carefully washed deposit is then converted to silver chloride by electrolysis in a chloride solution as described above. Electrodes prepared in this fashion from pure solutions have a lower bias potential than does the preceding type. The silver plating is carried out by first thoroughly cleaning the platinum electrodes as described in the preceding section on their preparation for platinization. They are then immersed in a solution containing 10 g/liter of pure KAg(CN)$_2$ (freed from excess cyanide by the addition of dilute AgNO$_3$ if necessary) and plated with silver for approximately 6 hrs at 0.4 mA/cm^2 (59). The anode and cathode compartments should be separated by a salt bridge or glass frit and adequate stirring should be provided. The silver plate obtained should be snowy white and velvet-like in appearance, and should

wet uniformly. After soaking in concentrated NH_4OH for 1 to 6 hr, it is washed with water frequently over a period of 1 to 2 days. The electrodes are then coated with AgCl as described above.

3. In the thermal-electrolytic type of AgCl electrode, silver is formed by heating a paste of silver oxide coated on a spiral of platinum wire. Part of the spongy silver mass is converted to AgCl by the electrolysis process described above. Because the silver is porous, the electrodes tend to be rather sluggish but they can be prepared with very small bias potentials ($\pm 20\,\mu V$). Bates (60) has described a convenient procedure for the preparation of this type of electrode.

Silver-silver chloride reference electrodes are available from all of the principal manufacturers of ion specific and pH electrodes. They usually are furnished with a saturated KCl (saturated with AgCl) filling solution, but some manufacturers will supply the electrodes in a "dry" condition so that the user may add his own electrolyte. This can be an advantage because a saturated KCl filling electrolyte is a source of difficulty in laboratories where the temperature fluctuates or goes much below 25°C. For the latter condition potassium chloride will precipitate in the junction and have to be dissolved with dilute KCl or water and flushed. For this reason most authorities (61) recommend the use of 3.5 F KCl (saturated with AgCl) for use in the silver chloride reference electrode. Junction potentials are not seriously altered so that pH measurements are unaffected.

4. Relating a Reference Electrode to the Standard Hydrogen Electrode

In using any reference electrode other than the hydrogen electrode there is the problem of knowing what potential on the hydrogen scale to ascribe to it. Direct comparison by means of the usual salt bridge or liquid junction presents two problems. First there is no direct way to determine the necessary single ion activities (a problem previously described in Section I). Second, the liquid-junction potential must be evaluated. There are two ways to deal with this. For the typical electrode system

$Ag/AgCl/KCl(m)$: liquid junction or salt bridge : test solution/test electrode

$$E^{0'} = E_{AgCl/Ag} \tag{2-44}$$

$$E^{0'} + E_j = E_{AgCl/Ag} + E_{\text{liquid junction}}$$

the potential $E^{0'}$ may be calculated from the known value of $E^0_{AgCl/Ag}$ by use of the equation

$$E^{0'} = E_{AgCl/Ag} = E^0_{AgCl/Ag} - \frac{RT}{F} \ln a_{Cl^-} \tag{2-45}$$

$$= E^0_{AgCl/Ag} - \frac{RT}{F} \ln m_{Cl^-} \gamma_{Cl^-} \tag{2-46}$$

Table 2-4 Standard Potentials $(E^{0'} + E_j)$ and Temperature Coefficients for Cells of the Type: $(Pt)/H_2, H^+(a=1):KCl/MCl(satd)/M$

MCl/M	KCl Molarity	$E^{0'} + E_j$ (V at °C)							$\dfrac{d(E^{0'} + E_j)}{dt}$ (mV/deg at 25°C)	Ref.
		10	15	20	25	30	35	40		
AgCl/Ag	3.5M (at 25°C)	0.215	0.212	0.208	0.205	0.201	0.197	0.193	−0.73	35, p. 7
	Satd	0.214	0.209	0.204	0.199	0.194	0.189	0.184	−1.01	35, p. 7
Hg$_2$Cl$_2$/Hg	0.1M (at 25°C)	0.336	0.336	0.336	0.336	0.335	0.334	0.334	−0.08	37, p. 161
	1.0M (at 25°C)	0.287	—	0.284	0.283	0.282	—	0.278	−0.29	37, p. 161
	3.5M (at 25°C)	0.256	0.254	0.252	0.250	0.248	0.246	0.244	−0.39	35, p. 7
	Satd	0.254	0.251	0.248	0.244	0.241	0.238	0.234	−0.67	35, p. 7
TlCl/Tl 40 wt% in Hg "Thalamid"	Satd	−0.565	−0.569	−0.573	−0.577	−0.581	−0.585	−0.589	−0.79	73

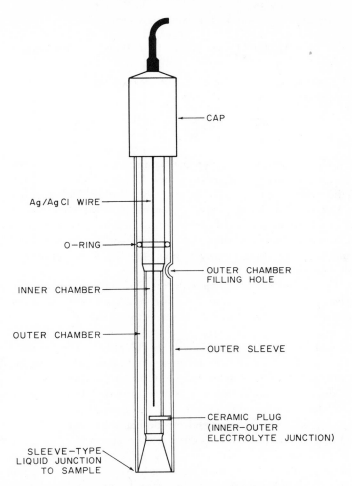

Figure 2-10. Silver-silver chloride reference electrode, commercial design.

The unknown term γ_{Cl^-} may be approximated by equating it to the corresponding mean ionic activity coefficient, $\gamma_\pm$, for the particular KCl solution concerned or by calculating the activity coefficient from the extended Debye-Huckel equation (62). The value of $(E^{0'} + E_j)$ may then be determined by adding to $E^{0'}$ the calculated liquid-junction potential, provided all of the parameters for its calculation are known or can be approximated (see Section I).

Alternatively, the silver-silver chloride half-cell can be directly compared with a hydrogen electrode, as for example in the cell

$$(Pt)/H_2, HCl\ (m)\ \text{or NBS buffer}:KCl\ (m)/AgCl/Ag \qquad (2\text{-}47)$$

$(E^{0'} + E_j)$ is then given by the observed cell potential minus that due to the hydrogen electrode, calculated from

$$E_H = \frac{RT}{F}\ln a_{H^+} = \frac{RT}{F}\ln m_{H^+}\gamma_{H^+} = -\frac{2.3RT}{F}pH \qquad (2\text{-}48)$$

For this calculation some arbitrary assumption must be made, such as $\gamma_{H^+} = \gamma_\pm$; $\gamma_\pm$ can be estimated from the Debye-Huckel equation. Alternatively, the known pH of an NBS primary standard buffer can be used in Equation 2-48. This latter procedure probably is the most practical and, if the pH of the solutions lies between 2 and 12 and they contain only simple ions in concentrations less than 0.2 M, good constancy of E_j is found. This value of $(E^{0'} + E_j)$ consequently relates specifically to solutions which are similarly constituted.

Values of $(E^{0'} + E_j)$ for silver chloride, calomel, and thalamid reference electrodes are listed in Table 2-4. For reasons described previously, the unsaturated 3.5 F KCl electrode appears to be the best choice for routine work.

Most commercially available silver chloride (or calomel) reference electrodes (see Figure 2-10) are not designed to pass substantial currents and should not be used for two-electrode voltammetry (where the current passes through the reference electrode). It is best to make your own silver chloride reference electrode for this purpose; a recommended design is shown in Figure 2-11 (63). This reference electrode has a low internal resistance and sufficient surface area (about 10 cm^2) to pass currents up to 20 μA without significant change of potential.

Figure 2-12 illustrates a small silver chloride reference electrode used by the authors that employs a dual glass junction (soft glass in Pyrex) with a low leak rate. This electrode is suitable for three-electrode voltammetry where the cell current does not pass through the reference electrode. The choice of filling electrolyte will depend on the solvent system used, although the authors have used an aqueous 1 M $(CH_3)_4N^+Cl^-$ filling solution connected to organic solvent solutions by a Luggin capillary bridge (see Chapter 3) that is filled with the sample solution. When used with organic solvents such as acetonitrile, dimethylsulfoxide, or dimethylformamide, the junction potentials certainly are not negligible, but they are reproducible to ± 5 mV or less. Although convenient, this practice is not recommended where it is necessary to exclude water completely.

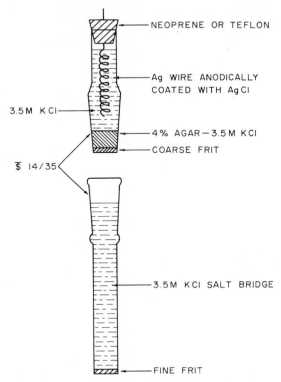

Figure 2-11. Design for a "home-made" silver-silver chloride reference electrode with low resistance.

5. The Mercury-Mercurous Chloride (Calomel) Electrode

The calomel electrode was used extensively as a chloride electrode, but it has been all but abandoned for this purpose in favor of the silver chloride electrode. The fixed-potential saturated or $3.5F$ KCl calomel electrode always has been popular for use with glass electrodes in pH measurements and in polarographic work; most of the vast compilations of aqueous polarographic half-wave potentials are referred to the aqueous saturated calomel electrode (SCE).

The electrode is based on the half-reaction

$$Hg_2Cl_2 + 2e^- \rightarrow 2Hg + 2Cl^- \tag{2-49}$$

and the "classical" electrode is prepared by grinding calomel (Hg_2Cl_2), mercury, and a little potassium chloride solution together and placing the

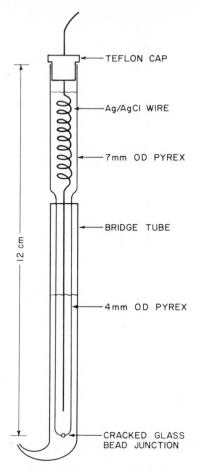

Figure 2-12. Design for a small, low leakage silver-silver chloride reference electrode and salt bridge.

resultant slurry in a layer about 1-cm thick on the surface of the mercury contained in a clean test tube. External contact to the mercury is usually made withh a platinum wire sealed into soft glass. Several forms of the calomel reference electrode are shown in Figure 2-13, and they are available from suppliers of pH and reference electrodes.

The preparation and properties of the calomel electrode have been reviewed (64); superior reproducibility is claimed for a calomel electrode prepared by shaking dry, finely divided $(0.1–0.5\,\mu)$ calomel with pure mercury. The calomel spreads over the mercury to form a "pearly" skin, and

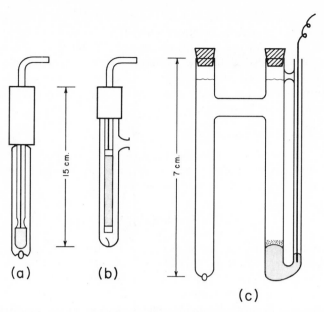

(a)   (b)

(c)

Figure 2-13. Calomel reference electrodes. (*a*), (*b*) Commercial designs with cracked-glass and fiber junctions, respectively. (*c*) "Home-made" design with cracked-glass junction on the bottom of left tube.

some of this skin is introduced on to a mercury surface where it spreads immediately. Solution is then added carefully to this "skin" electrode to give an electrode with substantial improvement in performance over that of paste electrodes.

The materials used to prepare the calomel electrode should be as pure as possible. Mercury may be purified by the methods described in Section IV. Finely divided calomel is prepared by chemical precipitation with HCl that has been rigorously freed of traces of bromide and iodide (64).

To prevent aqueous solution from creeping down between the mercury and the glass wall of the cell, the cell walls should be treated with silicone preparations to render the glass hydrophobic. Because this adversely affects the platinum contacts sealed into the glass, a design should be used that allows the mercury to form its own connection by filling a long-bore capillary tube; a removable platinum wire contact can then be used at the remote end of this tube, as in Figure 2-13.

To determine the potential of the saturated calomel electrode on the hydrogen scale, the same problems arise as discussed for the silver-silver chloride reference electrode, and $E^{0'}$ and $(E^{0'} + E_j)$ are determined in the same manner. The values recommended by Hills and Ives (65) and Bates (66) are tabulated for the calomel electrode in Table 2-4.

Although extensive data are available for the calomel electrode, the silver-silver chloride electrode appears to be superior to it because of its ease of preparation and use, lower sensitivity to the presence of oxygen, and smaller temperature hysteresis.

6. The Mercury-Mercuric Oxide Electrode

Ives (67) has reviewed the literature for the mercury-mercuric oxide electrode and has concluded that, among metal-metal oxide electrodes, it is uniquely well behaved. Investigations of the cell

$$(Pt)H_2/NaOH(aq)/HgO/Hg \qquad\qquad (2-50)$$

indicate that its potential, in agreement with theory, is constant within a mean deviation of $\pm 0.06\,mV$ for sodium hydroxide molalities between 0.001 and 0.3 mole/kg. The E^0 of the cell is 0.926 V at 25°C. Several other related fixed potential electrodes have been described (68), which in combination with a saturated calomel electrode (potential assumed to be 0. 2446 V versus the hydrogen electrode at 25°C) and a KCl salt bridge yield the values:

$$Hg/HgO/Ba(OH)_2(satd): \qquad E = 0.1462\,V \text{ versus NHE at } 25°C$$

$$Hg/HgO/Ca(OH)_2(satd): \qquad E = 0.1923\,V \text{ versus NHE at } 25°C$$

This electrode, which is available commercially, is well suited for use in alkaline solutions.

7. The Mercury-Mercurous Sulfate Electrode

Several commercial suppliers offer the mercury-mercurous sulfate electrode with a saturated potassium sulfate electrolyte. The potential $(E^{0'} + E_j)$ of this electrode system is 0.658 V on the hydrogen scale at 22°C (69). The electrode comprises one-half of the Weston standard cell (70), an international secondary voltage standard, and is outstanding in reproducibility (71), in spite of the slight tendency of mercurous sulfate to hydrolyze and its rather high solubility.

Because special preparative procedures are not necessary (except for the preparation of sulfate electrolytically), this reference electrode is recommended for use in place of the silver chloride or calomel electrodes when chloride ion must be rigorously excluded.

8. Thallium Amalgam-Thallous Chloride (Thalamid®) Electrode

The reference electrode system

$$Tl(Hg)/TlCl(s), TlCl(sat.), KCl \qquad (2\text{-}51)$$

has been reviewed (72) and is commercially available ("Thalamid" electrode, Jenaer Glaswerk Schott and Genossen, Mainz, Germany). It is claimed to be reproducible and usable at higher temperatures (135°C) than is the calomel electrode, and to be free of temperature hysteresis.

At present, this electrode is used primarily in symmetrical glass electrode cells (i.e., as the internal and external reference electrodes) with the principal advantage of a well-defined isothermal intersection point (isopotential point) that nearly coincides with the null point of the cell. (The family of straight lines that represent the variation of the cell potential with pH at various temperatures all intersect at approximately the same point, the isopotential point. The pH at which the cell potential is zero is called the zero point or null point.) This reduces the sensitivity of the pH measurements to small temperature changes and allows simpler instrumental compensation of temperature changes for pH measurements.

The standard potentials $(E^{0\prime} + E_j)$ of the Thalamid® reference electrode

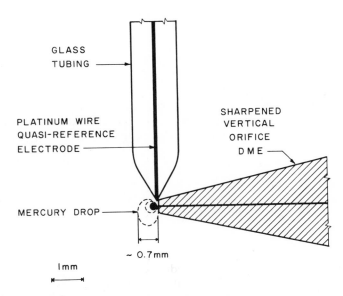

Figure 2-14. Platinum-wire quasi-reference electrode in combination with a dropping mercury electrode.

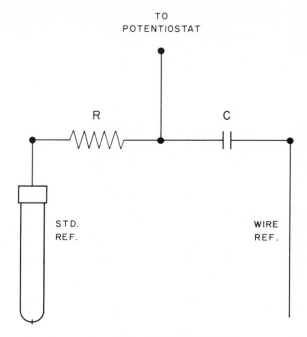

Figure 2-15. Circuit for dual reference electrode system.

in saturated aqueous KCl solution are tabulated in Table 2-4 (73).

9. Quasi-reference Electrodes

Precise potentiometric measurements require reference electrodes that are highly reproducible, but there are many applications where this is less essential. For example, in routine analytical polarographic or voltammetric measurements, the accurate measurement of the current is more important than the precise measurement of potential. For these purposes a simple quasi-reference electrode is suitable. In a halide-containing supporting electrolyte, a silver wire or mercury pool will adopt a reasonably steady potential that is reproducible to within ± 10 to 20 mV.

A platinum wire also will function satisfactorily as a quasi-reference electrode (74) in solutions of high resistance or in potentiostatic current measurements. Figure 2-14 illustrates a platinum wire for use in a high resistance medium (75); it minimizes the uncompensated resistance in the cell and replaces a bulky reference electrode and Luggin capillary. Wire quasi-reference electrodes also are used to measure fast electrode reactions by analysis of the system response to a current or potential step. Here the use of a standard reference electrode with its relative high resistance results in a long rise time (damped pulse form) as well as damped oscillations. In some experiments both fast response for meaningful information to be gathered in the first few microseconds and a long term stable reference potential are required. The use of a dual reference electrode system is recommended for this purpose (76) and is illustrated by Figure 2-15. At relatively short times (or high frequencies) the potentiostat responds to the potential at the wire reference electrode, to allow fast-pulse conditions to be met. At longer times, including D-C conditions, only the standard reference electrode, chosen for steady-state reference potential stability, is controlling. The values of R and C must be chosen to match the desired frequency response and input impedance of the potentiostat; the original paper should be consulted for details.

For potentiometric titrations, the potential changes at the end point are relatively large (100–200 mV) and a simple quasi-reference electrode may be quite adequate, even if it has an uncertainty or drift of a few millivolts. Also, for redox titrations which involve strong oxidizing agents with a platinum or noble metal indicator electrode, a silver wire or mercury pool cannot be used because they are easily oxidized. A platinum quasi-reference electrode cannot be used because its potential will shift with that of the indicator electrode. For such a situation, the glass electrode will serve as a highly satisfactory quasi-reference electrode if the hydrogen ion concentration remains essentially constant. A pH meter may be used to measure the potential changes, with the platinum indicator electrode connected into the terminal for the reference electrode in pH measurements and the glass electrode in its normal terminal.

10. Temperature Effects

The ordinary glass electrode-reference electrode pair that is used for pH measurements is not well suited to measurements far removed from room temperature. This is because the electrodes are immersed only partially, with the tips of the electrodes in the solution and the tops of the electrodes at ambient temperature. This creates a thermal gradient in the body of the electrode, which causes errors due to Seebeck (thermocouple) effects at junctions between dissimilar metals in the body of the electrodes. Whenever possible, the electrodes should be operated in an isothermal environment; thermal gradients should be limited to the external connecting leads which should be made of the same metal (e.g., copper).

Other difficulties may arise if the two electrodes are located in different thermal environments with the temperature of one electrode varying while the other remains constant. Although the true temperature coefficients of the potentials of single electrodes cannot be obtained by thermodynamic means,

Table 2-5   Reference Electrodes for Use in Dipolar Aprotic Solvents[a]

Reference Electrode	Acetonitrile	Propylene Carbonate	Dimethylformamide	Dimethylsulfoxide
H^+/H_2	S	S	S	NR
Glass (pH)	S		S	S
Glass (cationic)	S	S	S	S
Ag^+/Ag	S	S	S	S
$AgCl/Ag$	NR	NR	NR	NR
Hg_2Cl_2/Hg	NR	NR	NR	NR
Li^+/Li		S	S	S
$Li^+/Li(Hg)$		S		S
$TlCl/Tl(Hg)$				S
$Fe(Cp)_2^+/Fe(Cp)_2(Pt)$	S	S	S	S
Cp = cyclopentadienyl				
I_3^-/I^-		S		
Aqueous SCE + salt bridge	S	S	S	S
Aqueous AgCl/Ag + salt bridge	S	S	S	S
$CdCl_2/Cd(Hg)$			S	S

From Reference 57

[a] S = Satisfactory stability and reproducibility; NR = not recommended—unstable or not reproducible.

an indication of the changes in electrode potential can be obtained by observing the change in cell potential when nonisothermal conditions are created deliberately. Although these values have no thermodynamic meaning, they indicate the magnitude of errors that result from unavoidable thermal gradients in a cell. In Table 2-4 the change of the reference electrode potentials with temperature, $d(E^{0'} + E_j)/dT$, are tabulated relative to the hydrogen electrode at 25°C. The size of the temperature coefficients indicates that temperature control that is better than ±0.1°C is necessary for precise work.

C. REFERENCE ELECTRODES FOR USE IN POLAR APROTIC SOLVENTS

The increased use of polar aprotic solvents for electrochemical studies has inspired a search for suitable reference electrodes. Although the description of an aprotic solvent is somewhat ambiguous (see Chapter 4), we include in this class those solvents with dielectric constants greater than 30, which show weak acidic character and are difficult to oxidize or to reduce. Table 2-5 includes references to a number of reference electrode systems for use in such solvents and is drawn largely from the chapter by Butler (57); this is an excellent review of the extensive work since the earlier review by Hills (77). The recent monograph by Mann and Barnes (78) contains a brief discussion of nonaqueous reference electrodes. Strehlow (79) also has reviewed the problem of measuring electrode potentials in nonaqueous solvents and the comparison of potential scales in different solvents. Aspects of the problem related to pH measurements and acidity functions have been reviewed by Bates (80,81).

Until recently, the most popular reference half-cell for potentiometric titrations, polarography, and even kinetic studies has been the saturated aqueous calomel electrode (SCE), connected by means of a nonaqueous salt bridge (e.g., Et_4NClO_4) to the electrolyte under study. The choice of this paticular bridge electrolyte in conjunction with the SCE is not a good one because potassium perchlorate and potassium chloride have a limited solubility in many aprotic solvents. The junction is readily clogged, which leads to erratic junction potentials. For these practical reasons, a calomel or silver-silver chloride reference electrode with an aqueous lithium chloride or quaternary ammonium chloride filling solution is preferable if an aqueous electrode is used.

1. The Hydrogen Electrode

The hydrogen electrode has been used successfully in the formamides, anhydrous acetone, propylene carbonate, tetrahydrofuran, 1,1,3,3-

tetramethylguanidine, and acetonitrile (although there are conflicting reports about the reproducibility of the hydrogen electrode in acetonitrile). In DMSO solutions platinized platinum electrodes apparently decompose the solvent by catalytic reduction.

2. The Calomel Electrode

Calomel and other mercurous halides disproportionate in a number of organic solvents, and attempts to use the calomel electrode in polar aprotic solvents, have, for the most part, been unsuccessful. For this reason, it is not advisable to replace the aqueous electrolyte of an ordinary calomel reference electrode with an electrolyte dissolved in an aprotic solvent.

3. The Silver-Silver Chloride Electrode

The silver chloride reference electrode is not generally suitable as an electrode of the second kind because of the large solubility of AgCl in many aprotic solvents from formation of anionic complexes with chloride ion. In many cases the silver chloride solubility will essentially be that of the added chloride. This contributes significantly to the junction-potential in cells with liquid junction and makes the electrode unsuitable for precise potentiometric work.

4. The Silver-Silver Ion Electrode

One of the most satisfactory and widely used electrodes is the silver-silver ion electrode, which appears to be reversible in all aprotic solvents except those which are oxidized by silver ion. The electrode is easily made by putting a silver wire (or silver plated platinum) in a solution of 0.001 to 0.01 F $AgNO_3$ or $AgClO_4$. The polarizability of these electrodes (82) indicates that if they are to be used in voltammetric work, it should be with a three-electrode circuit (see Figure 2-2b) so that the cell current does not pass through the reference electrode.

Although $AgNO_3$ is nearly equitransferent in acetonitrile (83), junction potentials can be minimized further by use of a salt bridge that contains the same nonaqueous supporting electrolyte as the sample solution. By this approach, electrolytic connection is made with a compartment containing a silver electrode in the same electrolyte as the salt bridge, but with the addition of 0.001 to 0.01 F $AgNO_3$ (see Figures 2-16 and 2-17). This type of reference electrode requires more complex glassware than the conventional reference electrode, and the supporting electrolyte must not contain impurities that react with Ag^+, but the extra effort yields lower junction potentials and greater reproducibility of potential measurements.

The difficulties that can arise with aqueous-nonaqueous liquid junctions are illustrated by the values reported for the cell

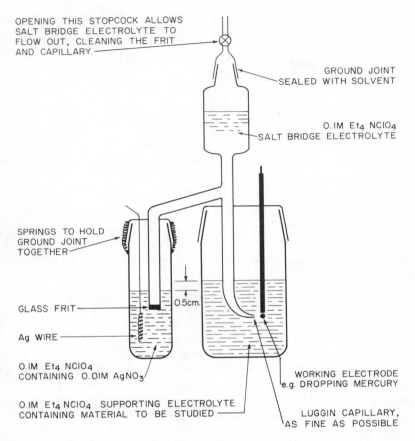

OPENING THIS STOPCOCK ALLOWS SALT BRIDGE ELECTROLYTE TO FLOW OUT, CLEANING THE FRIT AND CAPILLARY.

GROUND JOINT SEALED WITH SOLVENT

0.IM Et$_4$ NClO$_4$ SALT BRIDGE ELECTROLYTE

SPRINGS TO HOLD GROUND JOINT TOGETHER

GLASS FRIT

0.5cm.

Ag WIRE

0.IM Et$_4$ NClO$_4$ CONTAINING 0.0IM AgNO$_3$

0.IM Et$_4$ NClO$_4$ SUPPORTING ELECTROLYTE CONTAINING MATERIAL TO BE STUDIED

WORKING ELECTRODE e.g. DROPPING MERCURY

LUGGIN CAPILLARY, AS FINE AS POSSIBLE

Figure 2-16. Silver-silver ion reference electrode and salt bridge for use with nonaqueous solvents. Design allows junction potentials to be minimized.

aqueous SCE : junction or salt bridge : 0.01 F AgNO$_3$/Ag (acetonitrile)

$$(2\text{-}52)$$

These range (84) from 0.253 V for 0.1 F NaClO$_4$ in acetonitrile in the salt bridge to 0.300 V for a single junction with no salt bridge.

Silver-silver ion electrodes have been employed to study liquid-junction potentials between electrolyte solutions in different solvents by use of the cell

Ag/AgClO$_4$(0.01 F) (acetonitrile, $S1$) : TEA Pic (0.1 F) (bridge solvent, $S3$) :

AgClO$_4$(0.01 F)/Ag (solvent $S2$) $(2\text{-}53)$

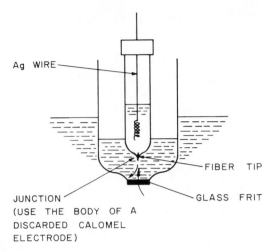

Ag WIRE

FIBER TIP

JUNCTION
(USE THE BODY OF A
DISCARDED CALOMEL
ELECTRODE)

GLASS FRIT

Figure 2-17. Design for simplified silver-silver ion reference electrode and salt bridge for use with nonaqueous solvents.

in which TEA Pic is tetraethylammonium picrate (28). When $S2$ is dimethylsulfoxide (DMSO), and the bridge solvent, $S3$, is varied over a representative group of solvents, the potentials are independent of the bridge solvent to within 5 mV. However, this is not true when the bridge solvent is formamide. The situation when the bridge solvent is water has not been tested because of the low solubility of TEA Pic in water. Furthermore, when $S2$ is formamide, water, or methanol, rather than DMSO, the potential of the cell varies by up to 100 mV as the bridge solvent, $S3$, is changed.

There is a linear correlation between the observed cell potential and the mutual heats of solution of the various solvents in the cell. Junction potentials up to 100 mV have been attributed to the free energy changes associated with the transport of solvent molecules across the junction; this is in addition to the junction potential due to the passage of ions across the junction. The liquid-junction potentials due to different solvents can be reduced by careful selection of bridge solvents and electrolytes. To minimize the liquid-junction potentials between two different solvents (as in cell 2-53), the bridge solvent must not interact strongly (small mutual heats of solution) with either of the other solvents, and the transport numbers of the cation and anion of the bridge electrolyte should be equal. Either tetraethylammonium picrate or tetrabutylammonium tetraphenylborate are reasonable choices as bridge electrolytes, but the latter is preferred for voltammetric work because the picrate ion is reduced easily.

5. Other Reference Electrodes for Use in Polar Aprotic Solvents

Emphasis has been given to the use of the silver-silver ion reference electrode because it is almost universally applicable, and because standardization on the use of one reference electrode system simplifies the comparison of data between different workers. However, a number of other reference electrodes have been used (see Table 2-5), particularly those which have resulted from the vast amount of battery research. These include the Li/Li^+ and other alkali metal electrodes which function reversibly in DMSO, propylene carbonate, and hexamethylphosphoramide. Thallium-thallous halide electrodes of the second kind also function reversibly in DMSO and propylene carbonate. The cadmium amalgam-cadmium chloride reference electrode also functions reversibly in dimethylformamide and may be a useful substitute for the silver-silver ion reference electrode, which may be unstable in dimethylformamide (85).

A number of redox couples at platinum electrodes (with both species soluble in the solvent) have been important to the establishment of a relation between potential scales in different solvents. These include ferrocene (dicyclopentadienyl iron (II))-ferrocinium$^+$ ion (3) as well as the iron (II)-(III) o-phenanthroline complexes (4). Although they are useful in the comparison of redox potentials, they appear to have no advantages over the silver-silver ion system in the preparation of practical reference electrodes.

6. Stability of Reference Electrodes Prepared in Organic Solvents

Reference electrodes in organic solvents ordinarily are not as stable with time as those with aqueous electrolytes. Most aprotic organic solvents undergo slow chemical change from reactions of impurities, hydrolysis of the solvent by traces of water, or reaction with a component of the half-cell reaction. The stability of reference electrodes should be checked frequently by comparison with fresh reference electrodes prepared with recently purified solvents and electrolytes.

D. REFERENCE ELECTRODES FOR USE IN NONPOLAR SOLVENTS

Solvents, such as dichloromethane, which are not highly polar present special problems. Their low dielectric constants promote extensive ion association, and cell resistances tend to be large. For this reason they are often used in mixtures with more polar solvents. Because dichloromethane and other nonpolar solvents are not miscible with water, use of an aqueous reference electrode with such solvents is not practical unless a salt bridge with some mutually miscible solvent is used. A better approach is to use a reference electrode of known reliability prepared in a solvent miscible with dichloromethane, or to use the reference electrode based on the half-cell in dichloromethane (86)

$$Ag/Ag_3I_4^-Bu_4N^+; \qquad (Bu = n\text{-butyl}) \tag{2-54}$$

At this time it is unclear whether there ultimately will be different references electrodes uniquely suited for each nonpolar solvent, or whether a single reference half-cell will come into wide use for a number of different solvents. This problem awaits further research and critical evaluation of the existing half-cell reactions that have been proposed.

E.   REFERENCE ELECTRODES IN FUSED SALT SYSTEMS

The kinds of investigations which require reliable reference electrodes in molten salts duplicate those in aqueous electrolytes. Specifically, these include (a) thermodynamic studies, such as the determination of the half-cell potentials of various electrodes relative to a particular reference electrode, or determination of thermodynamic data for molten salt mixtures; (b) kinetic studies, such as the measurement of exchange currents (important in battery and fuel cell technology; (c) mass transport and energy output studies, such as determination of transference numbers or measurement of thermoelectric power in pure or mixed electrolytes; (d) the study of electroactive chemical species in molten salts by use of voltammetric techniques (which may, of course, be used also for purposes of chemical analysis).

The literature on fused salt systems is too vast to survey here, but good access to the literature is provided in the review by Laity (87) and in the recent comprehensive handbook edited by Janz (88). Both of these sources contain useful information on solvent systems and suitable reference electrodes.

Alkali metal chloride and nitrate eutectic mixtures that melt in the range 200 to 500°C are widely used as fused salt solvents, because they have a large liquid range from their melting to their boiling points, and because they can be used at temperatures below the softening point of glass.

1. The Silver-Silver Ion Electrode

Of the reversible metal electrodes, silver has been most often employed. There is only one stable oxidation state of silver; above 300°C there is no danger of oxide formation because Ag_2O is unstable (89). The metal has no observable tendency to dissolve in molten silver salts and is highly reversible in mixed chloride and nitrate eutectics. The Ag(I) ion can be introduced into the melt by either adding silver nitrate to a nitrate melt (AgCl to a chloride melt) or by anodizing a silver electrode. The potentials of silver

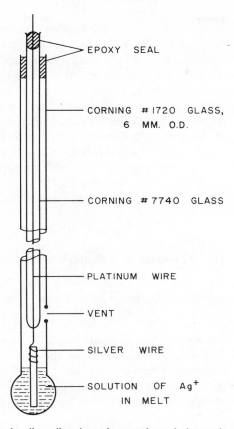

Figure 2-18. Design for silver-silver ion reference electrode for use in molten-salt systems.

nitrate concentration cells show ideal thermodynamic behavior up to 0.5 mole % in $(Na,K)NO_3$ eutectic and in $NaNO_3$ (90).

Silver metal usually is employed in the form of wires or foils. In contact with a melt containing Ag^+, silver continuously recrystallizes, so that a wire of small diameter may eventually be converted to a fragile string of loosely joined crystals. The rate of the process depends on temperature, and also appears to proceed more rapidly at a kinked or otherwise strained point in the wire.

The simplest reference electrode may be prepared by dipping a pure clean silver wire in a melt contained in a glass (or silica) tube. At higher temperatures, there is a possibility that glass will exchange sodium ions for silver ions; where this is a problem, silica is preferred over glass. The junction to the solution may be made by an asbestos fiber pinch-sealed into the bottom of the tube, by means of a porous ceramic plug sealed into the bottom of the tube, or through a thin glass bulb (see Figure 2-18).

Silver is rather difficult to seal into glass, and it is easier to seal a platinum wire into glass and then spot weld a sturdy silver wire to the platinum as shown in Figure 2-18. Corning No. 1720 glass is used to make the junction between the reference electrode compartment and the melt that contains the working electrode. This provides a barrier through which the transport of Ag^+ is insignificant, yet it has a resistance of less than 3000 Ω at 700°C (91).

Although a number of other reference electrodes are capable of reproducible behavior in molten salts, including the $Pt/PtCl_2$ electrode favored by Laitinen and coworkers (92) in chloride melts, none of these appear to be as widely applicable as the silver-silver ion electrode.

III. Voltammetric Indicator Electrodes

A. ELECTRODE MATERIALS AND THEIR ELECTROCHEMICAL BEHAVIOR

There is abundant evidence that the rate of electron transfer across an electrode/solution interface is dependent on the physical and chemical properties of the electrode material. The term electrocatalysis has been coined for this effect, and studies of the oxidation of hydrocarbons (93) and the reduction of water and hydrogen ion (94) have provided ample evidence for its existence.

Table 2-6 indicates the range of exchange current densities that have been observed for the hydrogen evolution reaction on various metals. Note that the value for mercury is about ten orders of magnitude smaller than that for platinum. This difference is consistent with the fact that mercury is a much more useful electrode material for the study of cathodic processes than is platinum or other noble metals.

Electrode reactions are analogous to any heterogeneous chemical reaction where reaction takes place on a catalytic surface, with one important difference (95). The heterogeneous chemical reaction does not involve a net charge transfer across the interface and is potential independent, while the electrode reaction involves a net charge transfer across the interface and therefore the reaction rate is potential dependent. In effect, the activation energy of the electrode reaction can be controlled by varying the potential.

The *net* current density, proportional to the moles of electrons transferred per square centimeter of electrode surface, is a measure of the rate of the

Table 2-6 The Exchange Current Density i_0 for the Hydrogen Evolution Reaction in 1 $M H_2SO_4$

Metal	$-\log i_0$ (A/cm^2)
Palladium	3.0
Platinum	3.1
Rhodium	3.6
Iridium	3.7
Nickel	5.2
Gold	5.4
Tungsten	5.9
Niobium	6.8
Titanium	8.2
Cadmium	10.8
Manganese	10.9
Thallium	11.0
Lead	12.0
Mercury	12.3

reaction. It may be thought of as the sum of a cathodic component corresponding to the reaction

$$O + e^- \rightarrow R \tag{2-55}$$

and an anodic component corresponding to the reverse action

$$R \rightarrow O + e^- \tag{2-56}$$

(Here, O and R represent oxidized and reduced forms, respectively, and their net charge is neglected.)

At equilibrium, there is no net current, and the cathodic and anodic components of the current are equal to one another and equal to the *exchange* current density. For a simple one-step transfer of a single electron, the relationship between current density and the potential is expressed by the Butler-Volmer equation

$$i = i_0[e^{-\alpha\eta F/RT} - e^{(1-\alpha)\eta F/RT}] = |i_c| - |i_a| \tag{2-57}$$

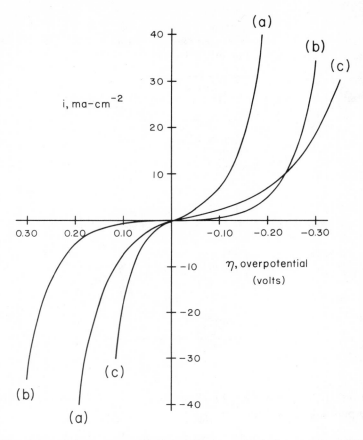

Figure 2-19. Current-voltage curves for three different sets of electron transfer kinetic parameters (i_0, exchange current density and α, transfer coefficient). (a) $i_0 = 1\,\text{mA-cm}^{-2}$, $\alpha = 0.50$; (b) $i_0 = 0.1\,\text{mA-cm}^{-2}$, $\alpha = 0.50$; (c) $i_0 = 1\,\text{mA-cm}^{-2}$, $\alpha = 0.25$.

where i is the net current density, i_o the exchange current density, i_c the cathodic current density, i_a the anodic current density, α the symmetry factor (the fraction of the potential difference, $\Delta\phi$, which assists the cathodic reaction; also known as the transfer coefficient), and η the overpotential (defined as $\eta = \Delta\phi - \Delta\phi_e$, where $\Delta\phi$ is the potential difference across the interface and $\Delta\phi_e$ is the potential difference across the interface at equilibrium).

Plots of current versus overpotential, for two different values of i_o and α

are shown in Figure 2-19. Note that when $\alpha = 0.5$ the curves are symmetric, and the slope of the curve as it crosses the zero current axis is determined largely by the value of i_o, the exchange current density. The cathodic and anodic components of the net current are indicated by the dotted lines for $\alpha = 0.5$.

Efforts have been made to correlate rates of electron transfer (as measured by exchange current densities or standard heterogeneous rate constants) with the physical properties of the electrode material. For example, a critical study has been made of correlations between the rate of the hydrogen evolution reaction and properties such as the heat of adsorption of hydrogen, work function of the metal, solubility of hydrogen in the metal, hardness of the metal, heat of fusion, and heat of sublimation (96). The results of another study (94) indicate that the work function of the metal is the key physical parameter in the hydrogen evolution reaction. Plots of the logarithm of the current density versus work function are linear but fall into two distinct groups—group *a* includes all transition metals and *sp* metals that are negatively charged at the potential where i_o is measured, while group *b* consists only of *sp* metals with negatively charged surfaces.

Considerable progress has been made in the development of theoretical approaches to the problem, but the existing general theory of electrocatalysis does not allow quantitative predictions. In the general theory described by Parsons (97), the kinetic equations do not contain any dependence on the work function of the metal. Two limiting cases are outlined: (*a*) for a simple electron transfer with the reactant only weakly bonded to the metal surface or for a two-step reaction with a single intermediate weakly adsorbed, there is no primary effect of the metal; (*b*) where the reactant or intermediate forms strong bonds with the metal surface, there is a strong dependence on the properties of the metal. For example, if the logarithm of the exchange current density for the hydrogen evolution reaction is plotted against the metal-hydrogen bond strength for a number of different metals, the familiar "volcano" plot is obtained, peaking near the heavier metals of the Group VIII triad.

Several groups (98–103) have developed semiquantitative theories of electron transfer in homogeneous solution, as well as at metal and semiconductor electrodes. A strong parallel is found to exist between electron transfer reactions in homogeneous solution and on an electrode surface, particularly if no chemical bonds are broken. In most of these theories, a strong dependence on the properties of the metal is not contained in the theory, and the electrode material is treated as an inert conductor with a Fermi distribution of electronic energies. The theories do discriminate, however, between the behavior of metal electrodes and semiconductor electrodes, which differ markedly in their distribution of electronic energies.

Although theoretical progress has been more rapid in the past 20 years, a completely satisfactory physical and chemical interpretation which accounts for the different behaviors of electrode materials still is not possible. We may feel intuitively that there ought to be some dependence on the geometric or crystal structure of the metal and on the electronic structure of the metal atoms but at this point we must be content with empirical observations. In the sections that follow, data are summarized for the most important electrode materials that are used in voltammetric work.

The voltage "window" or range of accessible potentials is limited on the negative side by reduction of the supporting electrolyte or solvent. In protic solvents, the limiting reaction will usually be hydrogen evolution. Therefore, for an electrode to be useful well into the negative potential region, it must have a low exchange current density for reduction of hydrogen ions. When the limit is reached, there is a characteristically exponential increase of current with potential consistent with Equation 2-57 and Figure 2-19. On the positive side, the potential range will be limited by oxidation of supporting electrolyte or solvent, by oxidation of the electrode material itself to form soluble metal ions or metal oxides, or by formation of molecular and chemisorbed oxygen in water or other oxygen containing solvents.

Figure 2-20 summarizes the positive and negative voltage limits for some commonly used electrode materials in several solvents. Wherever possible, the data for a particular solvent have been referred to a single reference electrode. Absolute values of the electrode potential for different solvent systems cannot be directly compared, however, because they are often referred to different reference electrodes and because of the uncertainty in our knowledge of junction potentials between different solvent systems.

The most commonly used electrode materials are mercury, various forms of carbon, gold, platinum and other noble metals, carbides, borides, and conducting tin oxide films on glass.

1. Mercury

Because mercury is a liquid down to $-39°C$, it can be used in dropping, streaming, or pool configurations that are impossible with solid electrodes. It also can be coated or plated in thin layers on other metals which then show properties similar to mercury electrodes. The dropping or streaming mercury electrodes have the special advantage of providing a continuously renewed surface which helps to minimize the effects from adsorption of solution impurities or from fouling of the electrode surface by films produced in the electrode reaction. In addition the surface is smooth and continuous and does not require the pretreatment and polishing that is common with solid electrodes.

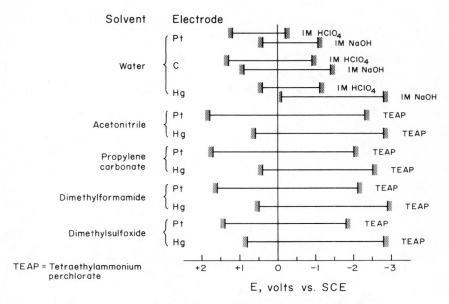

Figure 2-20. Voltage limits for various electrode materials in several solvents.

The advantages of the liquid surface and the large overpotential for hydrogen evolution make mercury the material of choice for cathodic processes, unless the use of mercury is specifically contraindicated by some incompatibility with the system. Incompatibility can arise from strong specific adsorption, as with some sulfur containing compounds, or in high temperature systems such as fused salts because of its low boiling point (356.6°C).

Cathodic Limits on Mercury. In aqueous or other protic solvents the reduction of hydrogen ion or solvent generally will limit the negative potential range. The nature of some electrode reactions at highly negative potentials on mercury has been examined recently (104). For example, K^+ and Na^+ ions are reduced reversibly in aqueous solutions, but the process is accompanied by a parallel irreversible reaction due to an amalgam dissolution reaction of the alkali metal with water which produces hydrogen.

In dipolar aprotic solvents, the proton availability is much lower and consequently the cathodic limit is extended. Reversible or nearly reversible waves can be readily observed for the reduction of Group I and some Group II metal ions (105).

In fused salt systems, reduction of traces of water (if present) or of the metal cations ordinarily will be the cathodic limiting process.

Anodic Limits on Mercury. Mercury is readily oxidized, particularly in the presence of anions which precipitate or complex mercury (I) or mercury (II) ions, such as the halides, cyanide, thiosulfate, hydroxide, or thiocyanate. For this reason, mercury is seldom used to study anodic processes except for those substances which are easily oxidized, for example, Cr(II), Cu(I), and Fe(II). Under carefully controlled conditions, mercury can be coated with a thin layer of mercury(I) chloride such that it does not interfere with electron transfer in the oxidation of a number of organic compounds, particularly amines (106).

2. Platinum, Gold, and Other Noble Metals (Pd, Rh, Ir)

Platinum and gold are the most commonly used metallic solid electrodes. These metals are readily obtained in high purity, are easy to machine, and can be fabricated readily into a variety of geometric configurations—wires, rods, flat sheets, and woven gauzes. They are resistant to oxidation, but are not totally inert as often assumed in early work.

Negative Potential Limit. Both platinum and palladium have extremely small overpotentials for hydrogen evolution, which is the basis for their use in the construction of reversible hydrogen electrodes. Gold has a significantly larger overpotential, but it is much smaller than for mercury. Platinum also adsorbs hydrogen readily and the amount of adsorbed hydrogen can be used to estimate the true surface area of the platinum. Palladium dissolves hydrogen readily in the bulk metal. Although this can be an asset in the construction of a palladium-hydrogen reference electrode, it makes palladium unsuited for use in most voltammetric work.

Gold does not appreciably adsorb hydrogen and this factor together with its larger overpotential for hydrogen evolution makes gold the metal of choice for the study of cathodic processes. Curiously, platinum is used more often than are all other metals combined, probably because of tradition and the fact that gold is much more difficult to seal into glass than is platinum.

Positive Potential Limit. At sufficiently positive potentials, all of the noble metals form an oxide film or layer of oxygen in aqueous solution with a fairly well defined stoichiometry which can be used to estimate true surface areas (see Section IIIB). The exact nature of this oxygen layer has been the subject of much investigation and some controversy, but a consensus seems to be emerging (107) that the oxygen film consists of chemisorbed oxygen with nucleation and growth of an oxide phase under severe anodic conditions.

The voltammetric behavior that is observed on platinum, palladium, rhodium, and gold in 1 F sulfuric acid is shown in Figure 2-21. The characteristic hydrogen adsorption-desorption peaks which appear on the first three metals are almost absent on gold. (Palladium was not taken into the hydrogen evolution region because of its capacity to dissolve hydrogen, which gives it almost unlimited capacity at ordinary voltammetric current levels.)

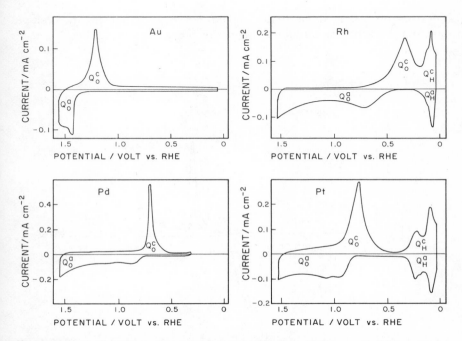

Figure 2-21. Cyclic voltammograms for platinum, palladium, rhodium, and gold electrodes in 1 F H$_2$SO$_4$. Q_0^c, current due to oxide reduction; Q_0^a, due to oxide formation; Q_H^c, due to H$_2$ formation; Q_H^a, due to H$_2$ oxidation.

The potential cycling illustrated by Figure 2-21 is a commonly used pretreatment procedure for attainment of a reproducible "active" surface (see Section IIIC on pretreatment procedures).

Less widely known is the fact that in aqueous solution this cycling procedure causes the dissolution of appreciable quantities of metal. The discrepancy between the integrated anodic and cathodic oxygen adsorption-desorption peaks has been shown to be due to dissolution of the metal.

Typical values are given in Ref 108 and indicate that platinum and gold dissolve to a much lesser extent than do palladium and rhodium.

Gold is more readily oxidized than platinum in the presence of complexing anions such as the halides or cyanide. The effect of chloride can be seen in Figure 2-22 (109). The enhancement of the peaks probably is due to the process

$$Au + 4Cl^- \rightarrow AuCl_4^- + 3e^- \qquad (2\text{-}58)$$

For this reason, gold electrodes should not be used at highly positive potentials in solutions containing halides or other complexing ions.

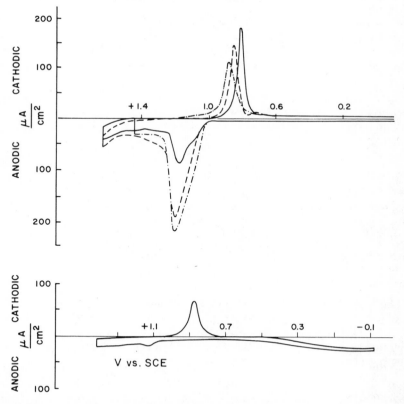

Figure 2-22. Cyclic voltammograms for a gold electrode. Upper curve for successive scans in the presence of chloride ion. Lower curve for sulfate electrolyte.

In polar aprotic solvents which are largely free of water, the formation of a chemisorbed oxygen layer is less pronounced and polished platinum shows a positive limit that is larger than any of the other commonly used electrode materials. Gold is almost as good, but it cannot be used with many complexing anions. On the negative side, gold is superior to other metals (except mercury) because of its larger overpotential for hydrogen evolution.

3. Carbon

Several different forms of carbon have been used to make satisfactory electrodes including spectroscopic grade graphite (usually impregnated with ceresin or paraffin wax), pyrolytic graphite (a high density, highly oriented form of graphite), carbon paste (spectroscopic grade graphite mulled in sufficient Nujol, bromonaphthalene, or bromobenzene to form a stiff paste), graphite dispersed in epoxy resin or silicone rubber, and vitreous or "glassy" carbon.

Vitreous or Glassy Carbon. The use of glassy carbon as an electrode material was first suggested in 1962 by Yamada and Sato (110). It is a proprietary preparation of the Tokai Electrode Manufacturing Company (111), but similar materials are available from other sources (112). Glassy carbon is an electrically conductive material, highly resistant to chemical attack and gas impermeable, and obtainable in a pure state. It has many properties in common with pyrolytic graphite, but does not need to be oriented as does pyrolytic graphite. Several groups (113–116) have described uses of glassy carbon. Some of its cited advantages relative to platinum are: (*a*) low cost; (*b*) pretreatment by polishing with metallographic paper; (*c*) larger overpotential for production of hydrogen and dissolved oxygen; and (*d*) increased reversibility for several redox couples and reactions which involve subsequent proton transfer. Relative to platinum the disadvantages are: (*a*) a high residual current in 1 F sulfuric acid, and (*b*) surface roughening as a result of recrystallization at high current densities.

Glassy carbon has a unique advantage in the determination of trace metals by stripping voltammetry. In one procedure (116) the trace metals are codeposited with mercury after addition of mercuric nitrate to the supporting electrolyte. As a result the trace metals are plated out into an extremely thin layer of mercury, and in the subsequent stripping step a single well resolved peak is obtained for each metal. After a determination the surface can be renewed by wiping with a cellulose tissue to remove all of the mercury.

Wax-Impregnated Graphite. Spectroscopic grade graphite is porous and penetration of the electrode by solution or by oxygen makes it unsuitable for use in voltammetry. Impregnation of the graphite under vacuum by molten ceresin wax or by paraffin gives much more satisfactory reproducibility. The surface is easily renewed by light sanding with fine sandpaper. The reproducibility in aqueous solution is improved by dipping the electrode in a surface active agent (0.001% Triton X-100), which increases the wettability of the electrode surface by the solution (117).

Pyrolytic Graphite. Pyrolytic graphite is produced by pyrolysis of hydrocarbons under reduced pressure to give a deposit of highly ordered carbon crystallites on a substrate maintained at 1000 to 2500°C. The graphite so formed is anisotropic with the planes of the hexagonal graphite rings parallel to the surface of the substrate. Pyrolytic graphite is highly impervious to liquids and gases, inert to chemical attack, and free of entrapped gases and metallic contaminants. Although relatively thin samples of pyrolytic graphite can be prepared, most of the data that have been reported were obtained with commercially available pyrolytic graphite in a thick coherent form prepared at about 2000°C. In fabricating an electrode, care must be taken to insure that only the planes parallel to the *ab* crystallographic axes are exposed. The edges of the planes are more vulnerable to penetration by solution, much as graphite behaves in the formation of intercalation compounds by diffusion of atoms between the planes of the graphite. The surface of pyrolytic graphite is renewed by light sanding or by cleavage. The fabrication and use of a pyrolytic graphite electrode for voltammetry in aqueous solution have been discussed (118); the behavior of a homemade pyrolytic carbon film electrode has been compared with that of wax-impregnated graphite (119). The reproducibility is as good and the carbon film electrode exhibits nearly reversible behavior for the hexacyanoferrate(III/II) couple.

At the low current densities (0.03 to 0.3 $\mu A/cm^2$) that are used in anodic stripping voltammetry, the reproducibility of pyrolytic graphite electrodes is improved by impregnation with ceresin wax (120). At higher current densities (30 $\mu A/cm^2$), wax impregnation is not necessary. Experience indicates that the wax-impregnated pyrolytic graphite electrode gives superior reproducibility relative to wax-impregnated spectroscopic graphite.

Because the residual current largely is a function of the differential capacitance of the electrode in solution, information about the latter is of practical significance. A differential capacitance study of the basal plane of stress annealed pyrolytic graphite indicates a surprisingly low minimum value of 3 $\mu F/cm^2$ which is attributed to a space charge component within the graphite (121). In contrast, for the edge orientation of pyrolytic graphite, the minimum capacitance value is 50 $\mu F/cm^2$; the value for glassy carbon is 13 $\mu F/cm^2$. The capacitance associated with the compact double layer should have a value of 15 to 20 $\mu F/cm^2$ in the absence of surface states on the basis of the values obtained with metal electrodes.

Dispersed Graphite Electrodes. A number of carbon electrodes have been prepared by dispersing graphite in ceresin wax (122), epoxy polymers (123), silicone rubber (124), or Nujol (125). The Nujol dispersion forms a thick paste and has been widely used by Adams and coworkers (126) who call it the carbon paste electrode (CPE). It is easy to prepare, shows fair reproducibility ($\pm 5\%$), and has a low residual current in the anodic region with a good anodic potential range. In this respect it is superior to platinum and gold because it shows none of the troublesome oxide film formation. In the negative region it shows a persistant residual current which Adams has attributed to the presence of oxygen dissolved in the paste or adsorbed on the surface of the graphite particles. The surface is renewed by removing about 3 mm of the paste and replacing it with fresh paste.

The carbon paste electrode prepared in the usual way tends to disintegrate in nonaqueous solvents. The addition of sodium lauryl sulfate (127) to the paste prevents wetting of the graphite by acetonitrile, nitromethane, or propylene carbonate; this makes the electrode more suitable for use in nonaqueous solvents.

A dispersion of graphite in ceresin wax has been suggested as a way of preparing an electrode suitable for use in nonaqueous solvents. The hot paste is tamped into a Teflon tube and allowed to solidify. No pretreatment is necessary and the surface can be renewed by wiping with cellulose tissue.

The use of a compressed acetylene-black electrode for coulometric analysis of adsorbed organic compounds has been proposed (128). A 1.2-cm-diameter glass tube that is closed with a frit is mounted in a rubber stopper and inserted into a vacuum flask. Next, about 0.1 g of acetylene black is added and tamped firmly; then a test solution containing about 0.5 mg of sample is added and drawn through the acetylene black at a controlled flow rate such that the compound to be determined is adsorbed quantitatively. A graphite collector electrode, which carries the current to the compressed acetylene black, is then introduced and held in position with a weight. A small amount of electrolyte is added above the collector, and several drops are drawn through the acetylene black. The electrode then is transferred to an electrolysis cell and a constant-current electrolysis is performed (typically with a 2-mA current); the potential between the graphite electrode and a calomel reference electrode is recorded. The potential-time curve usually shows a substantial break which indicates the coulometric endpoint. The utility of this technique appears to be greatest for semimicroanalysis of electroactive organic compounds.

The Carbon Cloth Electrode. When heated to high temperatures certain woven hydrocarbon polymer fabrics can be converted to carbon or graphite cloth that appears to have a voltage range in various solvents similar to that of other carbon electrodes. The use of this material (available from Carbon Products Division of Union Carbide Corporation) for anodic and cathodic work in aqueous and nonaqueous media has been reviewed (129). The electrochemical area of such electrodes is about three times their geometric area. The material has good mechanical properties and withstands current densities of 100 mA/cm^2. At current densities tenfold higher, the resistance of the cloth, normally just a few ohms per square centimeter, leads to uneven current distribution and poor potential control. This material appears to have its greatest utility in preparative scale electrosynthesis; it is not particularly suited to routine electroanalytical analysis. The cloth is sufficiently inexpensive that one can simply cut out a piece of the size needed and discard it after the electrolysis.

4. Metal and Semiconductor Materials (Borides, Carbides, Nitrides, and Silicides)

The use of boron carbide as an electrode material has been proposed (130), but it does not appear to have any advantage over vitreous carbon or pyrolytic graphite electrodes. In their comprehensive study of a range of borides, carbides, and nitrides as potentiometric indicator electrodes, Weser and Pungor (131) observed that most of the good indicator electrodes possess face-centered cubic crystal structures like gold and platinum. Although borides and silicides have low symmetry crystal structures and are extremely inert chemically, they are not suitable for redox electrodes.

Lanthanum hexaboride has been used as an electrode material and has a positive potential range that is about the same as that for mercury (132); the material displays the interesting possibility of generating La(III) with nearly 100% current efficiency.

Tin oxide coated glass has been used as an electrode material in electrochemical spectroscopy (133). By doping the tin oxide with antimony an n-type semiconductor is formed. The surface is chemically inert and is transparent in the visible region of the spectrum. However, it is more useful for its optical transparency than as an electrode material (see Section IIIE).

The current-voltage curves of semiconductor type electrodes deviate from the behavior of metals, mainly because of the internal resistance and nonuniform internal potential distribution of the electrode. Unlike solution resistance, it is not possible to compensate electronically for this deviation by positive feedback.

A comparative study of a number of metal and semiconductor materials as voltammetric electrodes has been made recently (134). All of the electrodes were examined in a similar manner and their characteristics recorded in terms of: (a) available positive potential limit versus SCE; (b) extent and reproducibility of residual currents; and (c) the reproducibility of

E_{peak} and i_{peak} using the hexacyanoferrate(II) ion as a model compound for the study of one-electron reversible oxidation. The results for a number of electrodes are summarized in Table 2-7.

Table 2-7   Anodic Potential Range of Electrode and Peak Potential for Oxidation of $Fe(CN)_6^{4-}$

| Electrode Material | Anodic Potential Range in Aqueous Solution (V vs. SCE)[b] | | | Peak Potential for Oxidation of $10^{-3} F$ $Fe(CN)_6^{4-}$ | |
	pH 1.0	pH 4.2	pH 10.0	pH 1.0	pH 4.2
Pt	1.1	1.1	0.9	0.35	0.23
C (vitreous)	1.3	1.4	0.95	0.33	0.28
C (pyrolytic graphite)	1.3	1.3	0.9	0.35	0.27
Pt + 10% Rh	1.4	1.1	0.6		0.26
Pd	1.2	1.0	0.15		0.47
Os	0.55	0.55	0.4	0.40	0.27
Ir	0.6	0.5–0.8	0.55		0.25
Re	0.6	0.4	0.1		0.37
Zr	1.5	1.3		NW	NW
Mo	0.2	−0.2	−0.1	NW	NW
Nb	NW[a]	NW	NW	NW	NW
Ta	NW	1.4	NW	NW	NW
Cr	0.9	0.7	0.35	0.34	0.27
Ti	1.6–1.9			0.45	NW
B_4C_3	1.3	0.9	0.8	0.35	0.27
TaC	0.5	0.4	0.1	0.45	
WC	0.95	0.8	0.7	0.47	

From Ref. 134.

[a]NW = poorly defined voltammogram.

[b]Anodic limit is defined as the potential at which the current becomes equal to one-half of the value of the peak current for the oxidation of $10^{-3} F$ $Fe(CN)_6^{4-}$.

The results obtained indicate that with the possible exception of chromium and tungsten carbide, none of the materials studied is comparable to vitreous carbon. There appears to be no reason why chromium cannot be used in place of platinum for measurements in the near-positive potential range; the voltammograms that are obtained are well-defined and reproducible. Despite having the largest positive potential limit, titanium is of little use as an electrode material; voltammograms are irreproducible and seriously influenced by the previous history of the electrode. Zirconium, tantalum, niobium, and molybdenum exhibit classic features of passivation; oxidation waves for Fe(II) are not obtained and the residual-current curves have significant slopes and are erratic.

Lead has been much used as a cathode material in organic electrosynthesis (135); it has a high hydrogen overvoltage and is easy to work mechanically. However, it does not appear to be as useful as vitreous carbon or platinum for general voltammetric work.

B. MEASUREMENT OF ELECTRODE AREA

Because of surface roughness, the real or true surface area of a solid electrode is greater than the projected or geometric area. However, if the electrode is polished to a smooth surface finish, this will be of no consequence in most voltammetric work. The depth of the depleted region around the electrode surface (the diffusion-layer thickness) is substantially larger than the characteristic dimensions of surface roughness for electrolysis times that are greater than 1 sec. (The diffusion-layer thickness may be crudely approximated by the term $(Dt)^{1/2}$, where D is the diffusion coefficient (cm^2/sec) and t is the time.)

However, in an adsorption measurement (where surface coverage is limited to a monolayer) or where the surface is rough (e.g., a platinized platinum electrode), an approximation to the true surface area may be needed. For a platinum electrode this can be obtained by measuring the amount of charge required to deposit a monolayer of adsorbed hydrogen atoms. Integration of the current-time curve for the hydrogen adsorption peaks (H_{C1} and H_{C2} of Figure 2-23) gives the integral shown as Q_H in Figure 2-24. The integration is carried out from about $+0.4$ to $+0.05$ V versus a hydrogen electrode. in 0.5 F H_2SO_4. The total number of coulombs is corrected for charging of the double layer. The double-layer charging current is assumed to be approximately constant and equal to the current measured at the minimum (about $+0.5$ V versus H in Figure 2-23) of the double-layer charging region. A monolayer of hydrogen corresponds to about $120\,\mu C/cm^2$ of real surface area (after correction for double-layer charging). The area also can be estimated by use of a constant current method, or by measurement of the coulombs required to oxidize a monolayer of adsorbed hydrogen (H_{A1}, H_{A2}, H_{A3} of Figure 2-23), or by measurement of the coulombs required to deposit a monolayer of oxygen (O_{A1}-O_{A2}, O_{A3} of Figure 2-23). The accuracy of these surface area measurements is such that the error limits probably are ± 10 to 20%. The relative merits of the methods have been discussed (49, 136).

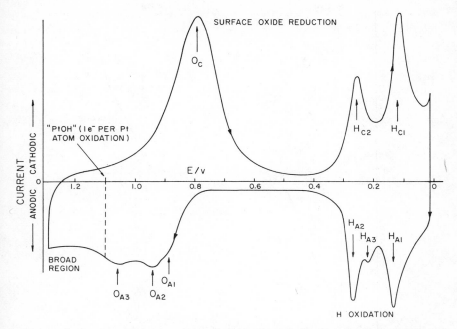

Figure 2-23. Cyclic voltammogram for a platinum electrode. Regions of oxide formation and reduction indicated as well as formation of H_2 (and atomic hydrogen) and its oxidation.

The area of a polished electrode (taken to be the projected or geometric area in most voltammetric experiments at times greater than 1 sec) usually is measured directly or electrochemically. If the electrode is of regular geometry, such as a disc, sphere, or wire of uniform diameter, its characteristic dimensions can be measured by use of a micrometer, optical comparator, or travelling microscope and the area calculated.

In practice two methods are used for stationary planar electrodes in quiescent solution—chronoamperometry and chronopotentiometry. By use of an electroactive species whose concentration, diffusion coefficient, and n value are known, the electrode area can be calculated from the experimental data. In chronoamperometry, the potential is stepped from a value where no reaction takes place to a value which ensures that the concentration of reactant species will be maintained at essentially zero concentration at the electrode surface. Under conditions of linear diffusion to a planar electrode the current is given by the Cottrell equation

$$i = nFAC \left[\frac{D}{\pi t} \right]^{1/2} \tag{2-58}$$

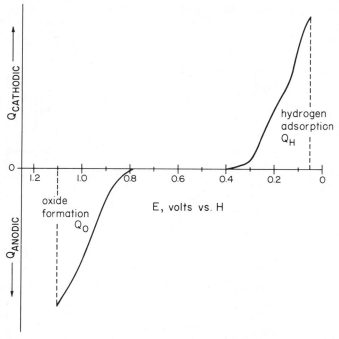

Figure 2-24. Current-time integral as a function of potential for a platinum electrode in $0.5\,F\,H_2SO_4$.

where i is the instantaneous current in amperes; F the faraday (96,490 C); n the number of electrons in the overall reaction; A the electrode area in cm^2; C the concentration in moles-cm^{-3}; D the diffusion coefficient in cm^2/sec; and t the time in seconds. The product $it^{1/2}$ should remain essentially constant at a shielded planar electrode for electrolysis periods from 1 to 30 sec or more. By use of a servorecorder accurate data usually can be obtained for times between 10 and 30 sec.

In the chronopotentiometric method the transition time is measured for a constant current and the electrode area is calculated by means of Equation 8-1. This equation is valid only for semiinfinite linear diffusion to a shielded planar electrode; the effects of electrode geometry (upward or downward diffusion) and electrode shielding on the transition time have been discussed (137). Lingane has described correction terms that can be applied to Equation 8-1 for diffusion to an unshielded planar electrode (138).

The step-potential method has an advantage over the constant-current method because the double-layer charging takes place very rapidly at the beginning of the measurement and afterward is negligible. Potential changes in the constant-current method cause the flow of double-layer charging current to change during the entire experiment, particularly at the transition time (137).

Both of these methods may be applied to nonplanar electrodes if the results are obtained at electrolysis times sufficiently short that the diffusion layer remains thin in comparison to the radius of curvature of the nonplanar electrode surface. For example, the spherical hanging mercury drop

Table 2-8   Selected Diffusion Coefficients at 25° in Water and Acetonitrile

Substance	Concentration (mM)	Solvent[a] Medium	$D \times 10^5 (cm^2/sec)$
Ag^+	2.5–4.0	0.1 F KNO_3	1.55 ± 0.02
$Fe(CN)_6^{3-}$	4.0	0.1 F KCl	0.762 ± 0.01
$Fe(CN)_6^{3-}$	4.0	1.0 F KCl	0.763 ± 0.01
$Fe(CN)_6^{4-}$	4.0	0.1 F KCl	0.650 ± 0.02
$Fe(CN)_6^{4-}$	4.0	1.0 F KCl	0.632 ± 0.02
Cd^{2+}	2.5	0.1 F KNO_3	0.69 ± 0.02
Cd^{2+}	2.5	1.0 F KNO_3	0.68 ± 0.02
Cd^{2+}	2.5	0.1 F KCl	0.716 ± 0.01
Cd^{2+}		0.1 F KCl	0.70 ± 0.013^b
Cd^{2+}	2.5	1.0 F KCl	0.80 ± 0.02
Cd metal		Mercury	1.6^c
Aniline	1–2	0.1 F TEAP[d] in acetonitrile	2.60 ± 0.15
Naphthalene	1–2	0.1 F TEAP[d] in acetonitrile	2.68 ± 0.1
Biphenyl	1–2	0.1 F TEAP[d] in acetonitrile	2.33 ± 0.2
9,10-Dihydro-9,10-dimethylphenazine	1–2	0.1 F TEAP[d] in acetonitrile	2.00 ± 0.05
Perylene	1–2	0.1 F TEAP[d] in acetonitrile	2.59 ± 0.15
9,10-Diphenylanthracene	1–2	0.1 F TEAP[d] in acetonitrile	1.97 ± 0.1

From Ref. 139 unless otherwise indicated.
[a] Solvent is water unless otherwise indicated.
[b] Weighted "best value"; Macero and Rulfs, *J. Electroanal. Chem.*, **7**, 328 (1964).
[c] Stevens and Shain, *Anal. Chem.*, **38**, 865 (1966).
[d] TEAP = tetraethylammonium perchlorate; average D calculated from data given by Bacon and Adams, *Anal. Chem.*, **42**, 524 (1970).

electrode (see Section IIID) provides chronoamperometric data that deviate less than 1 to 2% from the linear-diffusion Cottrell equation out to times of about 1 sec. With solid wire electrodes of cylindrical geometry, similar conclusions apply, but at short times surface roughness effects yield a real surface area that is larger than the geometric area.

Table 2-8 gives values for diffusion coefficients which can be used for determination of electrode areas (139). These values apply only to the specified temperature and supporting electrolyte composition. A critical evaluation of the measurement of electrode surface areas by the chronoamperometric method has been presented (140).

C.   ELECTRODE PRETREATMENT

There is ample evidence that the rate of electron transfer at a solid electrode is sensitive to the surface state and previous history of the electrode. An electrode surface that is not clean usually will manifest itself in a voltage sweep experiment to give a decrease in the peak current and a shift in the peak potential. Various pretreatment methods have been employed to clean or "activate" the surface of electrodes; the process is intended to produce an enhancement of the reversibility of the reaction (i.e., produce a greater rate of electron transfer) (141). This activation or cleaning process may function in two ways—by removing adsorbed materials which inhibit electron transfer and by altering the microstructure of the electrode surface.

The pretreatment process ordinarily begins by polishing the electrode surface until it is smooth and bright. Techniques similar to those used in the preparation of metallographic specimens can be used; the electrode is polished with successively finer grades of alumina, silicon carbide, or diamond dust. Simple polishing is often adequate for work in dipolar aprotic solvents, where adsorption may be less of a problem than in aqueous solutions. After thorough rinsing and drying with cellulose tissue, the electrode is ready for use. The polishing ordinarily is repeated daily or before every experiment.

More stringent cleaning is used in aqueous solutions. Adams recommends a chemical pretreatment with "cleaning solution" (chromic-sulfuric acid mixture) to oxidize the surface heavily, followed by thorough rinsing (142). Others recommend treatment with hot nitric acid (platinum and gold) or aqua regia (platinum only) to accomplish the same thing—the oxidation of the electrode surface to a reproducible state. After rinsing, the electrode is placed in the test solution and held at a potential that will reduce the oxide layer. This generally produces a clean and reproducible surface.

Sometimes the same result can be attained by heavily anodizing (e.g., 1 sec at 10 to 100 mA/cm^2) the electrode surface to remove adsorbed material.

The oxide layer then is reduced at a potential sufficient to remove the oxide layer but not reduce hydrogen ion or solvent. The pretreatment cycle is repeated before every experiment, usually in the test solution. For voltammetric work in aqueous solution, chemical pretreatment and electrochemical cycling is preferred to simple polishing or in addition to polishing.

The current-potential behavior shown in Figure 2-23 is claimed to be characteristic of a clean platinum surface in a clean test solution (143), and can be used as a criterion of solution and electrode cleanliness in aqueous 0.5 F H_2SO_4. The presence of organic material generally will cause a decrease in the hydrogen adsorption peaks and the appearance of new peaks (see Figure 4-4).

Less is known about electron-transfer kinetics on carbon electrodes, but one study indicates that electron transfer in the Fe(II)/Fe(III) system is slower on carbon electrodes than on platinum; the order of rate constants is $k_s(Pt) > k_s$ (glassy carbon) $> k_s$ (carbon paste) (144). The maximum rates are observed after treatment with chromic acid, except for the reduction of iodate ion where simple polishing gives the highest rate. The potentiometric response of some carbon electrodes after various pretreatments has been studied (145). In addition, the surfaces were examined by infrared spectroscopy, scanning electron microscopy, and direct titration of surface groups. The results indicate that the glassy carbon electrode has a negligible concentration of surface acid-base groups and comes closest to an ideal inert redox electrode.

D. CONSTRUCTION AND MASS TRANSPORT PROPERTIES OF VOLTAMMETRIC ELECTRODES

1. Mercury Electrodes in Unstirred Solution

Mercury is a widely used electrode material for the study of cathodic processes because of its high overpotential for hydrogen ion reduction. Its clean liquid surface eliminates the problems that are associated with solid electrodes (which must be polished or cleaned by various chemical pretreatments).

The purification of mercury and tests for its purity have been reviewed extensively (146, 147). Some of this work is summarized here because of its importance to electroanalytical methods that employ mercury. The normal impurities of mercury can be divided into three classes: (a) surface scum and oxides that can be removed by filtering the mercury through a pinhole; (b) dissolved base metals (such as zinc and cadmium) which are easily removed by chemical treatment; and (c) traces of dissolved noble metals which are not readily detectable. The latter often are not serious contaminants and can be removed by distillation of the mercury.

The usual methods of purification attempt to remove surface scum, base metals, and noble metals in turn. The mercury is first filtered through perforated filter paper ("pinholed"), and then treated by agitating under 2 F nitric acid by means of a stream of air drawn through it by a vacuum aspirator. The whole apparatus should be contained in a large polyethyelene basin in a hood to contain any spills, and a trap should be inserted between the flask and aspirator to catch any mercury that bumps over. The process can be carried out over a period of 1 to 3 days. As the metals are gradually removed, hollow bubbles of mercury (as large as 2 to 3 cm in diameter that float in the solution for a few seconds before collapsing) will signal that the base metals have been removed. If the mercury does not have a significant noble metal content, it probably is pure enough for general use at this stage, and will need only to be washed with water, dried, and pinholed. Mercury purified in this way contains less than a part per billion of zinc (often present at much higher concentrations in commercial distilled mercury), and also is free of other base metals. However, this procedure will not remove platinum or gold and even silver may be removed only in part. When the mercury contains these metals, it either should be distilled or exchanged for mercury that has been distilled.

The presence of noble metals in mercury will lead to a reduced overpotential for the evolution of hydrogen and should be reduced to trace levels (below 0.1 ppm). This is usually accomplished by vacuum distillation or by distillation in the presence of a stream of air. Base metals should be removed by chemical treatment before distillation and the still that is employed should be reserved for purifying mercury for electrochemical use. The use of commercial laboratory mercury stills is not recommended because they tend to become, in the course of time, reservoirs of impurity. Many still designs have been described in the literature, but those that employ surface evaporation avoid the problem of "bumping" which carries over impurities into the final product. The Bethlehem Apparatus Company, Inc., offers a cleaning service that employs three successive vacuum distillations of the mercury.

The individual who has occasional need for pure mercury is advised to obtain about 5 kg of mercury that is free of noble metals and further purify it, if necessary, to remove base metals. This mercury should then be reserved for electroanalytical work, and can be recovered from one experiment for use in the next. Redistillation of the mercury, once it has been freed of noble metals should not be necessary.

Pure mercury should retain a bright mirror surface indefinitely in contact with dry air and should not leave a ring if stored in a glass vessel. In fact the appearance of the mercury is a more sensitive test for base metals than is emission spectroscopy. Their presence at a concentration of 1 ppm is indicated by either the formation of a film or "tailing" of the mercury when it is shaken in a glass flask or rolled around in a clean porcelain dish (146). A simple test for purity consists of placing a few millimeters of mercury in a clean stoppered vial or flask along with about three times the volume of pure distilled water (148). When shaken vigorously, pure mercury will form a fairly stable foam which disappears gradually in from 5 to 15 sec. In contrast, mercury that contains a substantial amount of base metals does not foam at all and mercury that contains mere traces of base metals will form a foam that is stable for 1 or 2 sec and then collapse suddenly. For instance, an amalgam of 1 mg of copper per 1 kg of mercury will fail to give a stable foam when shaken with distilled water. The water that is used for the test must not contain any organic material and the container for the test must be scrupulously clean.

Although the preceding methods are satisfactory for purifying small quantities of mercury, electrolytic methods also are effective and convenient, and provide a purity equal to the purest reagent grade mercury (149). For purifying larger quantities (10 kg) of mercury, the Bethlehem mercury oxifier is recommended. It agitates the mercury in contact with air and will oxidize all of the base metal impurities in a period of 72 hr.

Mercury is toxic and every effort should be made to contain spills by placing trays under the apparatus. If a spill occurs, it should be immediately cleaned up with a suction apparatus or a "mercury magnet," which is a spiral of copper wire that has been amalgamated. Dry sweeping should be avoided because it disperses the mercury. It is almost inevitable that tiny droplets of mercury will become scattered about in a polarographic laboratory and these are almost impossible to clean up. The best insurance against mercury poisoning is a good ventilation system which changes the air several times per hour. Work with mercury should never be carried out in a closed, unventilated room. The greatest hazard exists when mercury is being distilled; stills should be located in an especially well-ventilated area. If reasonable precautions are observed there should be no significant hazard from work in a polarographic laboratory (150,151).

The disposal of mercury wastes is of some concern. Rather than being poured down the drain, soluble mercury salts can be reduced with sodium borohydride and the metallic mercury recovered and reused. Used metallic mercury can be stored under water in a tight container, preferably polyethylene to minimize the hazard of breakage. When a sufficient quantity has been accumulated it can be repurified.

Mercury most often is used in the form of a dropping electrode (Figure 2-25), a hanging mercury drop (Figure 2-26), or a mercury pool (Figure 2-27). It also may be plated on platinum to give an electrode with properties intermediate between those of platinum and mercury.

The Dropping Mercury Electrode. The usefulness of the dropping mercury

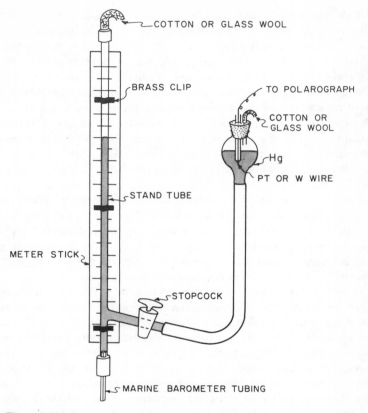

Figure 2-25. Dropping mercury electrode assembly.

electrode (DME) for analytical voltammetry was discovered by Heyrovsky, and the history of this discovery has been recounted recently (152). The DME usually is prepared from a 10 to 20-cm length of glass capillary tubing with an approximately 0.05-mm internal diameter (Corning marine barometer tubing frequently has the requisite diameter). The length of the capillary is selected so that a convenient head of mercury above the capillary will provide a drop time of 3 to 10 sec; such capillaries can be purchased from suppliers of polarographic instrumentation. Figure 2-25 illustrates a convenient arrangement for the DME. The meter stick adjacent to the stand tube is used to measure the height of the mercury column. The connections to the reservoir and capillary tubing can be made with neoprene rubber or Tygon tubing. They usually are wired on for safety.

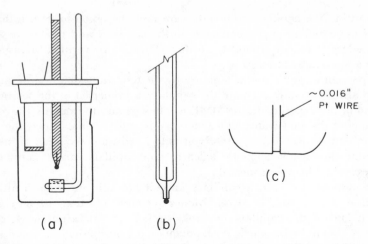

Figure 2-26. Hanging mercury drop electrode. (*a*) Electrode-cell assembly with scoop to transfer mercury drops from DME to the electrode (*b*) and (*c*) details of construction for amalgamated platinum electrode to which mercury drops are attached.

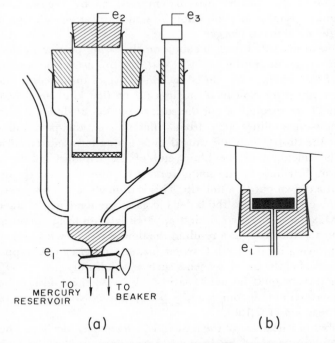

Figure 2-27. Mercury pool electrodes. (*a*) Cell with provision for pool replacement and (*b*) plastic cup for pool. e_1, working electrode; e_2, auxiliary electrode; and e_3, reference electrode.

In cutting the capillary to length, care must be taken to ensure that the end of the capillary is perpendicular to the capillary and that there are no rough edges or cracks around the orifice. This can be verified by inspection under a low-power microscope.

To prevent clogging of the capillary by the test solution, a flow of mercury should always be started before the capillary is immersed in the solution. At the end of the experiment, the DME is removed and thoroughly rinsed with solvent and distilled water while mercury is still flowing. When the water has evaporated, the mercury reservoir is lowered until the head of mercury above the capillary is below the sidearm. The capillary can be stored dry or in a test tube filled with mercury.

It is difficult to clean a capillary once it has become clogged although procedures are available (153). Meites recommends that 10 mm of the clogged end of the capillary be broken off. If this does not work, a new capillary should be installed. With adequate care, experienced workers have kept a given capillary in use for several years without any problems of clogging, particularly if the capillary is used daily. Most problems arise from improper rinsing or from infrequent use of the capillary. Clogging is readily detected by refusal of the mercury to flow or by ragged and erratic polarograms. Visual inspection under low magnification will usually reveal whether the capillary is clogged.

The appearance of current maxima on the rising portion of the current-voltage curve should not be confused with the problems that are created by clogging. Such maxima are caused by convective disturbances at the surface of the mercury drop. So-called maxima of the first kind involve streaming that results from inequalities of the surface tension at different parts of the drop. These inequalities arise from differences in the potential over the surface of the drop which are caused by a nonuniform current distribution (154). The shielding of the blunt end of the capillary accentuates this nonuniform distribution, and some workers recommend that the end of the capillary be drawn out to a fine tip. This is difficult to do, however, without making the orifice smaller; the latter increases the tendency of the capillary to clog. Maxima of the second kind apparently arise from vortex motion of the mercury in the drop (as a result of the flow of mercury from the capillary into the growing drop) (155). Current maxima often can be suppressed by the addition of surface active agents such as gelatin or Triton X-100 (156). However, these should be used sparingly because their adsorption on the electrode can eliminate completely a polarographic wave by inhibiting the electrode reaction (157,158).

A number of variations of the basic DME have been devised. The vertical orifice capillary, formed by bending a capillary into a right angle and cutting off the capillary near the bend, is claimed to eliminate problems of current maxima and depletion effects (growth of a drop in a solution depleted of electroactive material by the preceding drop), and to provide greater uniformity from drop to drop (159). The fabrication of a Teflon dropping mercury electrode for use in solutions of HF has been described (160). A number of mechanical devices have been developed that regulate the drop time by dislodging the drop at precisely controlled intervals (usually by a sharp rap on the capillary) (161,162). A regulated drop time often is used in experiments to synchronize current measurements with drop growth (see Chapter 5) or to synchronize electrodes in two different cells, as in differential or subtractive polarography (163).

The Hanging Mercury Drop Electrode. A number of techniques require a stationary electrode of fixed size—for example, chronocoulometry, anodic stripping voltammetry, and voltammetry with positive feedback iR compensation. This need is fulfilled, while still retaining the desirable properties of a mercury electrode, by the hanging mercury drop electrode (HMDE) shown in Figure 2-26. The electrode can be fabricated by sealing a small-diameter platinum wire in soft glass and etching the platinum wire back from the end in aqua regia. The wire then is amalgamated and one or more mercury drops are attached to its end by means of a scoop (Figure 2-26*a*). The latter is used to catch drops from a DME and transfer them to the HMDE (164). Kemula (165) uses a different style of hanging drop in which the drop hangs from a glass capillary tube connected to a reservoir containing mercury. A threaded screw advancing into the reservoir displaces mercury to form a drop on the end of the capillary. A more elaborate commercial version of this electrode is available (Metrohm).

The Mercury Pool Electrode. Mercury pools of sufficient diameter to approach a planar configuration obey the equations derived for linear diffusion to a planar electrode. This has certain theoretical advantages because of the large number of equations that have been derived for the planar electrode geometry, especially in terms of constant-current chronopotentiometry and linear-potential sweep chronoamperometry.

Compared to the dropping mercury electrode, the mercury pool has an enhanced analytical sensitivity; the ratio of the voltammetric peak current to the residual (charging) current is approximately 10 times as great for a mercury pool (166,167). In an effort to eliminate the rounding at the edge of the mercury pool that is created by the meniscus, three types of cells have been devised which use platinum rings to flatten the mercury pool (168). Although these are successful in providing a flatter surface, the exposure of platinum to the solution may decrease the overpotential for hydrogen evolution. Figure 2-27 illustrates two mercury pool electrode systems.

Mercury Plated Electrodes. Platinum can be plated with mercury to form an electrode with many of the properties of a mercury surface. This is done by cleaning a platinum electrode in nitric acid, connecting it as the cathode in perchloric acid solution over a pool of mercury, and then dipping it into the mercury while it is still connected in the circuit. The method works well for plating a planar platinum inlay electrode to give a flat mercury-coated surface, but it can be used for electrodes of any shape (169).

There is much evidence that mercury forms intermetallic compounds with platinum at the mercury-platinum interface. Apparently these do not diffuse to the surface of the mercury so that it retains many of the desired properties of mercury. However there is a slight reduction in the overpotential for hydrogen evolution at a plated electrode in comparison to a pure mercury electrode (170).

2. Solid Electrodes in Unstirred Solution

The most useful solid electrodes are platinum, gold, vitreous carbon, and carbon paste. The preferred configuration for theoretical work is a flat planar surface sealed in glass with epoxy or snuggly fitted into a Teflon shroud. The electrodes ordinarily are unshielded, although shielded designs have been described (137). Probably the most popular and widely used platinum electrode for voltammetry is the Beckman Instruments No. 39273 platinum-inlay electrode, which contains a disc of platinum directly sealed into glass. The electrode is easily polished and has a surface area of about 0.25 cm^2.

Platinum electrodes of smaller diameter may be made by sealing platinum wire into soft glass (Corning Code 0120 or 0088). The technique for doing this is shown by the sequence of drawings in Figure 2-28. A thin capillary of soft glass is first fused onto the platinum wire (avoid overheating which forms bubbles in the glass or melts the platinum); the glass bead that is formed on the wire is then sealed into a necked down soft glass tube. A planar voltammetric electrode may be made by cutting off the wire and polishing the platinum surface flush with the glass tube. Or a small sphere may be melted on the end of the wire to give a nearly spherical electrode. Glass that contains lead must be worked in an oxidizing flame to avoid reducing the lead (171).

Most other metals cannot be sealed into glass as easily as platinum because there is not a sufficiently good match between the coefficients of expansion of the glass and the metal (see Table 2-9).

For other metals, an electrode of planar configuration can be made by forcing a cylindrical billet of the metal into a slightly undersized hole that has been drilled in a Teflon rod. Shrinkable Teflon tubing also is available, which contracts about the metal wire or billet when it is heated with a hot

Table 2-9 Physical Properties of Selected Metals and Glasses

	Lattice Structure[a]	Density (20°C)	Melting or Softening Point (°C)	Linear Coefficient of Expansion (0–100°C) $\times 10^{-7}$
Ruthenium	HCP	12.45	2310	91
Osmium	HCP	22.61	3050	61
Rhodium	FCC	12.41	1960	83
Iridium	FCC	22.65	2443	68
Palladium	FCC	12.02	1552	111
Platinum	FCC	21.45	1769	91
Silver	FCC	10.5	961	188
Gold	FCC	19.3	1063	143
Pyrex (Corning No. 7740)		2.23	820	32.5[b]
Soft glass, Soda-lime glass (Corning 0088 flint glass)		2.47	700	92[b]
Vycor (Corning 7900)		2.18	1500	8[b]

[a] HCP = hexagonal close packed; FCC = face centered cubic (cubic close packed).
[b] 0–300°C

air dryer. This usually produces a tight fitting seal that is impervious to aqueous solutions.

When the electrode material is too soft or fragile to force into an undersized hole in Teflon, the holder shown in Figure 2-29 can be used. The electrode is inserted into the Teflon collet first. When tightly threaded onto the trunk, the Kel-F chuck squeezes the slip-fit Teflon collet to make a tight seal (172).

Thin-layer electrodes, which most commonly are made of platinum, require care and precision in fabrication (as shown in Figure 2-30). The spacing between the electrode and the precision capillary must be reproducible and is maintained by the precisely machined ridges or steps on the

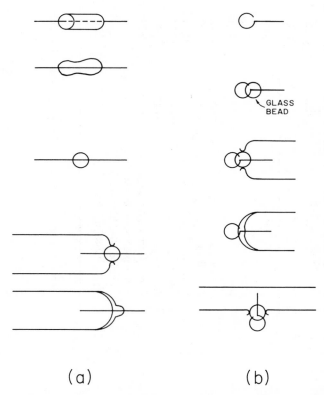

(a)   (b)

Figure 2-28. Techniques for sealing platinum in soft glass tubing. (*a*) Cylindrical wire electrode; (*b*) wire-loop electrode.

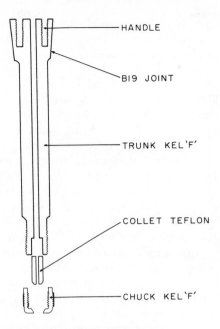

HANDLE

B19 JOINT

TRUNK KEL'F'

COLLET TEFLON

CHUCK KEL'F'

Figure 2-29.   Holder for soft or fragile electrode materials.

platinum rod. The cell is filled and emptied by controlling the pressure above the electrode with the upper Teflon stopcock. Several cycles of filling and emptying the cell are used before each experiment. The theoretical equations that apply to mass transport in a thin-layer cell have been reviewed (173–175). The primary advantage of the thin-layer cell is that it confines the reactant within 10^{-3} cm of the electrode surface so that each reactant particle has immediate access to the electrode surface. This simplifies the mass transport in the cell to the extent that the equations which govern thin-layer electrodes are simple combinations of Faraday's law and the Nernst equation for a reversible system. Surface active impurities are virtually excluded from the electrode surface unless they are present at high concentrations. Because the solution volume is so small, a low concentration of surface active material, even if it collects at the electrode surface, only covers a small fraction of the surface. For these reasons thin-layer cells promise to combine electrochemical techniques with electron and photon spectroscopy. The latter is necessary for studies of the relationship between the electrochemical behavior and the atomic and electronic structure of the electrode surface (175).

A simple thin-layer cell that is made by threading a platinum wire through a Teflon capillary has been described (176), as has a micrometer type thin-layer cell for estimating n values in electrochemical reactions (177) (see Figure 7-7).

Carbon Electrodes   The vitreous carbon electrode usually is fabricated by sealing a plug or disc, cut from a thick sheet of glassy carbon, into a glass tube with epoxy cement. The surface of the electrode is then polished until it is bright and smooth. No further treatment is necessary.

The carbon paste electrode is made by mulling together 15 g of graphite (Acheson, Grade 38 or equivalent) and 9 ml of Nujol until the entire mixture appears uniformly wetted and has the consistency of a stiff paste. The paste is packed into a Teflon cup like that shown in Figure 7-6*b* and smoothed with a spatula. Recently a more elaborate device has been described (178) which is shown in Figure 2-31. The carbon paste is packed into the cylindrical cavity; a new surface can be produced by advancing the threaded rod and slicing off the old surface with a thin wire stretched across a bow (pictured in Figure 2-31*b*). This saves the labor of digging out and repacking the carbon paste when a fresh electrode surface is desired. The thin-wire cutter also appears to produce a more even and reproducible surface. A recent report indicates that heating the graphite under vacuum before mixing with Nujol increases the anodic voltage range (179).

3. Hydrodynamic and Stirred Solution Electrodes

Certain advantages result when the electrode is moved past the solution or vice-versa. The increased mass transport increases the current and often increases the sensitivity (although not necessarily the signal-to-noise ratio). In addition, hydrodynamic electrodes such as the rotating platinum electrode and rotating disc electrode exhibit a current-potential behavior similar to that of the dropping mercury electrode. That is, they give the familiar plateau when the current is limited by mass transport to the electrode surface and the current is proportional to the solution concentration of the electroactive species.

The Vibrating Mercury Electrode.   A dropping mercury electrode which is connected by a shaft to an eccentric that is driven at about 200 Hz has been described (180–182). It produces a drop time of the order of 5 msec. A number of advantages are claimed over the conventional dropping mercury electrode: (*a*) the suppression of maxima without the addition of surfactants; (*b*) the elimination or minimization of kinetic and catalytic currents; and (*c*) the elimination of stirring effects upon the shape of current-voltage curves. The principle drawback to the electrode is that the surface-area increase

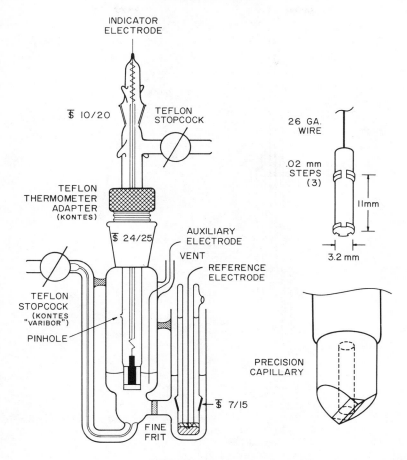

Figure 2-30. Thin-layer electrode assembly. Details of the platinum indicator electrode shown in upper right and of the precision ground glass capillary in lower right.

during a drop life is about 10 times that of the DME for its normal drop time. Therefore the charging current is increased by tenfold to produce a steeply increasing residual-current baseline.

The Rotating Platinum Electrode. Three designs of rotating platinum electrodes are shown in Figure 2-32. The first design (Figure 2-32*a*) produces the smoothest current-voltage curves because the flow is not as turbulent around the electrode surface as for the electrode of Figure 2-32*b*. The electrode of

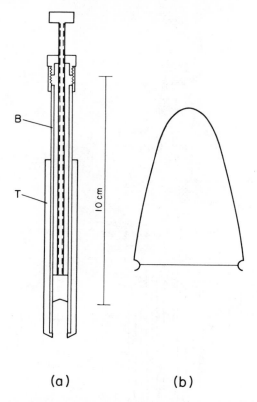

(a)   (b)

Figure 2-31. Carbon paste electrode with threaded screw to advance paste. (a) Electrode assembly with machined cylinder channel (*T*) and threaded piston (*B*); (b) bow with stretched wire for slicing off old surface after the electrode plug is advanced.

Figure 2-32c must be rotated smoothly with little eccentricity or wobble to obtain reproducible currents; this method is not recommended by the authors. The electrodes usually are rotated at about 600 rpm; a rotator designed for this purpose is commercially available (Sargent-Welch). Contact to the platinum wire is made internally by filling the electrode with mercury. A stationary wire dips into the mercury pool at the top to make external contact to the potential control circuitry. The platinum-glass seals are prone to crack, which causes erratic currents that are associated with the leaking of mercury to the electrode surface. Once cracked, the electrodes are not easily repaired and should be discarded. Table 2-10 indicates the dependence of the current on the rotational speed of the electrode.

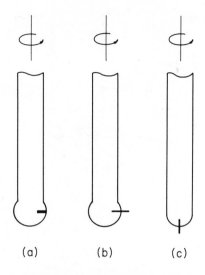

Figure 2-32. Rotating platinum electrodes.

Table 2-10   Limiting Currents at Electrodes in Stirred Solution

Electrode	Expression for the Limiting Current
Rotating platinum electrode (RPE)	$i_L = knA(R)^{0.3-0.6}C$
Rotating disc electrode (RDE)	$i_L = 1.88 \times 10^5 nD^{2/3}\nu^{-1/6}r^2\omega^{1/2}C$
Platinum tubular electrode (PtTE)	$i_L = 5.31 \times 10^5 nD^{2/3}L^{2/3}V_f^{1/3}C$
Multiple mesh flow through electrode (MME)	$i_L = kNV_f^{1/3}C$

i_L = limiting current (A)
n = number of electrons transferred per mole in the overall reaction
D = diffusion coefficient (cm^2/sec)
R = Rotational speed (rps)
ω = angular velocity (rad/sec, $\omega = 2\pi R$)
r = disc radius (cm)
A = electrode area (cm^2)
ν = kinematic viscosity (cm^2/sec) (viscosity/density)
L = length of tubular electrode (cm)
V_f = solution flow rate (cm^3/sec)
N = Number of meshes in the multiple mesh electrode
C = concentration (moles/cm^3)
k = constant

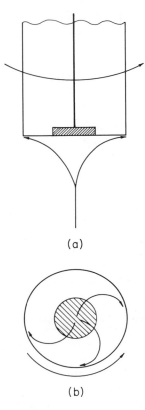

(a)

(b)

Figure 2-33. Rotating disc electrode.

The rotating disc and ring-disc electrodes that are shown in Figures 2-33 and 2-34 are the most useful of the hydrodynamic electrodes, from both the practical and theoretical viewpoints. With such electrodes the equations for mass transport are in close agreement with experimental results. Furthermore, the ring-disc electrode allows the products that are produced at the disc to be detected as they are swept past the ring. The ring and disc are insulated from one another so that their potentials can be controlled independently by means of a circuit such as that of Figure 5-19.

The construction of these electrodes requires precise machining of the disc, the Teflon shroud, and the spacers. Ring-disc electrodes now are available commercially (Pine Instrument Company, Grove City, Penn.), and most workers will prefer to buy them rather than to attempt their fabrication. The expression for the limiting current at a disc electrode is shown in Table 2-10.

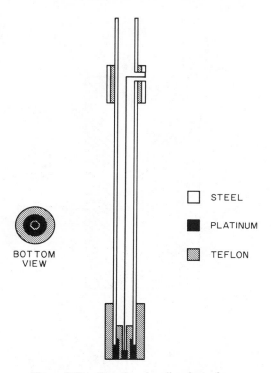

STEEL

PLATINUM

TEFLON

BOTTOM
VIEW

Figure 2-34. Rotating ring-disc electrode.

A small monograph on the theory and applications of the ring-disc electrode recently has been published (183). It confirms that this electrode system is a versatile tool for the study of electrochemical reactions.

4. Flow Electrodes

Rather than move the electrode past the solution, the sample solution can be flowed past a stationary electrode. The tubular platinum electrode (Figure 2-35) and the gold micromesh flow-through electrode (Figure 2-36) are both ingenious attempts to produce electrodes that are useful for the measurement of electroactive materials in a continuously flowing stream. Applications of these electrodes to the analysis of flowing streams have been described (184,185), and their mass transport properties are summarized in Table 2-10. The manner of their construction is fairly obvious from the figures; the references can be consulted for more detail.

The flow electrode of Figure 2-37 is designed for external generation of a coulometric intermediate. This is most useful in situations where the sample

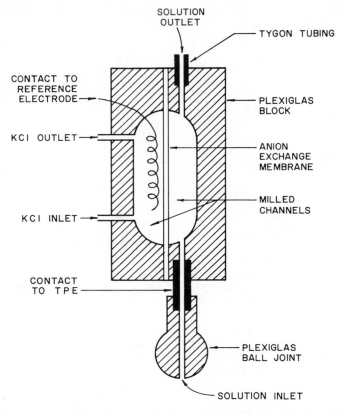

Figure 2-35. Tubular platinum electrode (TPE) for flowing sample solutions.

contains some species which would interfere with the electrode reaction that generates the coulometric reactant. A substantial flow is required through the electrode to sweep the products of the reaction into the sample, and a design which is claimed to be an improvement over that shown in Figure 2-37 has been recently described (186).

E.   OPTICALLY TRANSPARENT ELECTRODES

Several methodologies have been developed to permit spectral observations by use of optically transparent electrodes. An internal reflectance spectrophotometry (IRS) cell system is depicted in Figure 2-38. Such measurements can be made by use of electrodes that are composed of solid

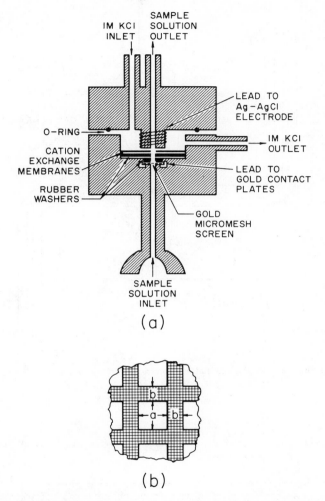

Figure 2-36. Gold micromesh flow-through electrode. (*a*) Cell system; (*b*) detail of gold micromesh screen.

germanium (187,188), or of tin oxide (189,190) or metal films coated on glass (191,192). The electromagnetic radiation strikes the electrode-solution interface at an angle that exceeds the critical angle and penetrates the solution to a depth of less than one wavelength of the light. Therefore only a thin layer of solution next to the metal-film electrode interacts with the beam of light. Hence this type of experiment provides a means to study

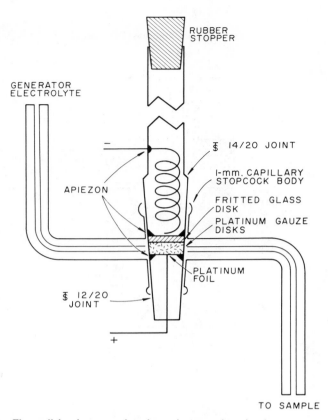

Figure 2-37. Flow cell for the external coulometric generation of a titrant.

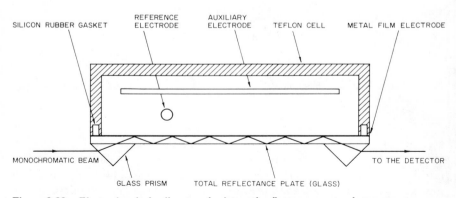

Figure 2-38. Electrochemical cell system for internal reflectance spectrophotometry.

98

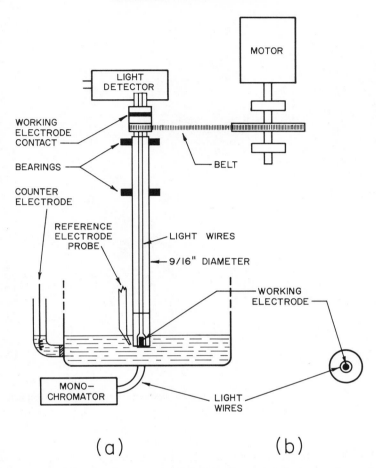

Figure 2-39. Rotating disc electrode surrounded by light wires for spectroelectrochemical measurements.

processes that occur at or near the electrode, with little or no interference from bulk solution components. The provision for multiple reflections enhances the sensitivity of the measurement.

A second technique is the measurement of spectral transmission through optically transparent tin oxide coated on glass or quartz (193, 194), through thin metal films of platinum, gold, or mercury on glass (195, 196), or through gold minigrid electrodes that are produced from thin gold foils by a photoetching process (197, 198). These electrodes are not too difficult to fabricate and provide a way to measure the spectral properties of the

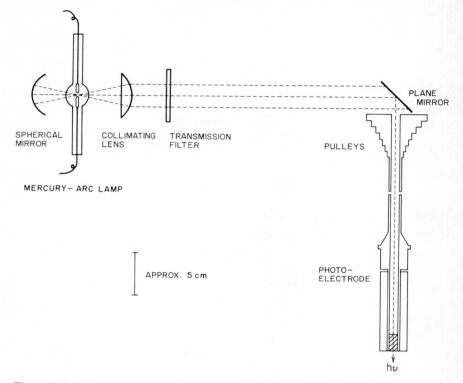

Figure 2-40. Transparent rotating disc electrode system for producing and electrochemically monitoring photo-chemical reaction products.

products of an electrode reaction. The construction of optically transparent platinum and gold electrodes by standard vapor deposition techniques has been discussed (199). These same electrodes also can be plated with thin films of mercury (about 10^{-6} cm) to provide a 0.4-V extension in the negative direction without seriously degrading the optical properties.

Figure 2-39 illustrates a rotating disc electrode that is surrounded by a concentric shell of light wires sealed in epoxy cement. Products produced electrochemically at the disc electrode are swept past the light wires, and their spectrum is recorded by slowly scanning with the monochromator (200).

Figure 2-40 depicts a rotating electrode system that reverses the roles of the light and the electrode from that of preceding assembly. Here an intense beam of light shines through the transparent disc to produce a photochemical reaction. The products of the photochemical reaction are swept past the disc electrode where they may be monitored for their electroactivity (201).

IV. Potentiometric Indicator Electrodes

A. MERCURY INDICATOR ELECTRODES

Although mercury seldom is used as an indicator electrode in redox titrations (because it is so readily oxidized), it is used extensively for potentiometric titrations with complexing agents such as ethylenediaminetetraacetic acid (EDTA). In this application the M^{2+} cation that is to be titrated forms a soluble EDTA complex which is appreciably less stable than the HgEDTA complex (202). The mercury electrode is constructed in a form like that of Figure 2-41a or b, and is used in conjunction with a saturated calomel reference electrode. The electrode is immersed in a solution that contains the ion to be titrated (M^{2+}), and a few drops of 0.01 M HgY^{2-} are added (where Y^{4-} represents the EDTA anion). This establishes the potential of the mercury electrode according to the half-cell reaction

$$Hg^{2+} + 2e^- = Hg \qquad (2\text{-}59)$$

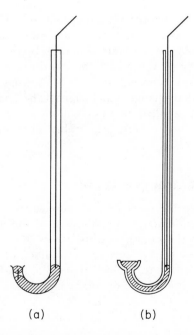

(a)   (b)

Figure 2-41. Potentiometric mercury indicator electrodes.

so that E_{cell} is given by

$$E_{cell} = E_{Hg}{}^0 - E_{sce} + \frac{2.3RT}{2F} \log[Hg^{2+}] \qquad (2\text{-}60)$$

Both the Hg^{2+} and M^{2+} ions form complexes with EDTA by the general reaction

$$M^{2+} + Y^{4-} = MY^{2-} \qquad (2\text{-}61)$$

with equilibrium formation constants

$$K_{Hg} = \frac{[HgY^{2-}]}{[Hg^{2+}][Y^{4-}]} \qquad (2\text{-}62)$$

$$K_M = \frac{[MY^{2-}]}{[M^{2+}][Y^{4-}]}, \qquad K_{Hg} \gg K_M \qquad (2\text{-}63)$$

Substitution of Equations 2-62 and 2-63 in Equation 2-60 yields

$$E_{cell} = E_{Hg}{}^0 - E_{SCE} + \frac{2.3RT}{2F} \log \frac{[HgY^{2-}]K_M[M^{2+}]}{K_{Hg}[MY^{2-}]} \qquad (2\text{-}64)$$

During the course of the titration the concentration of HgY^{2-} remains essentially constant because it is so much more stable than the MY^{2-} complex. Therefore E_{cell} is determined mainly by the ratio $[M^{2+}]/[MY^{2-}]$, which changes slowly in the middle of the titration but rapidly near the equivalence point as the concentration of M^{2+} drops to a small value. This gives a sharp potential change which signals the endpoint. The method is general and can be applied to most cations which form soluble EDTA complexes that are appreciably less stable then the mercury(II)-EDTA complex.

B.   SOLID INDICATOR ELECTRODES

The number of reversible metal/metal ion electrodes is limited so that the accurate direct potentiometric measurement of the activity of a metal ion with an electrode of the same metal usually is not feasible, except perhaps with the Ag/Ag^+ system. However, a number of metal ion-metal half reactions are sufficiently reversible to give a satisfactory potentiometric titration with a precipitating ion or complexing agent. These couples include Cu^{2+}/Cu, Pb^{2+}/Pb, Cd^{2+}/Cd, and Zn^{2+}/Zn. However, all of these metals can be determined by EDTA titration and the mercury electrode that is described in the preceding section.

Platinum and gold often are used as inert redox indicator electrodes in titrations where both the oxidized and reduced species are soluble in solution (e.g., Ce(IV)/Ce(III), Fe(III)/Fe(II), $Cr_2O_7{}^{2-}$/Cr(III), and I_2/I^-). Here the electrode functions as an inert substrate which can mediate electron transfer and respond to the potential determined by the relative concentrations of the soluble redox species. Platinum electrodes are used most widely because the metal is easily sealed into soft glass to make a convenient electrode, and electron transfer reactions seem to be as fast on platinum as any other substrate. When used in this manner, the electrode should have a reasonably large surface area (1 to 2 cm^2). This can be achieved by spot-welding a platinum foil to a short length of sturdy platinum wire that has been sealed into glass tubing. If preferred, a tightly wound spiral of platinum wire can be used to obtain a reasonable surface area. A larger surface area makes the electrode more resistant to polarization, so that redox measurements can be made with a low impedance voltmeter or potentiometer.

1. pH-Sensitive Solid Indicator Electrodes

Although platinum and iridium under certain conditions will respond to changes in pH, they have not been widely used for this purpose and do not appear to be as reliable as the antimony electrode. The latter apparently functions as a pH-sensitive electrode according to the reaction

$$Sb_2O_3 + 6H^+ + 6e^- \rightarrow 2Sb + 3H_2O \qquad (2\text{-}65)$$

The electrode can be made by casting pure antimony in a glass tube, then cracking off the glass tube and remounting the antimony stick in a glass or plastic tube with epoxy cement. The electrode is not as reproducible as a glass electrode, but gives close to a 59-mV per pH unit response at 25°C, and it can be used in slurries or solutions that contain fluoride ion (which would ruin glass electrodes). The electrode is available commercially (Leeds and Northrup) in a form that is intended for industrial applications. It also has been used in a micro form for pH titrations of nanoliter samples of biological fluids. In this application the antimony that is contained in a glass capillary is drawn out with a microelectrode puller into a 5-μ- diameter tip which can be immersed in a 5-nl sample under oil (203).

C.   SELECTIVE ION ELECTRODES

In the past 10 years there has been a renaissance in the development of electrodes which show good selectivity and sensitivity for the measurement of ion activities, and indirectly, the concentration of enzymes, enzyme sub-

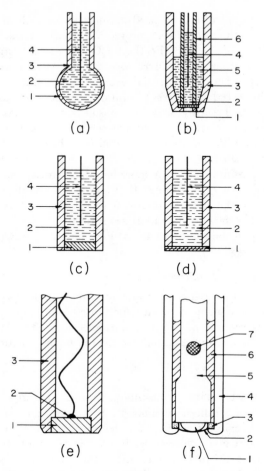

Figure 2-42. Selective ion electrodes. (*a*) Glass membrane; (*b*) liquid ion exchange; (*c*) homogeneous solid membrane; (*d*) heterogeneous solid membrane; (*e*) solid membrane without reference electrode; and (*f*) gas permeable membrane. 1, sensing electrode; 2, electrolyte, 2(*e*) ohmic contact, 2(*f*) gas permeable membrane; 3, membrane support; 4, reference electrode, 4(*f*) outer electrode body; 5(*b*) liquid ion exchanger, 5(*f*) electrode body; 6(*b*) reference electrode body, 6(*f*) electrolyte; and 7, liquid junction.

Table 2-11 Commercial Ion-Selective Electrodes

Electrode Designated For	Type	pH Range	Principal Interferences
Ammonia	Permeable membrane		Volatile amines
Bromide	Solid state	0–14	CN^-, I^-, S^{2-}
Cadmium	Solid state	1–14	Ag^+, Hg^{2+}, Cu^{2+}, Fe^{2+}, Pb^{2+}
Calcium	Liquid ion exchanger	5.5–11	Zn^{2+}, Fe^{2+}, Cu^{2+}, Ni^{2+}
Chloride	Solid state	0–14	S^{2-}, Br^-, I^-, CN^-
Chloride	Liquid ion exchanger	2–10	I^-, NO_3^-, Br^-, HCO_3^-, SO_4^{2-}, F^-
Cupric	Solid state	0–14	S^{2-}, Ag^+, Hg^{2+}, Fe^{3+}
Cyanide	Solid state	0–14	S^{2-}, I^-
Fluoride	Solid state	0–8.5	OH^-
Fluoroborate	Liquid ion exchanger	2–12	NO_3^-, Br^-, OAc^-
Iodide	Solid state	0–14	S^{2-}
Lead	Solid state	2–14	Ag^+, Hg^{2+}, Cu^{2+}; high levels of Cd^{2+}, Fe^{3+}
Nitrate	Liquid ion exchanger	2–12	ClO_4^-, I^-, Br^-, NO_2^-, Cl^-
Perchlorate	Liquid ion exchanger	4–10	I^-, NO_3^-, Br^-
pH (H^+)	Glass	0–14	Na^+ at high pH
Potassium	Liquid ion exchanger		Cs^+, NH_4^+, H^+, Ag^+

Table 2-11(*Continued*)

Electrode Designated For	Type	pH Range	Principal Interferences
Silver/Sulfide	Solid state	0–14	Hg^{2+}
Silver	Glass	4–8	H^+
Sodium	Solid state		Ag^+, H^+
Sodium	Glass	3–12	Ag^+, H^+, Li^+, K^+
Sulfur dioxide	Permeable membrane	1.7	HF, acetic acid, HCl
Thiocyanate	Solid state		S^{2-}, CN^-, $S_2O_3^{2-}$, Cl^-, OH^-
Water hardness (divalent cation)	Liquid ion exchanger	5.5–11	Zn^{2+}, Fe^{2+}, Cu^{2+}, Ni^{2+}, Ba^{2+}

strates, and neutral gaseous molecules like CO_2 and NH_3. Selective ion electrodes can be classified roughly into four types as shown in Figure 2-42. These are the glass electrodes, which can be made with good selectivity for H^+ and other cations (Figure 2-42a); liquid ion exchanger electrodes, which can be made selective for both cations and anions (Figure 2-42b); homogeneous and heterogeneous solid membrane electrodes, which can be made selective for both cations and anions (Figure 2-42c,d,e); and permeable membrane electrodes, which allow diffusion of gaseous molecules (CO_2 and NH_3) that by hydrolysis or some other reaction produce OH^- or H^+ ions. The latter are sensed by a solid-state electrode inside the membrane. Table 2-11 lists a number of commercially available electrodes by type.

The mechanism of operation of some of these electrodes is still imperfectly understood, although great progress has been made by means of theoretical models in recent years. For recent reviews of the theory and application of selective ion electrodes the reader should consult references (19,204,205). One of the interesting sidelights is the fact that the internal reference electrode may be replaced by an apparent ohmic contact in many instances, as illustrated by Figure 2-42e for the solid membrane electrode. Thus the glass electrode can be filled with mercury in place of the internal reference electrode (206) or a gold contact that is plated over with copper can be used (207). Likewise a selective ion electrode for calcium ion has been described which is coated on a platinum electrode (208); the contact appears to be mainly ohmic. This practice has been criticized (209) in the case of the calcium-sensitive electrode that is coated on platinum, with the contention that the internal reference must involve some clearly defined half-reaction. However, this claim seems unwarranted in view of the fact that a number of electrodes operate satisfactorily with an internal ohmic contact.

Durst (210) has discussed the methodology of selective ion electrodes and their many applications. The simplest method of measurement is direct potentiometry by use of the Nernst equation. However this makes extreme demands on the reproducibility of the junction potential and there is the problem of variation of activity with ionic strength. Concentration cell techniques have proved to be very precise, especially in terms of null-point potentiometry. The usual procedure involves adjustment of the composition of one of the half-cell solutions to match the other. This is indicated by a potential null between identical indicator electrodes that are specific for the ion. The method has been employed at the National Bureau of Standards to measure fluoride ion at the level of 0.4 to 190 ng in volumes that range from 5 to 100 μl (211); the relative error is less than 1%.

Calibration techniques that use buffers to adjust the ionic strength and pH of the solution are effective for certain kinds of samples. For example, the detection of fluoride in public water supplies often is carried out by dilution of both standards and samples with a buffer that contains acetic acid, sodium chloride, and sodium citrate (with the pH adjusted to pH 5.0 to 5.5 by use of sodium hydroxide). This buffer performs three functions: (*a*) it fixes the ionic strength of the standards and samples to the same level, principally determined by the buffer; (*b*) the solution is buffered in a region where OH^- ion does not interfere; and (*c*) any Fe^{3+} or Al^{3+} ions are complexed by citrate to release the fluoride ion that is bound by these ions.

The electrodes also can be used effectively to determine the endpoints in potentiometric titrations, a technique that generally provides better accuracy than direct potentiometry. Reference 210 summarizes some selected applications.

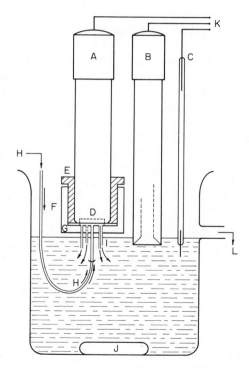

Figure 2-43. Flow cell for a selective ion electrode. *A*, sensor electrode; *B*, reference electrode; *C*, solution ground; *D*, sensing membrane; *E*, Teflon sleeve; *F*, Plexiglas cap; *G*, washer; *H*, sample inlet flow; *I*, sample outlet flow; *J*, magnetic stirring bar; *K*, potentiometer; *L*, solution outlet.

Table 2-12   Biological Materials That Can Be Assayed by Ion-Specific
Electrodes

Enzyme	Substrate	Sensing Electrode
β-glucosidase	Amygdalin	CN^-
Rhodanese	$S_2O_3{}^{2-}/CN^-$	CN^-
β-cyanoalanine synthase	L-Cysteine/CN^-	S^{2-}/CN^-
α-chymotrypsin	Diphenylcarbamyl fluoride	F^-
Urease	Urea	Glass/antibiotic
Deaminase enzymes		Glass/antibiotic
Cholinesterase	Acetylcholine	Acetylcholine
Peroxidase	H_2O_2	I^-
Catalase	H_2O_2	I^-
Glucose oxidase	β-D-glucose	I^-
Hypoxanthine oxidase	Hypoxanthine	I^-
Uricase	Uric acid	I^-

From G. A. Rechnitz, *Research / Development*, **18** (August 1973).

Undoubtedly the area of greatest potential application is in the field of
clinical chemistry and the study of biological systems. Electrodes can be
fabricated which can penetrate into volumes as small as a single cell
(35, 212). The potential of the electrodes for continuous analysis of enzymes
has been demonstrated by use of flow cells of the type shown in Figure 2-43
(213). Table 2-12 lists a few of the enzymes which can be determined by use
of selective ion electrodes (214).

References

1. J. O'M. Bockris and A. K. N. Reddy, *Modern Electrochemistry*, Vol. 2, Plenum Press, New York, 1970, pp. 644–678.
2. V. A. Pleskov, *Usp. Khim.*, **16**, 254 (1947).
3. H. M. Koepp, H. Wendt, and H. Strehlow, *Z. Elektrochem.*, **64**, 483 (1960).
4. I. V. Nelson and R. T. Iwamoto, *Anal. Chem.*, **33**, 1795 (1961).
5. O. Popovych and A. J. Dill, *Anal. Chem.*, **41**, 456 (1969).
6. O. Popovych, *Crit. Rev. Anal. Chem.*, **1**, 73 (1970).
7. E. Grunwald, G. Baughman, and G. Kohnstam, *J. Am. Chem. Soc.*, **82**, 5801 (1960).
8. O. Popovych, *Anal. Chem.*, **38**, 558 (1966).
9. A. J. Parker and R. Alexander, *J. Am. Chem. Soc.*, **90**, 3313 (1968).
10. I. M. Kolthoff and M. K. Chantooni, Jr., *J. Phys. Chem.*, **76**, 2024 (1972).

11. D. A. MacInnes, *The Principles of Electrochemistry*, Dover Publications, Inc., New York, 1961, p. 231 [an unabridged and corrected version of the second (1947) printing of the work first published by the Reinhold Publishing Corp. in 1939].

12. J. J. Campion, *J. Chem. Educ.*, **49**, 827 (1972).

13. P. Henderson, *Z. Physik. Chem.*, **59**, 118 (1907); **63**, 325 (1908).

14. Ref. 11, p. 85, 226.

15. P. A. Rock, *Electrochim. Acta*, **12**, 1531 (1967).

16. G. N. Lewis and L. W. Sargent, *J. Am. Chem. Soc.*, **31**, 363 (1909).

17. D. A. MacInnes and Y. L. Yeh, *J. Am. Chem. Soc.*, **43**, 2563 (921); Ref. 11, p. 236.

18. W. H. Smyrl and J. Newman, *J. Phys. Chem.*, **72**, 4660 (1968).

19. A. K. Covington, "Reference Electrodes," in *Ion Selective Electrodes*, R. A. Durst, ed., National Bureau of Standards Special Publication 314, U. S. Govt. Printing Office, Washington, D. C., 1969, Chap. 4.

20. P. A. Rock, *J. Chem. Educ.*, **47**, 683 (1970).

21. N. P. Finkelstein and E. T. Verdier, *Trans. Faraday Soc.*, **53**, 1618 (1957).

22. E. A. Guggenheim, *J. Am. Chem. Soc.*, **52**, 1315 (1930); *J. Phys. Chem.*, **36**, 1758 (1930).

23. A. J. Zielen, *J. Phys. Chem.*, **67**, 1474 (1963).

24. E. W. Baumann, *J. Electroanal. Chem.*, **34**, 238 (1972).

25. *Orion Research Inc. Newsletter*, **1**, 21 (1969).

26. E. L. Eckfeldt and W. E. Proctor, Jr., *Anal. Chem.*, **43**, 332 (1971).

27. "Lazaran Reference Electrode," Bulletin 7202, Beckman Instruments, Inc., Fullerton, Calif. 92634.

28. B. G. Cox, A. J. Parker, and W. E. Waghorne, *J. Am. Chem. Soc.*, **95**, 1010 (1973).

29. J. Jackson, *Chem. Ind.*, 272 (1969).

30. M. F. Ryan, *Science*, **165**, 851 (1969); letter.

31. C. C. Westcott and T. Johns, "pH Measurements in Tris Buffers," Applications Research Technical Report No. 542, Beckman Instruments, Inc., Fullerton, Calif. 92634.

32. N. Salling and O. Siggaard-Andersen, *Scand. J. Clin. Lab. Invest.*, **28**, 33 (1971).

33. P. C. Caldwell, *Int. Rev. Cytol.*, **24**, 345 (1968).

34. L. A. Geddes, *Electrodes and the Measurement of Bioelectric Events*, Wiley-Interscience, New York, 1972.

35. M. Lavallee, O. F. Schanne, and N. C. Hebert, ed., *Glass Microelectrodes*, John Wiley and Sons, Inc., New York, 1969.

36. R. N. Khuri, "Ion-Selective Electrodes in Biomedical Research," in *Ion Selective Electrodes*, R. A. Durst, ed., National Bureau of Standards Special Publication 314, U. S. Govt. Printing Office, Washington, D. C., 1969, Chap. 8.

37. D. B. Cater and I. A. Silver, "Microelectrodes and Electrodes Used in Biology," in *Reference Electrodes*, D. J. G. Ives and G. J. Janz, eds., Academic Press, New York, 1961, Chap. 11.

38. K. Tsuji and P. J. Elving, *Anal. Chem.*, **41**, 216 (1969).

39. W. R. Turner and P. J. Elving, *Anal. Chem.*, **37**, 467 (1965).

40. R. G. Clem, F. Jakob, and D. Anderberg, *Anal. Chem.*, **43**, 292 (1971).

41. W. N. Carson, C. E. Michelson, and K. Koyama, *Anal. Chem.*, **27**, 472 (1955).

42. R. A. Durst, *J. Chem. Educ.*, **43**, 437 (1966).

43. H. P. Raaen, *Anal. Chem.*, **37**, 1355 (1965).

44. R. G. Bates, *Determination of pH*, 2nd ed., Wiley-Interscience, New York, 1973, Chap. 10.

45. G. J. Hills and D. J. G. Ives, "The Hydrogen Electrode," in *Reference Electrodes*, D. J. G. Ives and G. J. Janz, eds., Academic Press, New York, 1961, Chap. 2, p. 102.

46. T. C. Franklin, M. Naito, T. Itoh, and D. H. McClelland, *J. Electroanal. Chem.*, **27**, 303 (1970).

47. Ref. 45, p. 100.

48. Ref. 44, p. 290.

49. A. M. Feltham and M. Spiro, *Chem. Rev.*, **71**, 177 (1971).

50. Ref. 44, p. 294; Ref. 45, p. 92.

51. Ref. 44, p. 282; Ref. 45, p. 95.

52. M. Fleischmann and J. N. Hiddleston, *J. Sci. Instrum. Series 2*, **1**, 667 (1968).

53. J. V. Dobson, *J. Electroanal. Chem.*, **35**, 129 (1972).

54. T. S. Light, Ref. 19, p. 421.

55. Ref. 19, p. 115.

56. G. J. Janz, "Silver-Silver Halide Electrodes," in *Reference Electrodes*, D. J. G. Ives and G. J. Janz, eds., Academic Press, New York, 1961, Chap. 4.

57. J. N. Butler, "Reference Electrodes in Aprotic Organic Solvents," in *Advances in Electrochemistry and Electrochemical Engineering*, Vol. 7, P. Delahay, ed., Interscience Publishers, New York, 1970, pp. 106–114.

58. G. D. Pinching and R. G. Bates, *J. Res. Nat. Bur. Stand.*, **37**, 311 (1946); Ref. 44, p. 329.

59. Ref. 56, p. 205.

60. Ref. 44, p. 330.

61. R. G. Bates, Ref. 19, p. 421.

62. Ref. 44, p. 48.

63. L. Meites and S. A. Moros, *Anal. Chem.*, **31**, 23 (1959).

64. Ref. 45, Chap. 3.

65. Ref. 45, pp. 160–161.

66. R. G. Bates, Ref. 35, p. 7.

67. D. J. G. Ives, Ref. 45, Chap. 7.

68. G. J. Samuelson and D. J. Brown, *J. Am. Chem. Soc.*, **57**, 2711 (1935).

69. G. Mattock, *pH Measurement and Titration*, Heywood, London, 1961, p. 153; Ref. 19, p. 122.

70. Ref. 44, p. 395.

71. Ref. 44, p. 398.

72. Ref. 19, p. 124.

73. F. G. K. Baucke, *J. Electroanal. Chem.*, **33**, 135 (1971).

74. D. J. Fisher, W. L. Belew, and M. T. Kelley, in *Polarography 1964*, Vol. 2, G. J. Hills, ed., Interscience Publishers, 1966, p. 1043.

75. W. L. Belew, D. J. Fisher, M. T. Kelley, and J. A. Dean, *Chem. Instrum.* **2**, 297 (1970).

76. C. C. Herrmann, G. P. Perrault, and A. A. Pilla, *Anal. Chem.*, **40**, 1173 (1968).

77. G. J. Hills, Ref. 45, p. 433.

78. C. K. Mann and K. K. Barnes, *Electrochemical Reactions in Nonaqueous Systems*, Marcel Dekker, Inc., New York, 1970, Chap. 1.

79. H. Strehlow, "Electrode Potentials in Non-Aqueous Solvents," in *The Chemistry of Non-aqueous Solvents*, Vol. II, J. J. Lagowski, ed., Academic Press, New York, 1967, Chap. 4.

80. R. G. Bates, "Acidity Functions for Amphiprotic Media," *ibid.*, Chap. 3.

81. R. G. Bates, "Medium Effects and pH in Nonaqueous Solvents," in *Solute-Solvent Interactions*, J. F. Coetzee and C. D. Ritchie, eds., Marcel Dekker, Inc., New York, 1969, Chap. 2.

82. B. Kratochvil, E. Lorah, and C. Garber, *Anal. Chem.*, **41**, 1793 (1969).

83. W. Ward, quoted by Ref. 57, p. 115.

84. R. C. Larson, R. T. Iwamoto, and R. N. Adams, *Anal. Chim. Acta*, **25**, 371 (1961); quoted in Ref. 57, p. 135.

85. Ref. 78, p. 18.

86. D. Coutagne, *Bull. Soc. Chim. France*, 1940 (1971).

87. R. W. Laity, "Electrodes in Fused Salt Systems," in Ref. 45, Chap. 12.

88. G. J. Janz, ed., *Molten Salts Handbook*, Academic Press, New York, 1967.

89. S. Senderoff and A. Brenner, *J. Electrochem. Soc.*, **101**, 31 (1954); quoted in Ref. 87, p. 585.

90. L. G. Boxall and K. E. Johnson, *Anal. Chem.*, **40**, 831 (1968).

91. R. D. Caton, Jr., and C. R. Wolfe, *Anal. Chem.*, **43**, 660 (1971).

92. H. A. Laitinen and W. S. Ferguson, *Anal. Chem.*, **29**, 4 (1957).

93. Ref. 1, pp. 1156–1170.

94. S. Trasatti, *J. Electroanal. Chem.*, **39**, 163 (1972).

95. Ref. 1, pp. 986–989, 1141–1146.

96. A. T. Kuhn, C. J. Mortimer, G. C. Bond, and J. Lindley, *J. Electroanal. Chem.*, **34**, 1, (1972).

97. R. Parsons, *Surface Sci.*, **2**, 418 (1964); **18**, 28 (1969).

98. R. A. Marcus, *Electrochim. Acta*, **13**, 995 (1968).

99. N. S. Hush, *Electrochim. Acta*, **13**, 1005 (1968).

100. V. G. Levich, "Oxidation-Reduction in Solution," in *Advances in Electrochemistry and Electrochemical Engineering*, Vol. 4, P. Delahay and C. W. Tobias, eds., Interscience, New York, 1966, p. 249.

101. H. Gerischer, "Semiconductor Electrode Reactions," *ibid.*, Vol. 1, 1961, p. 139.

102. W. Mehl, "Reactions at Organic Semiconductor Electrodes," in *Reactions of Molecules at Electrodes*, N. S. Hush, ed., Wiley-Interscience, New York, 1971, p. 305.

103. R. R. Dogonadze, "Theory of Molecular Electrode Kinetics," *ibid.*, p. 135.

104. R. M. Reeves, M. Sluyters-Rehbach, and J. H. Sluyters, *J. Electroanal. Chem.*, **34**, 55, 69 (1972); **36**, 101, 287 (1972).

105. J. N. Butler, *J. Electroanal. Chem.*, **14**, 89 (1967); J. B. Headridge, M. Ashraf, and H. L. H. Dodds, *ibid.*, **16**, 116 (1968).

106. T. Kuwana and R. N. Adams, *J. Am. Chem. Soc.*, **79**, 3609 (1957); *Anal. Chim. Acta*, **20**, 51, 60 (1959).

107. D. A. J. Rand and R. Woods, *J. Electroanal. Chem.*, **31**, 29 (1971).

108. D. A. J. Rand and R. Woods, *J. Electroanal. Chem.*, **35**, 209 (1972).

109. J. N. Gaur and G. M. Schmid, *J. Electroanal. Chem.*, **24**, 279 (1970).

110. S. Yamada and H. Sato, *Nature*, **193**, 261 (1962).

111. C. E. Plock, *J. Electroanal. Chem.*, **18**, 289 (1968).

112. H. E. Zittel and F. J. Miller, *Anal. Chem.*, **37**, 200 (1965).

113. V. J. Jennings, T. E. Forster, and J. Williams, *Analyst*, **95**, 718 (1970).

114. M. Kopanica and F. Vydra, *J. Electroanal. Chem.*, **31**, 175 (1971).

115. G. Cauquis and D. Serve, *J. Electroanal. Chem.*, **34**, Appl. 1 (1972).

116. T. M. Florence, *J. Electroanal. Chem.*, **27**, 273 (1970).

117. P. J. Elving and D. L. Smith, *Anal. Chem.*, **32**, 1849 (1960).

118. F. J. Miller and H. E. Zittel, *Anal. Chem.*, **35**, 1866 (1963).

119. A. L. Beilby, W. Brooks, Jr., and G. L. Lawrence, *Anal. Chem.*, **36**, 22 (1964).

120. U. Eisner and H. B. Mark, Jr., *J. Electroanal. Chem.*, **24**, 345 (1970).

121. J. P. Randin and E. Yeager, *J. Electroanal. Chem.*, **36**, 257 (1972).

122. J. R. Covington and R. J. Lacoste, *Anal. Chem.*, **37**, 420 (1965).

123. H. S. Swofford, Jr. and R. L. Carman III, *Anal. Chem.*, **38**, 966 (1966).

124. E. Pungor, E. Szepesuary, and J. Havas, *Anal. Letters*, **1**, 213 (1968).

125. R. N. Adams, *Anal. Chem.*, **30**, 1576 (1958).

126. R. N. Adams, *Electrochemistry at Solid Electrodes*, Marcel Dekker, Inc., New York, 1969, p. 26.

127. L. S. Marcoux, K. B. Prater, B. G. Prater, and R. N. Adams, *Anal. Chem.*, **37**, 1446 (1965).

128. J. D. Voorhies and S. M. Davis, *Anal. Chem.*, **32**, 1855 (1960).

129. R. Hand, A. K. Carpenter, C. J. O'Brien, and R. F. Nelson, *J. Electrochem. Soc.*, **119**, 74 (1972).

130. T. R. Mueller and R. N. Adams, *Anal. Chim. Acta*, **23**, 467 (1960).

131. A. Wesor and E. Pungor, *Proc. IMEKO Symp., Budapest*, 1969.

132. D. J. Curran and K. S. Fletcher III, *Anal. Chem.*, **40**, 78 (1968).

133. J. W. Strojek and T. Kuwana, *J. Electroanal. Chem.*, **16**, 471 (1968).

134. J. F. Alder, B. Fleet, and P. O. Kane, *J. Electroanal. Chem.*, **30**, 427 (1971).

135. H. Lund and P. Iversen, "Practical Problems in Electrolysis," in *Organic Electochemistry*, M. M. Baizer, ed., Marcel Dekker, Inc., New York, 1973, p. 199.

136. T. Biegler, D. A. J. Rand, and R. Woods, *J. Electroanal. Chem.*, **29**, 269 (1971).

137. A. J. Bard, *Anal. Chem.*, **33**, 11 (1961).

138. P. J. Lingane, *Anal. Chem.*, **36**, 1723 (1964).

139. M. von Stackelberg, M. Pilgram, and W. Toome, *Z. Elektrochem.*, **57**, 342 (1953).

140. Ref. 126, pp. 45–61.

141. W. G. French and T. Kuwana, *J. Phys. Chem.*, **68**, 1279 (1964).

142. Ref. 126, pp. 206–208.

143. H. Angerstein-Kozlowska, B. E. Conway, and W. B. A. Sharp, *J. Electroanal. Chem.*, **43**, 9 (1973).

144. R. J. Taylor and A. A. Humffray, *J. Electroanal. Chem.*, **42**, 347 (1973).

145. V. Majer, J. Vesely, and K. Stulik, *J. Electroanal. Chem.*, **45**, 113 (1973).

146. C. L. Gordon and E. Wichers, *Annals N.Y. Acad. Sci.*, **65**, 369 (1957).

147. J. F. Coetzee, "Mercury," in *Treatise on Analytical Chemistry*, I. M. Kolthoff and P. J. Elving, eds., Part II, Vol. 3, Interscience Publishers, New York, 1967, pp. 235–249.

148. O. H. Müller, *Chem. Eng. News*, 1528 (1942).

149. M. E. Hanke and M. Johnson, *Science*, **78**, 44 (1933).

150. "How Poisonous Is Mercury?" Bethlehem Apparatus Company, Hellertown, Pa.

151. G. D. Christian, *J. Electroanal. Chem.*, **23**, 172 (1969).

152. J. Koryta, *J. Chem. Educ.*, **49**, 183 (1972).

153. F. L. Lambert, *Chemist-Analyst*, **46**, 10 (1957).

154. H. H. Bauer, *Electrochim. Acta*, **18**, 427 (1973).

155. S. Lal, A. Kumar, and H. H. Bauer, *J. Electroanal. Chem.*, **42**, 423 (1973).

156. L. Meites, *Polarographic Techniques*, 2nd ed., Interscience Publishers, New York, 1965, p. 320.

157. C. N. Reilley, W. G. Scribner, and C. Temple, *Anal. Chem.*, **28**, 450 (1956).

158. R. W. Schmid and C. N. Reilley, *J. Am. Chem. Soc.*, **80**, 2087 (1958).

159. I. Smoler, *Coll. Czech. Chem. Communs.*, **19**, 238 (1954).

160. A. M. Bond, T. A. O'Donnell, and A. B. Waugh, *J. Electroanal. Chem.*, **39**, 137 (1972).

161. W. L. Belew, D. J. Fisher, H. C. Jones, and M. T. Kelley, *Anal. Chem.*, **41**, 779 (1969).

162. D. K. Means and H. B. Mark, Jr., *Anal. Letters*, **4**, 23 (1971).

163. W. D. Shults, D. J. Fisher, and W. B. Schaap, *Chem. Instrum.*, **1**, 7 (1968).

164. R. D. DeMars and I. Shain, *Anal. Chem.*, **29**, 1825 (1957).

165. W. Kemula, "Voltammetry with the Hanging Mercury Drop Electrode," in *Advances in Polarography*, Vol. 1, I. S. Longmuir, ed., Pergamon Press Ltd., 1960, pp. 105–143.

166. C. A. Streuli and W. D. Cooke, *Anal. Chem.*, **25**, 1691 (1953).

167. C. A. Streuli and W. D. Cooke, *Anal. Chem.*, **26**, 963 (1954).

168. J. R. Kuempel and W. B. Schaap, *Anal. Chem.*, **38**, 664 (1966).

169. L. Ramaley, R. L. Brubaker, and C. G. Enke, *Anal. Chem.*, **35**, 1088 (1963).

170. M. Z. Hassan, D. F. Untereckler, and S. Bruckenstein, *J. Electroanal. Chem.*, **42**, 161 (1973).

171. E. L. Wheeler, *Scientific Glassblowing*, Interscience Publishers, New York, 1958, p. 14.

172. A. Capon and R. Parsons, *J. Electroanal. Chem.*, **39**, 275 (1972).

173. A. T. Hubbard and F. C. Anson, "The Theory and Practice of Electrochemistry with Thin Layer Cells," in *Electroanalytical Chemistry*, Vol. 4, A. J. Bard, ed., Marcel Dekker, Inc., New York, 1970, p. 129.

174. C. N. Reilley, *Rev. Pure Appl. Chem.*, **18**, 137 (1968).

175. A. Hubbard, *Crit. Rev. Anal. Chem.*, **3**, 201 (1973).

176. J. C. Sheaffer and D. G. Peters, *Anal. Chem.*, **42**, 430 (1970).

177. J. E. McClure and D. L. Maricle, *Anal. Chem.*, **39**, 236 (1967).

178. J. Lindquist, *J. Electroanal. Chem.*, **18**, 204 (1968).

179. J. Lindquist, *Anal. Chem.*, **45**, 1006 (1973).

180. R. E. Cover and J. G. Connery, *Anal. Chem.*, **41**, 918 (1969).

181. R. E. Cover and J. G. Connery, *Anal. Chem.*, **41**, 1191 (1969).

182. R. E. Cover and J. G. Connery, *Anal. Chem.*, **41**, 1797 (1969).

183. W. J. Albery and M. L. Hitchmann, *Ring-Disc Electrodes*, Clarendon Press, Oxford, 1971.

184. W. J. Blaedel and S. L. Boyer, *Anal. Chem.*, **43**, 1538 (1971).

185. W. J. Blaedel and S. L. Boyer, *Anal. Chem.*, **45**, 258 (1973).

186. R. B. Kesler, *Anal. Chem.*, **35**, 963 (1963).

187. H. B. Mark, Jr. and B. S. Pons, *Anal. Chem.*, **38**, 110 (1966).

188. D. R. Tallent and D. Evans, *Anal. Chem.*, **41**, 835 (1969).

189. W. N. Hansen, T. Kuwana, and R. A. Osteryoung, *Anal. Chem.*, **38**, 1810 (1966).

190. N. Winograd and T. Kuwana, *J. Am. Chem. Soc.*, **92**, 224 (1970).

191. B. S. Pons, J. S. Mattson, L. O. Winstrom, H. B. Mark, Jr., *Anal. Chem.*, **39**, 685 (1967).

192. S. Gottesfeld and M. Ariel, *J. Electroanal. Chem.*, **34**, 327 (1972).

193. T. Kuwana, R. K. Darlington, and D. W. Leedy, *Anal. Chem.*, **36**, 2023 (1964).

194. T. Osa and T. Kuwana, *J. Electroanal. Chem.*, **22**, 389 (1969).

195. A. Yildiz, P. T. Kissinger, and C. N. Reilley, *Anal. Chem.*, **40**, 1018 (1968).

196. W. R. Heineman and T. Kuwana, *Anal. Chem.*, **43**, 1075 (1971).

197. W. R. Heineman, J. N. Burnett, and R. W. Murray, *Anal. Chem.*, **40**, 1974 (1968).

198. M. Petek, T. E. Neal, and R. W. Murray, *Anal. Chem.*, **43**, 1069 (1971).

199. W. Von Benken and T. Kuwana, *Anal. Chem.*, **42**, 1114 (1970).

200. J. E. McClure, *Anal. Chem.*, **42**, 551 (1970).

201. D. C. Johnson and E. W. Resnick, *Anal. Chem.*, **44**, 637 (1972).

202. C. N. Reilley and R. W. Schmid, *Anal. Chem.*, **30**, 947 (1958).

203. B. Karlmark, *Anal. Biochem.*, **52**, 69 (1973).

204. G. A. Rechnitz, *Accounts Chem. Res.*, **3**, 69 (1970).

205. G. Eisenman, ed., *Glass Electrodes for Hydrogen and Other Cations*, Marcel Dekker, Inc., New York, 1967.

206. C. D. Ritchie and R. E. Uschold, *J. Am. Chem. Soc.*, **89**, 1721 (1967).

207. Ref. 205, p. 446.

208. R. W. Cattrall and H. Freiser, *Anal. Chem.*, **43**, 1905 (1971).

209. M. D. Smith, M. A. Genshaw, and J. Greyson, *Anal. Chem.*, **45**, 1782 (1973).

210. Ref. 19, Chap. 11.

211. Ref. 19, p. 378.

212. Ref. 205, Chap. 18.

213. R. A. Llenado and G. A. Rechnitz, *Anal. Chem.*, **45**, 826 (1973).

214. G. A. Rechnitz, *Research / Development*, **24**, 18 (1973).

Three

ELECTROCHEMICAL CELLS

I. Introduction

A. GENERAL REQUIREMENTS

Electrochemical techniques are used under a variety of conditions. No matter what the purpose of the procedure, the design and construction of a suitable cell must meet certain basic requirements related to the desired level of precision in the measurement, optimum electrode geometry, the chemical reactivity of the system, the working temperature and pressure of the system, the scale of the system (micro or macro), the need to exclude contaminants from the laboratory atmosphere, and reasonable convenience in changing solutions and cleaning the cell between measurements. The cell also may need to be designed to be compatible with auxiliary equipment such as rotating electrodes, spectroscopic systems, or systems for controlling temperature and pressure.

There are several factors to consider in optimizing cell geometry. In laboratory-scale experiments the working solution volume may range from 10 μl of a dilute solution (as in a thin-layer cell) to a solution volume as large as a liter (as in preparative electrochemistry on the mole scale). Currents may range from the submicroampere region to several amperes. The specific resistance of the cell electrolyte may be as low as 2 Ω-cm (1 F aqueous HCl) or as large as 10^5 Ω-cm (0.1 F sodium acetate in glacial acetic acid).

1. Effects of Solution Resistance

The electrode geometry becomes a crucial factor whenever the ohmic (iR) drop in a cell becomes large. First, the ohmic drop imposes a natural limit on the current that can pass through the cell because the product of the total cell resistance and the current cannot exceed the output voltage of the potential source. This limit may be encountered in solvent systems of high specific resistance or in solutions of moderate resistance if the current is large, as in preparative electrochemistry. Second, the ohmic drop results in an error in the measured potential of the working electrode using either two electrode or three electrode circuits (see Figure 2-2).

2. Ohmic Drop in Two-Electrode Circuits

In a simple two-electrode cell of the type often employed in the polarography of aqueous solutions, the cell current passes through both the working electrode and the reference electrode. As discussed in Chapter 2, it is possible to design a reference electrode of low resistance and sufficiently large area so that its potential is not seriously affected by microampere polarographic currents. However, if accurate half-wave potential measurements are to be made, the product of the total cell resistance and the current must not exceed 1 to 2 mV. This requires that the cell resistance not exceed 200 Ω for currents of 10 μA. The approximate cell resistance can be conveniently measured with a simple A-C conductivity bridge. The value of the measured resistance will depend on the surface area of the electrodes, so the measurement is best made with the actual electrodes to be used. Because an electrochemical cell does not behave as a pure resistance, a D-C ohmmeter should not be used because the cell potential will cause a false reading.

3. Ohmic Drop in Three-Electrode Circuits

In modern coulometry and voltammetry the use of a potentiostat and a three-electrode configuration is the routine practice. The three electrodes are usually called the working, reference, and counter (or auxiliary) electrodes (see Figure 2-2). The cell current passes between the working electrode immersed in the test solution and the counter electrode which may be in the test solution but is usually isolated from it by a single- or double-junction glass frit.

The reference electrode normally is located in a salt bridge and connected to the solution by a Luggin capillary whose tip is placed close to the working electrode. This minimizes the error in the measured potential associated with the iR drop in the solution between the tip of the Luggin capillary and the working electrode. If the cell currents are large, the solution resistance high, or fast potential changes necessary, careful attention must be paid to optimizing the electrode geometry. To achieve the greatest accuracy of potential control, the characteristics of the potentiostat and electrolysis cell must be carefully matched as components of a feedback control system.

The external leads from the potentiostat to the electrodes may also contribute significant resistance and capacitance that must be taken into account if the cell currents are large and if fast response is desired. Most metallic working electrodes will have very low resistance, but a typical dropping mercury electrode may have a resistance as large as 100 Ω because the mercury-filled lumen of the capillary is so small (about 0.005-cm diameter). This resistance makes a contribution to the total cell resistance and to the uncompensated resistance in a three-electrode circuit.

The extent to which a three-electrode circuit can compensate for ohmic drop in the cell has been the subject of a number of investigations and is well understood in principle although exact equations may not be available for

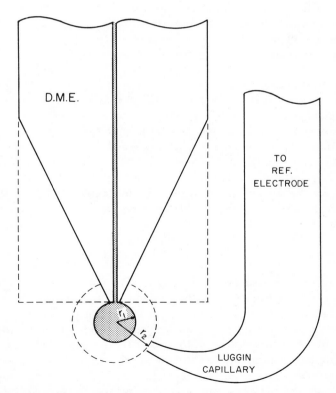

Figure 3-1. Luggin capillary and its placement relative to a dropping mercury electrode

every electrode geometry. The effects are illustrated by reference to Figure 3-1, which shows a dropping mercury electrode and the tip of a Luggin capillary which makes contact with the reference electrode. The dropping mercury electrode (DME) may either be the normal blunt tip glass capillary (dashed lines of Figure 3-1), or a tapered capillary may be used to minimize the shielding effect which distorts the symmetry of the current flux at the mercury drop.

Experimental measurements (1–3) of the uncompensated solution resistance indicate that it changes very rapidly with distance in the immediate vicinity of the mercury drop, and attains an approximately constant limiting value at a distance greater than about 0.5 cm. The latter corresponds to a distance roughly 10 times the maximum radius of the mercury drop. This is illustrated in Figure 3-2, where the potential of a reference electrode in a movable Luggin capillary probe at different distances from the mercury drop has been measured with respect to an identical reference electrode located several centimeters away from the drop on the side opposite the counter electrode. An important feature to note is the nearly spherical symmetry of the gradients at points close to the DME. (The slight gradients on the side of the DME in the direction of the counter electrode indicate that the symmetry probably would be seriously distorted if the counter electrode were placed closer than 1 cm from the DME, or if the cell diameter were smaller than 1 cm.)

The ohmic iR drop at the DME is that expected for a spherically symmetric radial current flux between the surface of an inner sphere of radius r_1 centimeters (the mercury drop in this case) and the surface of an (imaginary) outer sphere of radius r_2 (dashed circle of Figure 3-1 at the tip of the Luggin capillary). Using the model of concentric spherical electrodes of radii r_1 and r_2 separated by a medium of specific resistance ρ (Ω-cm), the expression for the resistance of the solution occupying the volume between the two spheres, assuming a spherically symmetric current flow, is given by

$$R = \frac{\rho}{4\pi}\left(\frac{1}{r_1} - \frac{1}{r_2}\right) \qquad (3\text{-}1)$$

This equation apparently was first derived by Ilkovic (4) and Nemec (5) has derived an equivalent form of the equation. In the case under consideration, the solution iR drop "seen" at the tip of the Luggin capillary will be that between r_1 (the surface of the mercury drop) and r_2 (the distance from the center of the mercury drop to the tip of the Luggin capillary). If the glass capillary has a blunt end, the shielding of the drop requires the introduction of a correction factor. An empirical factor of $3/2$ has been suggested (6), and a value of $16/9$ has been proposed on theoretical grounds (7). Another group (2) uses the value of $3/2$ because it gives better agreement with their experimental results. Assuming this approximation and that the tip of the Luggin capillary is more than 0.5 cm from the drop surface, the uncompensated solution resistance, R, of a blunt-end dropping mercury electrode is given by

$$R \approx \frac{3}{2} \frac{\rho}{4\pi r_1} \qquad (r_2 \gg r_1) \qquad (3\text{-}2)$$

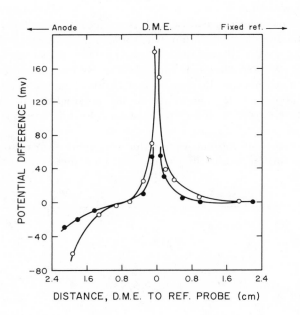

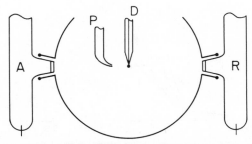

Figure 3-2. Potential difference between a reference electrode in a movable Luggin capillary probe (p) and an identical fixed reference electrode (R) placed opposite to the counter electrode (A). The dropping mercury electrode (D) is placed between the fixed reference electrode and the counter electrode; the probe electrode is on side of DME opposite to the fixed reference electrode.

For $r_1 = 0.05$ cm, and a potential error not to exceed 2 mV, the specific resistance of the electrolyte must be less than 200 Ω-cm for currents of 10 μA. This approximate calculation indicates that the errors in potential due to iR drop will be 200 mV when the specific resistance of the electrolyte is 10^4 Ω-cm.

The solution iR drop at the DME will also be time dependent because r_1, the drop radius, is a function of time. For this reason a stationary hanging mercury-drop electrode is to be preferred or the vertical orifice (Smoler) dropping mercury electrode can be used (see Figure 2-14). Belew et al. (8) claim that the tip of a platinium wire quasi-reference electrode can be placed as close as 0.1 drop-diameter (about 0.003 cm) because the drop grows in the downward direction. This gives nearly complete compensation in an electrolyte with a specific resistance of 15,000 Ω-cm for a cell with total resistance of about 10^5 Ω. The effect on the polarograms of placing the quasi-reference electrode at different distances from the electrode surface is shown in Figure 3-3.

Several instrumental approaches to reduce the effect of iR drop have been described (see further details in Chapter 5). One method is to introduce

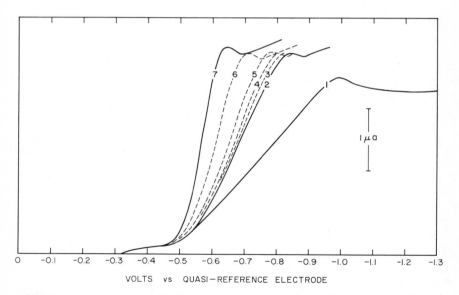

VOLTS vs QUASI–REFERENCE ELECTRODE

Figure 3-3. Polarograms as a function of reference electrode placement in a solution system with a specific resistance of 15,000 Ωcm. (1) Two electrode polarogram; (2)–(6) Quasi-reference electrode placed increasingly close to DME; (7) Quasi-reference electrode less then 0.1 drop radius from DME.

electronic positive feedback compensation in the potentiostat that is proportional to the cell current and thereby approximately cancel the iR drop error. A second method reduces the effects of uncompensated resistance by measuring periodically the ohmic potential drop by means of the current-interrupt method. Bezman (9) has described a simple instrument which stores the measured ohmic potential drop in an analog sample-and-hold memory which is returned as a correction to the potentiostat input. A third method (10) uses the potential difference between two probes located at different distances from the working electrode to provide a correction signal to the potentiostat input.

4. Optimum Geometry in Voltammetry with Microelectrodes

In electrolytes of specific resistance of 100 Ω-cm or less, the geometry is not very critical. The working electrode should be placed between the counter and reference electrodes and is ordinarily an appreciable distance (greater than 1 cm) from both. At small, nearly spherical electrodes, all points on the surface of the electrode will be essentially equidistant from the counter electrode and the current density will be almost uniform over the surface of the microelectrode.

The reference electrode is usually connected to the input of a follower amplifier (see Chapter 5) of high input impedance, so that the current and iR drop through the Luggin capillary and reference electrode are negligible. The Luggin capillary and reference electrode may have a rather high resistance (10–100 kΩ) unless fast response is desired, in which case the reference electrode must have a low resistance.

As the specific resistance of the solution increases, the geometry becomes more important as has been shown in the previous sections. The uncompensated resistance will remain large unless the tip of the reference electrode is located very close to the working electrode surface. This tip must be quite small, otherwise the current density will be nonuniform over the electrode surface because of distortion of the equipotential lines by the tip of the reference electrode.

5. Electrode Geometry in Controlled Potential Electrolysis

When fast response and accuracy of potential control are desired, considerable attention must be paid to the design of the cell-potentiostat system, and several recent papers (11–14) have discussed the critical parameters and made recommendations for optimum cell design. In general, to achieve stability and an optimum potentiostat rise time for a fast potential change, the total cell impedance should be as small as possible, and the uncompensated resistance should be adjusted to an optimum (nonzero) value which depends on the characteristics of the cell and potentiostat (12,15). The electrode geometry also should provide for a low-resistance reference electrode and a uniform current distribution over the surface of the working electrode, which requires that the equipotential lines in the solution near the electrode surface be parallel to the electrode surface.

A cell system has been designed for the potentiostatic transient investigation of fast electrode reactions (12). The main novel feature of the cell is the elimination of the classical Luggin capillary, as shown in Figure 3-4. The design provides a low ohmic resistance reference electrode with low stray capaitances.

In controlled potential coulometry, fast response is usually not necessary and some design compromises can be made for the sake of convenience. Here, the most important thing is to arrange the counter and working electrodes so that a uniform current density will be obtained. If this is not possible, the positioning of the reference electrode tip becomes further restricted. This can be understood by reference to Figure 3-5 which shows a cross-section of an idealized potential and current distribution on the inside

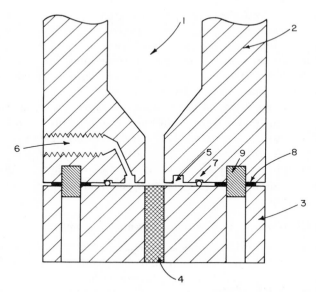

Figure 3-4. Cell design for the study of fast electrode reactions. System provides low-resistance reference electrode and low stray capacitances. (1) counter electrode chamber, (2) Kel-F top, (3) Teflon bottom, (4) working electrode, (5) reference electrode groove, (6) reference electrode connection, (7) Viton O-ring, (8) stainless steel spacer, (9) stainless steel locating pin.

of a cylindrical gauze working electrode with an eccentric, cylindrical counter electrode, under concentration polarization conditions. As in all current distribution systems, the current and potential lines are orthogonal, and the current distribution is indicated by the convention of spacing the current lines to indicate current flux per unit of area (the more closely spaced the current lines, the higher the current density). The metal of the gauze electrode is assumed to be an equipotential, and the boundary (for mathematical purposes) is the adjacent layer of solution (the electrical double layer). The potential that is represented by the dashed contour lines at the electrode surface is equivalent to the usual electrochemical potential difference between the metal and this adjacent layer of solution measured with respect to some reference half-cell. The terms electrode potential or potential of the surface of the electrode refer to this potential difference at specific points along the surface of the electrode.

When polarization occurs at an electrode with nonideal geometry (e.g., when the current is limited by rate of electron transfer or by mass transport),

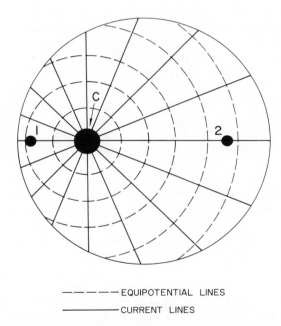

——————EQUIPOTENTIAL LINES
————————CURRENT LINES

Figure 3-5. Idealized potential and current distribution diagram for the inside of a cylindrical gauze electrode and an eccentric interior cylindrical counter electrode (C). Positions 1 and 2 represent other placements of the counter electrode.

there is a gradient in potential in the solution adjacent to the electrode and associated with this is a tangential as well as normal component of the current at the electrode surface (16). This causes the equipotential lines to intersect the electrode and the current lines to enter the electrode at angles other than 90°. (In the absence of polarization, or in a polarized electrode with ideal geometry, the equipotential lines would be parallel to the electrode surface and the current lines would intersect the electrode at an angle of 90°.)

Under the effects of electrode polarization, proper placement of the reference electrode in the cylindrical gauze cell is dictated by the potential distribution. If the reference electrode is placed at location 2, the electrode potential exceeds the control potential in the region of location 1 (i.e., it is more positive in the case of an oxidation or more negative in the case of a reduction). To avoid exceeding the control potential, the reference electrode tip must be positioned on an equipotential which does not intersect the working electrode (such as location 1). There will inevitably be some uncompensated iR drop in the solution, and to minimize uncompensated iR drop errors, the tip should be located on an equipotential which approaches the working electrode closely. If the distance from the working electrode is made greater, more uncompensated iR drop is introduced, but there is no inherent potential control difficulty in this arrangement and the larger iR drop error is not detrimental in most coulometric procedures. Simply stated, the best practice is to *locate the reference electrode tip on the line of minimum separation between the counter and working electrode* (11). This will achieve satisfactory potential control (for analytical purposes) for all types of geometrical arrangements.

B. MATERIALS FOR THE CONSTRUCTION OF CELLS AND ELECTRODES

Materials used for the construction of cells should be usable over a wide temperature range; dimensionally stable; inert to aqueous and organic solvents and reagents; durable; and easy to machine or fabricate. Another highly desirable property is for the material to be transparent so that the electrodes and the solution can be observed.

1. Glass

Glass comes closest to fulfilling all of the requirements above and is the most widely used material. It is attacked by hydrofluoric acid, concentrated alkalies, and basic fused salts but has outstanding chemical resistance to almost all other chemical reagents. The chemical and physical properties of the most commonly used laboratory glasses are summarized in Table 3-1.

Borosilicate glasses are the easiest to work with and can be used at temperatures up to 600°C. Borosilicate fritted glass discs also are widely used as cathode-anode separators or to isolate reference electrode and working electrode compartments. The pore sizes for glass frits of standard porosities are listed in Table 3-2.

Table 3-1   Chemical Composition and Thermal Properties of Common Laboratory Glasses

Component	Chemical Composition (%)[a]		
	Soda-Lime "Soft Glass" Standard Flint	Pyrex Corning No. 7740	Vycor Corning No. 7900
SiO_2	71–73.5	80.5	96
B_2O_3	—	12.9	
Na_2O	14–17	3.8	
K_2O	0–1.5	0.4	
CaO	5–6	—	
MgO	3.5–4.5	—	
Al_2O_3	—	2.2	
Fe_2O_3	0.006–0.1	—	
Thermal properties [b]			
Linear coefficient of expansion (°C, 0–300°C) $\times 10^{-7}$	92	32.5	8
Softening point (°C)	700	820	1500
Annealing point (°C)	521	565	910
Strain point (°C)	480	515	820

[a]E. L. Wheeler, *Scientific Glassblowing*, Interscience Publishers, Inc., New York, 1958, p. 12.
[b]*Laboratory Glass Blowing with Corning's Glasses*, Corning Glass Works, Corning, N. Y.,   1961.

Table 3-2   Porosity of Fritted Glass

Porosity	Type	Nominal Pore Size (μ)
Coarse	C	40–60
Medium	M	10–15
Fine	F	4–5.5
Very fine	VF	2–2.5
Ultra fine	UF	0.9–1.4

Type C: Filtering coarse materials; mercury
      filtration.
Type M: Gas dispersion; salt bridges
Type F: Salt bridges.

The large variety of commercially available ground glass joints, glass frits, and stopcocks, and the services of a glassblower make it easy to fabricate cells for special purposes, and a selected few of these are described in a later section. Most companies that sell electrochemical instrumentation offer a number of ceils of different styles, as well as electrodes and other accessories.

2. Synthetic Polymers

A number of plastics are used in electrochemical cells, and some of their pertinent physical and chemical properties are described here and in Table 3-3.

Teflon.   The Teflon TFE polymer is prepared from tetrafluoroethylene by Du Pont; it is white and nearly opaque except in thin sheets. Teflon FEP is a copolymer of tetrafluoroethylene and hexafluoropropylene, which is more readily molded and is translucent. The FEP polymer has more limited chemical resistance, particularly to halogenated solvents and a lower working temperature limit of 205°C. The TFE polymer is the most chemically resistant plastic available. It is inert, even at elevated temperatures, to practically all chemicals except fused alkali metals, chlorine trifluoride, and fluorine at elevated temperature and pressure. Teflon tape and molded stirring bars have been reported to react vigorously with sodium/potassium alloy at room temperature (17).

Teflon is relatively soft and very light finishing cuts must be used to machine it to a smooth finish and precise tolerances. Because it will flow under pressure, it frequently is used to fabricate electrodes by force fitting rods or wires into a slightly undersized hole that has been drilled in the

Teflon, or by using heat shrinkable Teflon tubing. Using either of these techniques tight seals can be made between the Teflon and the electrode material, with little danger that solution will creep between the Teflon and the electrode. Seals to Teflon TFE are impractical with ordinary adhesives, although special adhesive kits are available for use with Teflon FEP.

Although Teflon TFE is used at temperatures up to 300°C as insulation for wiring, it may give off traces of the monomer when heated much above 250°C, and there is a report (18) that it may give off traces of fluoride at temperatures as low as 90°C. It should probably not be used above 200°C on a continuous basis.

Kel-F. This is a chlorotrifluoroethylene polymer manufactured by Minnesota Mining and Manufacturing Company. The translucent plastic has negligible water vapor transmission and is nonflammable. Kel-F is suitable for use from -200 to $+200$°C, and is unaffected by concentrated alkalies and strong acids, including aqua regia. It is widely used in handling liquid hydrofluoric acid and is resistant to most organic solvents. Kel-F is swelled slightly by highly halogenated compounds and some aromatics. It is more rigid than Teflon and easier to machine with good dimensional stability. The material can be molded under heat and pressure and this technique has been used to seal electrodes (19).

Nylon. The various polyamides are referred to as Nylon and are characterized by toughness and good abrasion resistance. These materials are readily machined and molded and can be used to 150°C. Although Nylon will withstand boiling water, it absorbs water with resultant swelling and loss of flexural strength. Nylon is resistant to weak acids, weak and strong alkalies, petroleum oils, and many common solvents. It is attacked by strong acids, oxidizing agents, phenols, and formic acid. Because of its tendency to absorb liquids Nylon is not recommended for uses where it will have prolonged contact with solutions.

Acrylates. Polymethylmethacrylate (Lucite, Plexiglas, Perspex) is easily machined and is widely used with dilute aqueous solutions. Acrylates are resistant to nonoxidizing acids and weak alkalies, but are attacked by concentrated oxidizing acids and strong alkalies. They will withstand petroleum oils and most alcohols, but are generally unsuitable for use in contact with organic solvents. They are dissolved by ketones, esters, and aromatic and chlorinated hydrocarbons. They are thermoplastic and cannot be used continuously above 75°C. The acrylates are perfectly clear and transparent and often are used to make shields or inert atmosphere enclosures. The material burns slowly when ignited.

Table 3-3 Physical Properties and Chemical Resistance of Plastics

	Teflon TFE	Teflon FEP	Kel-F 81	Methyl Methacrylate	Polyamide (Nylon-Zytel)	Polystyrene	Polypropylene	Linear Polyethylene	Conventional Polyethylene
Temperature limit (°C) (continuous)	250	200	140	50	140	70	130	110	80
Density (g/cm³)	2.15	2.15	2.13	1.19	1.14	1.05	0.90	0.94–0.96	0.91–0.93
Linear coefficient of expansion (°C×10⁻⁵)	10	8.3–10.5	4.8		10–15	6–8		11–13	16–18
Dielectric constant	2.2	2.2	2.5	3.5	4.0–7.6	2.5		2.3	2.3
Dielectric strength (V/mil)	400–600	500–600	400–600	400	385	500–700		450–500	460–700
Water absorption (%/24 hr)	0.01	0.01	0.01	0.3	1.5	0.03–0.05	0.02	0.02	0.03
Relative O₂ permeability	0.6	0.6				0.11	0.11	0.08	0.40
Thermal conductivity [cal/(cm²)(sec)/°C/cm ×10⁻⁴]	6.0	6.0	6.3		5.5	2.9		12	8
CHEMICAL RESISTANCE									
Acids, inorganic	E	E	E	X	X	NR	E	E	E
Acids, organic	E	E	E	X	X	G	E	E	E
Alcohols	E	E	E			G	E	E	E
Aldehydes	E	E	E			NR	G	G	G
Amines	E	E	E			G	G	G	G
Bases	E	E	E			E	E	E	E
Esters	E	E	G	NR		NR	E	E	E
Ethers	E	E	G			X	G	G	G
Glycols	E	E	E		E	G	E	E	E
Hydrocarbons, aliphatic	E	E	E			NR	G	G	G
Hydrocarbons, aromatic	E	E	X	NR		NR	G	G	G
Hydrocarbons, halogenated	E	E	X	NR		NR	G	G	G
Ketones	E	E	G	NR		NR	G	G	G

E: excellent—No effect to 1 year   X: fair—Short exposures causes some effect at room temp.

130

Epoxy resins. The Epoxy resins possess excellent adhesive properties and are often used to seal electrode materials into glass and plastic tubes (except Teflon). They will generally withstand temperatures to 200°C and are resistant to weak acids and alkalies and to organic solvents generally. They are affected by strong alkalies and attacked by certain strong acids.

Polyolefins. Polyethylene (conventional and linear) and polypropylene have excellent chemical resistance and are readily molded and machined, although they are rather soft. Conventional polyethylene adheres well to metals, and polyethylene tubing can be readily sealed around metal rods and wires to make simple electrodes suitable for use at temperatures below 60°C. The material is resistant to mineral acids and bases (except concentrated sulfuric and perchloric acids) and most organic solvents except halogenated or aromatic hydrocarbons.

Polyvinyl Chloride. This material (PVC) is most commonly encountered in the form of flexible tubing (Tygon and other trade names). It has good transparency and chemical resistance, and is most commonly used for short gas connections between the cell and the gas supply. Its heat resistance varies with composition, but most formulations should not be used above 80°C. It is resistant to salt solutions, alkalies, weak acids, alcohols, and aliphatic hydrocarbons. PVC is subject to attack by strong acids and oxidizing agents, and dissolves in ketones and esters; it also is attacked by aromatic hydrocarbons. The tubing frequently is compounded with plasticizers, such as dioctylphthalate, and these may leach out and be a source of contamination. Although polyvinyl chloride tubing often is used for gases, it has a high gas permeability. Hence at modest flow rates over moderate distances considerable oxygen contamination of a nitrogen or argon supply occurs when PVC tubing is used.

3. Changes in Solution Composition Caused by Structural Materials

Materials used for the construction of electrodes and cells may alter the composition of solutions either by surface adsorption of solutes or by dissolution of adventitious impurities in the structural material. Several studies have indicated that polyethylene, polypropylene, Teflon, stainless steel, borosilicate glass (Pyrex), Vycor, quartz, "soft glass," and silicone-coated surfaces all adsorb appreciable quantities of certain ions. For most of the elements investigated, the total adsorption on plastic or glassware is negligible at millimolar concentrations of the solute. However, in work at trace levels (below $10^{-5}\ M$), such as those encountered in the sampling and analysis of natural waters, considerable attention must be paid to the choice of materials used in sampling containers and the various chemical procedures.

Cleaning procedures adopted often alter the surface and vary the degree of adsorption, and the problem is complicated by the fact that new glass and plastic containers, often from the same batch or supplier, show differences in adsorption properties even when handled and pretreated in an identical manner. Yoe and Koch (20) have discussed these problems in some detail.

Others (21) have studied the adsorption from dilute aqueous solutions on glass and plastic surfaces of Cs, Ce(III), ^{90}Sr-Y, ^{140}Ba-La, ^{95}Zr-Nb, ^{106}Ru-Rh, and I$^-$ at the tracer level ($10^{-10}\ M$). The results indicate that for most of the elements studied use of borosilicate glassware is preferable rather than polypropylene, but that for Cs, Ru, and Zr, there is a lower loss on polypropylene surfaces. The protection of laboratory glassware with a silicone containing hydrophobic agent (Desicote, Siliclad) results in some decrease in adsorption, but probably not enough to justify the trouble and expense. They also found that an 18-hr presoaking of the samples in the test solution did not effectively reduce the adsorption of ^{144}Ce. However, the radiotracer method used cannot distinguish net adsorption from an exchange process in which an unlabeled ion is exchanged for a radioactive ion. Other workers (22) have suggested that presoaking is a useful way to diminish adsorption losses, although the evidence is limited.

The adsorption of trace silver ion ($2 \times 10^{-6}\ M$) has been studied by use of a silver ion-selective electrode (23); the results are shown in Figure 3-6. For periods of up to a day, silver losses on Teflon were less than 2%; at the end of 30 days the losses (28%) were least for Vycor. The study indicates that none of the materials is suitable for long-term storage of solutions containing low levels of silver unless a complexing ligand such as thiosulfate is present.

Mercury is particularly liable to loss on polyethylene. Coyne and Collins have reported that the loss of mercury at 0.05 ppm concentration (added as mercuric chloride) is such that in four hours 95% is lost from filtered creek water samples and 40% is lost from distilled water samples, both stored in polyethylene (24). This loss rate is so rapid that some mechanism other than adsorption may be responsible; one suggestion is that traces of reducing agents react with Hg(II) to form small amounts of Hg(I) which disproportionate to Hg(II) and Hg. The mercury formed is presumed to be lost by volatilization.

A wide variety of solvents, reagents, and structural materials encountered normally in the trace element analysis of seawater have been analyzed for trace element impurities by neutron activation analysis and gamma-ray spectrometry (25). Some of the results obtained for 10 trace elements are shown in Table 3-4 and indicate that many substances contain high impurity levels of various elements. Particular note should be made of the high concentrations of zinc in neoprene and natural rubber tubing and in Kimwipe cellulose tissue; zinc, antimony, and iron in polyvinyl chloride

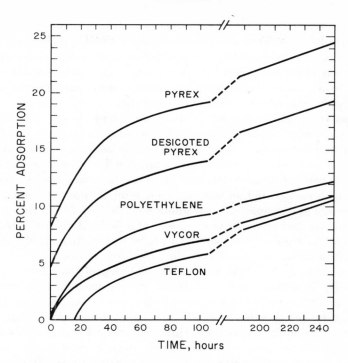

Figure 3-6. Adsorption of silver ion as a function of time.

tubing; antimony in one source of polyethylene tubing, and iron in one source of polyethylene bottles; iron and antimony in borosilicate glass; a large concentration of antimony in a sample of Vycor glass; and cobalt in a sample of Nylon. Teflon appeared to be relatively free of trace metal impurities, and the purest material of those studied was a sample of Plexiglas.

Some samples of plastic tubing contain plasticizers (particularly polyvinyl chloride tubing), and fillers and mold release agents often are employed. In contact with a solution these materials may be leached out and introduce contaminates to the sample.

C. THE MAINTENANCE OF AN INERT ATMOSPHERE

Because oxygen is chemically reactive with many substances and electrochemically reducible, electrochemical work requires exclusion of oxygen. This is generally done by purging the system with a flow of inert gas (nitrogen, argon, helium, etc.), by enclosing the system in an inert atmosphere glove box, or by using a vacuum line. A properly designed vacuum line system is considered the most effective way of rigorously excluding oxygen and water, but it is possible to maintain an atmosphere below 1 ppm of oxygen and water in a glove box, and a glove box allows the use of simpler cells and more convenient manipulation.

For polarographic and voltammetric work where the presence of oxygen interferes merely because of its electrochemical reducibility, purging the solution and cell with a flow of inert gas is entirely satisfactory and is capable of reducing the oxygen concentration in the cell atmosphere to something between 20 and 200 ppm. However, there are species which react rapidly with oxygen and an inert gas containing 20 ppm of oxygen may be intolerable. For example, a 50-ml solution of 10^{-4} M Cr(II) that is purged with nitrogen containing 20 ppm oxygen at a flow rate of 200 ml/min will be completely oxidized in about 15 min, assuming that all of the oxygen flowing into the system reacts with the Cr(II) ion.

The choice of inert gas to some extent depends on local costs and availability. In addition to the cost of the gas, transportation and demurrage costs of gas cylinders can be substantial. Nitrogen is most often used because it is cheapest and is widely available in a "prepurified" grade that contains not more than 20 ppm of oxygen. However, nitrogen reacts with lithium at room temperature, forms molecular complexes with certain transition metals (26), and may be reactive in fused salt systems at high temperature. In these circumstances the use of argon or helium is preferred. Helium can be obtained in very high purity by diffusion through quartz (27) or by use of adsorption columns (28). Hydrogen also can be obtained in an extremely pure state by electrolysis and/or diffusion through palladium-silver alloys, but its flammability, low density, and potential reactivity with some systems make it less suitable for routine use. The use of propane (29), natural gas (methane), carbon dioxide (30), Freon (31), and N_2-H_2 generated from hydrazine (32) have all been described for use in special circumstances, but they are not widely used. The use of a gas that is denser than air, such as argon, is preferred so that the solution will be covered with a blanket of inert gas.

Residual oxygen can be removed from nitrogen or other inert gases by bubbling through aqueous solutions containing vanadous ion or chromous ion which are continuously regenerated through contact with amalgamated zinc. Meites (33) gives a convenient recipe for the use of vanadous ion. Gases treated by bubbling through aqueous solutions will not be suitable for use with nonaqueous solvents unless there is provision for drying the gas, and the continual carryover of water will require an inconveniently large drying capacity.

Table 3-4 Trace Element Concentrations in Structural Materials

Sample	Source	Concentration (ppb)									
		Zn	Fe	Sb	Co	Cr	Sc	Cs	Ag	Cu	Hf
Construction materials											
Teflon	a	9.3	35	0.4	1.7	<30	<0.004	<0.01	<0.3	22	NM
Plexiglas	b	<10	<140	<0.01	<0.05	<10	<0.002	<0.06	<0.03	<9.5	NM
Polyvinyl chloride	c	7120	270,000	2690	45	2	4.5	<1	<5	630	NM
Surgical rubber tubing	d	3×10^6	UM	<100	<30	UM	<8	<100	1240	<6	NM
	e	5×10^6	UM			UM					
Neoprene rubber	f	1.8×10^7	UM	290	2300	UM	3090	UM	<1000	NM	NM
Nylon (block)	g	UM	UM	UM	1.4×10^6	UM	UM	UM	UM	UM	UM
Polyethylene tubing	h	55	7.4	9000	140	254	11	<100	<300	NM	<100
Container and related material											
Borosilicate glass	i	730	280,000	2900	81	UM	106	<100	<0.001	NM	597
Vycor glass	i	UM	UM	1.1×10^6	UM	UM	UM	UM	UM	UM	UM
Quartz tubing	j	1.5	395	0.05	0.44	6.5	0.03	1.1	0.05	2	<0.005
Quartz tubing	k	<1	NM	<0.01	12	2.5	0.39	<0.1	<0.01	0.04	<0.005
Quartz tubing	l	UM	UM	1940	0.89	UM	UM	1390	<0.1	0.09	<0.01

Table 3-4, (continued)

Sample	Source	Zn	Fe	Sb	Concentration (ppb)						
					Co	Cr	Sc	Cs	Ag	Cu	Hf
Container and related material-											
Quartz tubing	m	21	NM	43	0.64	230	0.18	0.30	<0.1	0.05	27
Quartz tubing	n	33	NM	38	1.1	602	0.16	<0.1	<0.1	0.03	26
Polyethylene	o	28	10,400	0.18	0.07	76	0.008	<0.05	<0.1	6.6	<0.5
Polyethylene	p	25	10,600	0.83	0.31	19	0.36	<0.15	<0.1	15	<0.5
Kimwipe tissue	q	48,800	1000	16	24	500	14	<0.1	~0.8	NM	NM

[a] Du Pont.
[b] Manufacturer unknown.
[c] Manufacturer unknown.
[d] Kent Latex Products, Inc.
[e] Rubber Latex Products, Inc.
[f] Atlantic India Rubber Works.
[g] Manufacturer unknown.
[h] Interstate Plastics Company.
[i] Corning Glass Company.
[j] "Spectrosil," Thermal American Fused Quartz Company.
[k] "Suprasil," Engelhard Industries, Inc.
[l] Quartz Products.
[m] United States Quartz.
[n] General Electric Company.
[o] Nalgene, No. 6250 container.
[p] The Chemical Rubber Co., 2/5 dram polyethylene, flip-top vial.
[q] Kimberly-Clark.
NM: not measured.
UM: unable to measure because of interfering radionuclides.

Perhaps the most efficient and convenient method for providing a supply of dry oxygen-free inert gas employs BTS catalyst (Badische Anilin und Soda-Fabrik) in conjunction with molecular sieves. The catalyst is a pelleted form of finely divided copper on an inert support and is used in a system like that shown in Figure 3-7. The nitrogen passes through molecular sieves, BTS catalyst, and another column of molecular sieves. The output gas will have an oxygen and water content each below 1 ppm in a carefully built system. The three-way stopcocks in the inlets are used to bleed the regulators and inlet lines when the cylinders are changed. During the regeneration cycle, the BTS catalyst is heated to about 150°C by means of a heating tape while gradually replacing the inert gas with hydrogen. During the regeneration

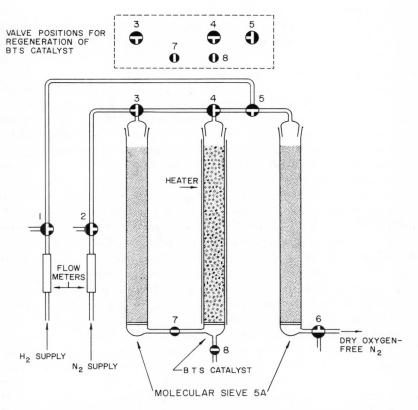

Figure 3-7. Gas purification system for removal of oxygen (BTS catalyst) and water (molecular sieves).

cycle the stopcock at the bottom of the BTS column is opened to allow water produced in the regeneration step to bleed off while the stopcock between the BTS column and the left molecular sieve column is shut off. The catalyst may be regenerated many times; the extent of exhaustion of the catalyst is marked by the progress of a yellowish band as the reduced catalyst is oxidized (34). This method appears to be superior to the use of heated copper gauze; the latter has a much lower surface area which is rapidly coated with an oxide layer.

A whole laboratory can be readily served by one purification system if the purified gas is fed into a manifold with several outlets. Copper or aluminum tubing with Swagelok fittings and needle valves are convenient for this purpose.

Where possible gas connections should be made with metal or glass tubing. Where short flexible connections are necessary polyvinyl chloride (Tygon) tubing is recommended. Most plastic materials have an appreciable permeability (35); a 1-m length of 1/4-in. ID polyvinyl chloride tubing with 1/16-in. wall thickness will allow the diffusion of about 0.8 ppm of oxygen at a flow rate of 150 ml/min.

A common practice is to saturate the purified, oxygen-free gas with solvent in a gas-dispersion cylinder that is filled with a solution identical to the test solution. This avoids the evaporation of the solvent or the differential evaporation of a volatile constituent of the test solution that would otherwise accompany the continuous purging of the system. Such a practice is particularly important for volatile solvents like acetonitrile or for solutions that contain volatile constituents like ammonia.

When exclusion of both water and oxygen is necessary, glove boxes are commercially available that can control water and oxygen levels below 1 ppm (36). To achieve this level requires a carefully designed system that provides for continuous circulation and purification of the inert atmosphere. Most of the commercial glove boxes provide for positive circulation of the atmosphere through a purification train that removes water with molecular sieves and oxygen with BTS or a similar copper catalyst. Some systems incorporate a small percentage of hydrogen in the atmosphere so that the BTS catalyst will be reduced continually.

Although expensive, these units are convenient to use after a little practice. Their continued use with relatively volatile organic solvents may pose a problem, however. Organic solvents are adsorbed by the molecular sieve columns, causing premature exhaustion of their capacity to adsorb water. Sulfur containing compounds also poison the BTS or copper catalyst. For these reasons, solvents should be freed of oxygen before they are put in the glove box and should be capped. Even though the atmosphere in a glove box is maintained below 1 ppm, the water level in a solvent will not be reduced below the level present when the solvent was introduced into the glove box. The glove box is of little use if the solvents have not been carefully purified and protected from moisture during all transfers.

The rate of circulation in the purification train is important; a throughput not less than one box volume per minute is recommended. The time required to achieve a specific reduction in the level of oxygen is approximated by the relation

$$t = 2.3 \frac{V}{FE} \log \frac{C^0}{C} \qquad (3\text{-}3)$$

where C^0 is the initial concentration of oxygen; C the concentration after time t; V the box-volume; F the flow rate; and E the efficiency (fraction of O_2 removed upon passage through the purification train). The introduction of 0.5 liter of air into a 500-liter glove box produces a concentration of 200 ppm of oxygen. This level can be reduced to about 1 ppm in 5 to 10 min at a throughput of one box-volume per minute, which implies that the purification system has an efficiency of about 50%. To reduce oxygen and water below 20 to 50 ppm is difficult unless the circulation rate exceeds 10 box-volumes/hr.

Oxygen in the glove box atmosphere may be monitored with a Hersch galvanic cell (37) as pictured in Figure 3-8. The sensor cathode is a fleece of graphite or porous silver exposed on one side to the sample gas stream while the other side is lined by a separator soaked with sodium hydroxide. The separator in turn contacts a cadmium anode. The two electrodes are bridged by a microammeter. The maximum readout current is about 10 μA for each ppm of oxygen in the inert gas at a flow rate of 37.4 ml of inert gas per minute (flow being measured at 20°C and 1 atm).

The lifetime of a tungsten filament exposed to the glove box atmosphere may also be used as a rough indication of the oxygen content of the inert gas (38).

Water in the glove box atmosphere can be monitored by direct amperometric measurement in a cell (Moisture Monitor) first described by Keidel (39) and available commercially (Du Pont Instruments). The cell is shown in Figure 3-9 and consists of two platinum wires closely wound in a spiral and jacketed in Teflon or glass. The approximately meter-long element is in turn wound into a helix and potted in a metal housing. The platinum wires are bridged by a high resistance P_2O_5 electrolyte; in use absorbed water is quantitatively electroyzed to hydrogen and oxygen at the platinum electrodes by the application of a D-C voltage greater than the

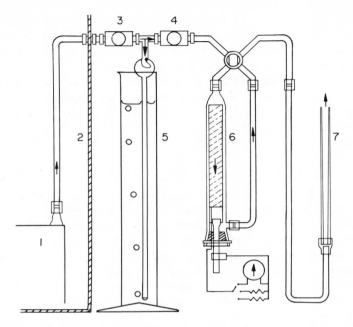

Figure 3-8. Hersch galvanic cell for the determination of oxygen at trace levels in gases.

decomposition potential of the water. This not only provides continuous indication of water content, but also maintains the film in an absorbent condition. The current output is proportional to the water concentration in the gas flowing through the cell and can be used to measure concentrations from less than 1 ppm up to about 1000 ppm.

If a glove box is not available, Schlenk type glassware used with medium vacuum and inert gas is a relatively inexpensive and convenient approach. The use of this glassware is described in a general reference (40). Dissolution, recrystallization, solution transfers, and other simple operations can be carried out on the bench top. Syringe techniques using apparatus closed with rubber septa also have been developed (41).

The most rigorous exclusion of water and oxygen can be accomplished on a high vacuum line. There appears to be a significant psychological barrier that has to be overcome before most workers will go to the trouble of building a suitable vacuum system, so that this approach is regarded as a technique of last resort.

The capital investment in a vacuum line may be considerably less than the cost of a good glove box and with well designed cells and accessories a

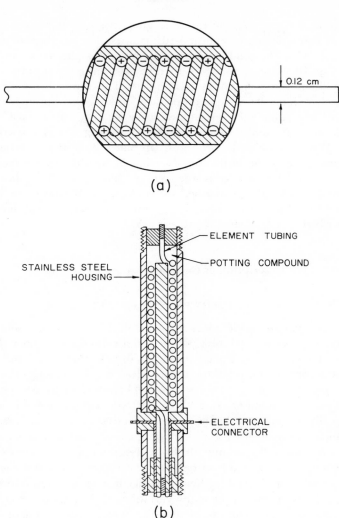

Figure 3-9. Keidel coulometric moisture analyzer for the determination of water in gases. At a flow rate of 100 ml/min a moisture content of 1 ppm gives a current response of 13.2 μA.

vacuum line can be convenient to use. A simple manifold with a diffusion pump backed by a mechanical pump can serve as the core of the system (42,43). One or more nitrogen traps are placed between the pumps and the manifold leading to the cell. In addition to a supply of liquid nitrogen for the traps, the system requires a supply of pure, dry inert gas for backfilling and flushing the electrochemical cells after evacuation. A versatile and convenient cell is shown in Figure 3-10; its operation has been described in great detail (44). The solid supporting electrolyte is placed in ampoule B' and dried under vacuum; solvent is then vapor transferred into the ampoule until the desired concentration is attained. The same is done with the reference electrode electrolyte in ampoule A' and for the sample in ampoule C'. For polarographic work, enough mercury is added to section C to cover the platinum wire and a small Teflon-clad stirring bar is floated on the mercury. With ampoules A', B', and C' in place, the cell is attached to the vacuum line at the joint on the top of section B. By tilting the ampoules and manipulation of the stopcocks the cell is filled and the polarographic operations carried out as described by Anderson and coworkers. The cell also may be used for cyclic voltammetry and coulometry. A similar cell has been described for use in electrosynthesis (45).

II. Description of General-Purpose Cells

A.   CELLS FOR VOLTAMMETRY AND POLAROGRAPHY

For ordinary voltammetric measurements in low resistance solutions, a simple cell employing a spoutless beaker and a close fitting rubber stopper is adequate. The stopper is drilled to accept the electrodes and the purging tubes, and to provide for addition of reagents or other necessary functions. A reference electrode of the style shown in Figure 2-11 is convenient for two-electrode voltammetry. If a three-electrode circuit is used, a higher resistance silver-silver chloride or calomel electrode of the type used in pH measurements may be used and a platinum wire, carbon rod, or mercury pool may be used as the counter electrode. Isolation of the counter electrode usually is not necessary in a polarographic measurement because the current is so small that it produces a negligible change in solution composition.

Figure 3-11 shows one of the many variations of the popular H-cell widely used for polarography. In this design the sample compartment may be detached for easy cleaning. The side-arm of the reference electrode compartment is closed with a glass frit, and when the cell is detached the side-arm is capped so that the reference electrode will not dry out. The stopcock allows the solution and accumulated mercury to be drained before rinsing and filling with the next sample. Strong oxidants may attack mercury, and it may be desirable to protect the accumulated mercury pool from a dropping mercury electrode under a layer of some inert immiscible solvent that is denser than the electrolyte solution, such as chloroform (46,47). Nitrogen or other inert gas is bubbled through the solution by

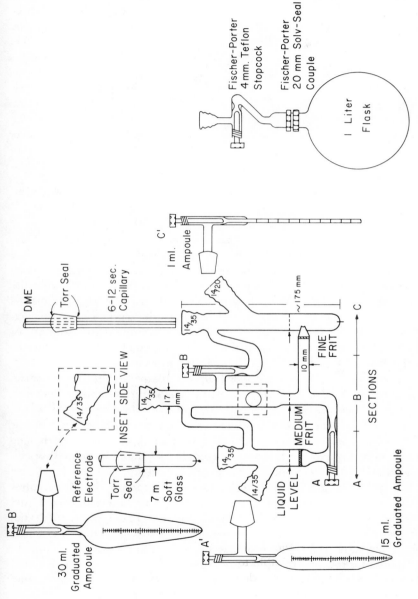

Figure 3-10. Electrochemical cell for use on a vacuum line.

Fischer–Porter 4 mm. Teflon Stopcock

Fischer–Porter 20 mm Solv–Seal Couple

I Liter Flask

DME

Torr Seal

6-12 sec. Capillary

C'

I ml. Ampoule

$^{14}/_{35}$

$^{14}/_{20}$

~175 mm

FINE FRIT

10 mm

B

INSET SIDE VIEW

$^{14}/_{35}$

$^{14}/_{35}$

17 mm

MEDIUM FRIT

Reference Electrode

Torr Seal

7 m Soft Glass

$^{14}/_{35}$

$^{14}/_{35}$

LIQUID LEVEL

A

SECTIONS

A   B   C

B'

30 ml. Graduated Ampoule

A'

15 ml. Graduated Ampoule

143

Electrochemical Cells

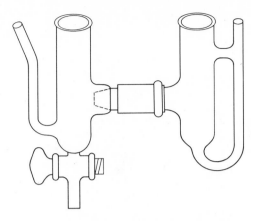

Figure 3-11. H-cell for polarography with demountable sample compartment on left side.

means of the tube sealed near the bottom of the cell. Removing 99% of the oxygen ordinarily takes about 5 min; a glass frit dispersion tube with a flow rate of 20 to 50 ml/min will cut the time to 2 to 3 min. A two-way stopcock (not shown) is used to pass nitrogen over the top of the solution while the polarogram is recorded. The cell top is sealed with a rubber stopper drilled to loosely accommodate the DME and purging tube used to pass inert gas over the surface of the solution; the inert gas exits around the capillary of the DME. This cell and a number of other polarographic cells are available from the Sargent-Welch Scientific Company (United States).

We have found the Leeds and Northrup Company No. 7961 coulometric cell pictured in Figure 3-12 convenient for voltammetric and coulometric work except where there is the need to rigorously exclude oxygen or water. The glass cell is a standard 100-ml spoutless beaker (Corning No. 1040 or equivalent) and may be quickly changed by unsnapping the polyethylene retaining ring. The polyethylene top contains six molded holes that accommodate various electrodes, purging tubes, and salt bridges. In Figure 3-12 the cell is set up for three electrode polarography with the counter electrode isolated by a glass frit. (For coulometric work, a second glass fritted tube may be inserted inside the first to provide greater isolation of the counter electrode.) The reference electrode may be located in a Luggin probe as shown, or in a bridge tube like that shown in Figure 2-6d. This cell is well suited for routine voltammetry in organic solvents using a bridge tube filled with the organic solvent in conjunction with an aqueous or organic solvent reference electrode. The cell may be used for coulometry by inserting

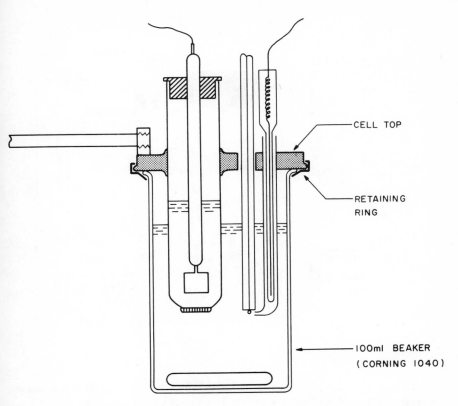

Figure 3-12. Versatile electrochemical cell with polyethylene top and retaining ring. The system includes a magnetic stirring bar, an isolated counter electrode, and a Luggin capillary reference electrode.

a platinum working electrode concentric with the glass fritted tube holding the counter electrode; it may also be used with a mercury pool, if a platinum or tungsten contact is sealed into the bottom of the cell. A magnetic stirring bar, resting on the bottom of the cell or floating on top of the mercury pool, provides adequate stirring. Cells of comparable versatility are commercially available from Metrohm (available through Brinkmann Instruments or Princeton Applied Research Corp.,) or from Beckman Instruments, Inc., both in the United States.

Figure 3-13 illustrates a cell which can be used for two- or three-electrode polarography. It has an integral salt bridge with provision for changing the bridge solution without disrupting or disassembling the reference electrode.

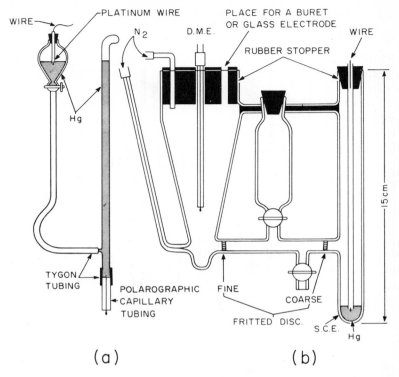

Figure 3-13. Cell for voltammetry and amperometric titrations. (*a*) Dropping mercury electrode assembly. (*b*) Cell with provision to flush the salt bridge.

This allows the use of an indifferent salt-bridge electrolyte solution which prevents cross contamination of the electrolytes in the sample and reference electrode compartments. The tapered cell can accommodate sample sizes from a few up to 50 ml; furthermore, it has a sufficiently large top that a second reference electrode may be introduced for three-electrode voltammetry, using the first reference electrode as the counter electrode. The large top also can accommodate other electrodes and a buret for specialized applications—for example, it is convenient to use with a glass electrode for studies of complexation equilibria; both the polarographic response and the solution pH can be monitored simultaneously. A slightly different cell has been described which is particularly suited to the determination of stability constants by the polarographic method (48).

B. CELLS FOR COULOMETRY AND PREPARATIVE ELECTROCHEMISTRY

The controlled current coulometric titration technique is inherently accurate and provides the basis for the precise determination of the Faraday constant to about 1 part per 100,000 (49). Two important factors in the design of cells for coulometric titrations are: (*a*) the need to maintain the cell resistance as low as possible; and (*b*) the necessity to provide adequate isolation of the counter electrode. Errors can arise from outflow of sample from the working electrode compartment or by inflow of reactive products produced at the counter electrode. Some loss through glass-frit or membrane separators is inevitable because the flow of current necessarily involves transport of material. The transport may be by migration, diffusion, and hydrodynamic flow. Migration can be reduced by using a supporting electrolyte in large excess over the sample concentration. The effect of the remaining two transport mechanisms may be reduced by special cell designs. Accurate results have been obtained with a four-compartment cell (see Figure 3-14) in which the two central bridge compartments are periodically emptied into the sample compartment by applying nitrogen pressure (49). They are refilled by applying vacuum.

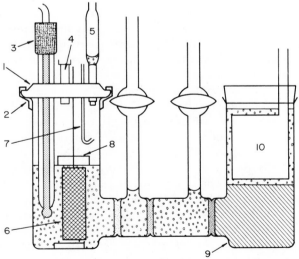

Figure 3-14. Four-compartment cell for controlled potential electrolysis and coulometric titrations. Two central bridge compartments can be emptied to sample compartment by application of nitrogen pressure and refilled by vacuum. 1, polyethylene top; 2, lock ring; 3, combination glass-calomel electrode; 4, N_2 inlet tube; 5, N_2 outlet tube; 6, Pt gauge electrode; 7, cell rinse assembly; 8, polyethylene spray shield; 9, 0.1 F KCl in 3% agar gel; 10, Ag anode.

A bridge compartment with sufficient hydrostatic head to ensure that fresh electrolyte flows continuously into both the working electrode and counter electrode compartments is another useful approach (50). However, the volumes of the solutions in both compartments increase inconveniently during a titration. This method also suffers if the background electrolyte contains reactive impurities (e.g., oxygen).

The problem of adequate isolation has been studied theoretically and practically in the context of coulometric titrations by Lindberg (51), who employed mathematical models of a four compartment cell like that of Figure 3-14. This work indicates that the loss of sample by diffusion will not be greater than 0.1% if at least two fine frit separators are used with at least 4 to 5 cm of electrolyte between the frits and if the time of electrolysis is short (3–15 min). A tube closed with a fine frit containing a 12-mm layer of fine silica gel above the frit has been suggested.

Ion exchange membranes have been used as separators (52,53), particularly in coulometric titrations. The fixed negative charge of a cation exchange membrane tends to exclude anions, so that current through the membrane is primarily due to cation transport with the reverse true for anion exchange membranes.

Other devices may be useful. For example, if the counter electrode is the anode, a silver anode may be used with a halide-containing supporting electrolyte; oxidation of the silver produces a layer of silver halide on the anode which can be periodically removed (54). Hydrazinium chloride (55) also has been used as an anodic depolarizer (particularly in hydrochloric acid solutions to prevent formation of chlorine at a platinum anode); it is quantitatively oxidized to nitrogen gas.

In both controlled potential and controlled current coulometry as low a cell resistance as possible is desirable. This obviously is affected by cell design (narrow passages between working electrode and counter electrode compartments should be avoided), but it is mainly determined by the specific resistance of the electrolyte. The latter typically will range from 5 Ω-cm for 0.5 F H_2SO_4 in water to 220 Ω-cm for 0.1 F tetra-n-butylammonium perchlorate in dimethylformamide. Typical cell resistances with these two electrolytes might range from 20 to 1000 Ω. A cell resistance as large as 1000 Ω may produce some difficulties. For example, to pass a current of 100 mA through the cell would require a potentiostat output of 100 V. Few potentiostats have an output voltage much larger than this, and many solid-state potentiostats are limited to 10 to 20 V. (Despite their large current outputs, the latter are useless for preparative-scale electrochemistry in cells with total resistances of 1000 Ω or more because the current is limited to 10 to 20 mA.)

Another factor to consider is the resistive heating of the electrolyte. In the example cited, 10 W of power are dissipated by passing 100 mA through a cell with 1000 Ω of resistance. This is sufficient to raise the temperature of the electrolyte significantly unless provision for cooling is made. Finally, large potential gradients in the cell may produce electrophoretic transport of solvent through membrane or glass-frit separators, producing loss or contamination of the sample, depending on the direction of flow.

1. Optimization of Cell Design for Controlled Potential Coulometry

Using concepts largely based on the work of Booman and others (56–59) a detailed study of the characteristics of three-electrode controlled-potential electrolysis cells as components of control systems has been made (14). The results indicate that excessive phase shift in a poorly designed cell can be a prime contributor to instability and oscillation of the potential control circuit. Ideally, a minimum phase shift cell will have a completely uniform potential and current distribution at the working electrode. If this cannot be realized, the reference electrode tip should be positioned on the line of minimum separation between the working electrode and the counter-electrode separator (for reasons previously described). The reference electrode resistance is more significant than had been realized previously, and it is important to keep this resistance as low as possible, along with the cell resistance, especially in organic solvents where resistances tend to be larger. The bridging capacitance between the counter electrode and the follower input of the reference electrode also should be minimized by use of cables in which the reference electrode conductor is shielded or guarded separately from that of the counter electrode (see Chapter 5).

Figure 3-15 shows a mercury-pool cell that has a virtually uniform potential and current distribution because of the large parallel spaced glass-frit separator. A somewhat more convenient cell with a less ideal current distribution is shown in Figure 3-16. Note the position of the reference electrode tip between the counter-electrode separator and the mercury pool. Both of these cells have approximately 10-cm^2 mercury-pool working electrodes with 6-ml sample solution volumes.

A platinum working electrode cell is shown in Figure 3-17. The projected area of the gauze is about 70 cm^2 with a solution volume of about 25 ml. Except for minor edge effects, this cell has a nearly uniform potential and current distribution because of the cylindrical symmetry of the platinum gauze and porous Vycor counter-electrode separator. The tip of the reference electrode is positioned close to the platinum gauze on a line representing the shortest distance between the gauze and the Vycor tube.

The time required for a simple controlled potential coulometric determination is determined by the efficiency of mass transport in the cell. The current decays exponentially according to the equation

$$i_t = i_0 e^{-kt} \qquad (3\text{-}4)$$

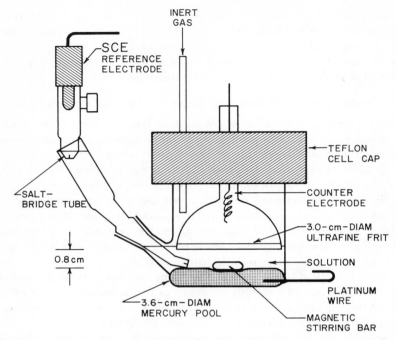

Figure 3-15. Mercury-pool cell for controlled-potential electrolysis with uniform potential and current distribution.

where i_t is the current at time t, i_0 is the initial current, and k is a function of the rate of mass transport. For a simple Nernst diffusion-layer model of mass transport k is given by the expression

$$k = \frac{DA}{\delta V} \tag{3-5}$$

where D is the diffusion coefficient of the electroactive species, A the electrode area, V the solution volume, and δ the thickness of the diffusion layer near the electrode surface. The actual functional dependence of k on the variables is more complex than indicated by Equation 3-5 (60), but as a rule of thumb, the ratio of solution volume to electrode area should be as small as possible and δ should be made as small as possible by efficient stirring.

Completion of electrolysis is generally taken at the time when the current has decayed to 0.1% of its initial value. For this condition

$$t = \frac{6.9}{k} \tag{3-6}$$

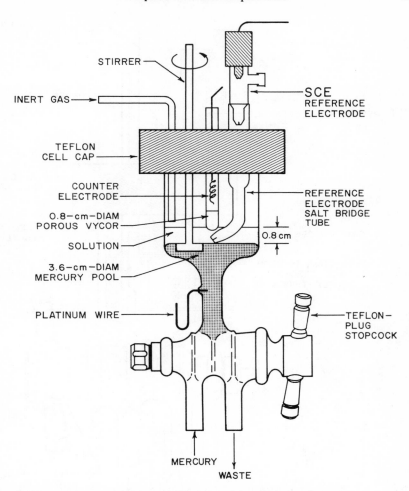

Figure 3-16. Mercury-pool cell for controlled-potential analytical coulometry.

which means that it is desirable to have k as large as possible. Cells like those depicted in Figures 3-15, 3-16, and 3-17 typically have k values of 0.008 $\sec^{-1}$. By increasing the efficiency of stirring and decreasing the cell volume values of k in the range 0.02 to 0.01 $\sec^{-1}$ can be achieved, which greatly shortens the time for a coulometric determination. Ultrasonic stirring (60) has been employed as have highly efficient rotated cells in which the solution is maintained in a thin layer over the electrode surface by centrifugal force (61).

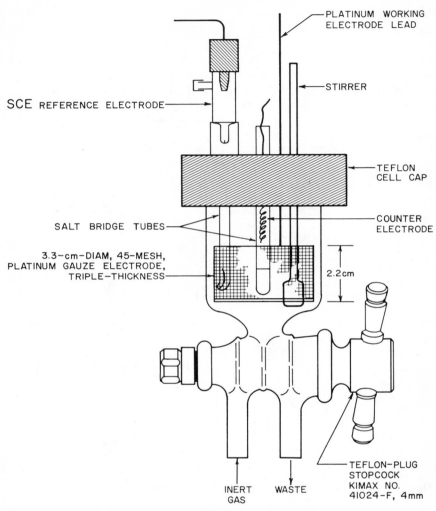

Figure 3-17. Platinum-gauze cell for controlled-potential analytical coulometry.

2. Cells for Preparative Electrochemistry

Organic and inorganic preparative electrochemistry requires that solution volumes and electrode areas be scaled up. Particular attention must be paid to optimizing the geometry of these cells in order to obtain uniform current and potential distribution over the working electrode. This is particularly important in organic electrochemistry, where products may depend critically on the potential and where the solvent-electrolyte systems used tend to have large resistances. The latter lead to large potential-control errors in a poorly designed cell. The limitations of potentiostat output and resistive heating also require that the cell be designed to obtain low resistance, and that provision for cooling be supplied. To avoid contamination of the working-electrode solution by the products of reaction at the counter electrode some designs provide for a continuous flow of fresh electrolyte through the counter electrode compartment.

Several successful cell designs and references to earlier literature are given in the chapter by Lund and Iversen in the recent monograph edited by Baizer (62) and in the shorter reviews by Eberson and Schäfer (63) and by Chang, Large, and Popp (64).

C. CONTROL OF TEMPERATURE AND PRESSURE

Diffusion coefficients in aqueous solution have a temperature coefficient of about $+2\%/\text{deg}$ (65) which means that polarographic diffusion currents or voltammetric peak currents increase about 1 to $2\%/\text{deg}$. The rates of follow-up chemical reactions of reactive species produced at the electrode surface depend even more strongly on temperature. For these reasons, temperature control to $\pm 0.1\,^\circ\text{C}$ or better is required in all careful quantitative work. This is accomplished by immersing the cell in a temperature controlled oil or water bath, or by circulating water in a jacketed cell (66). Temperature control in coulometry is less critical unless there is substantial cell heating by the current that passes through the cell. Coulometry usually is carried out at the ambient temperature of the laboratory.

Van Duyne and Reilley (67) have designed low-temperature electrochemical cells to study reactive intermediates at temperatures down to $-130\,^\circ\text{C}$. The variable temperature cell (Figure 3-18) is designed with opposed, coaxial working and Luggin-probe electrodes surrounded by a platinum-coil counter electrode. The cylindrical symmetry provides uniform current and potential distribution across the platinum working electrode. The Luggin probe is connected to an external aqueous reference electrode maintained at room temperature. Butyronitrile/0.1 F tetra-n-butylammonium perchlorate proved to be the most useful low-temperature medium; the 4-ml volume of solution is deaerated by nitrogen bubbled in through a long syringe needle. Cooling of the cryostat is achieved by passing compressed dry nitrogen through a copper coil immersed in a 4-liter dewar of liquid nitrogen. A vacuum jacketed heater also is located in the inlet line. The temperature of the cooling gas, and, in turn, the cryostat is automatically regulated with a recorder-controller equipped with a sensing thermocouple located in the gas space surrounding the cell. The actual solution temperature is measured

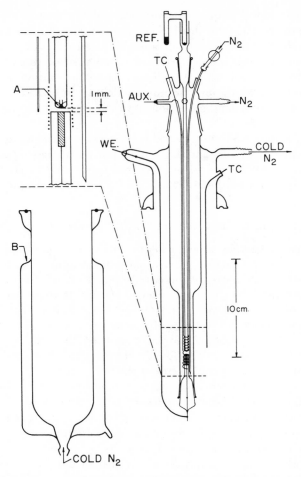

Figure 3-18. Electrochemical cell for low-temperature studies.

with a second thermocouple. The cell has been used to study radical-ion decay mechanisms, and for the evaluation of rate constants and activation parameters for homogeneous chemical reactions coupled to charge transfer. Other workers (68–70) also have reported stabilization of reactive intermediates by means of low temperature electrochemistry.

A special cell has been described (71) which employs a bimetal thermistor electrode to measure the exotherms and endotherms associated with electrochemical reactions. This is done concurrently with cyclic voltammetric measurements.

Fused salt systems and organic solvents with melting points above room temperature (such as dimethylsulfone) require higher temperatures. These may range from simple systems using beakers wrapped with heating tapes to carefully designed cells that are contained in high temperature furnaces with temperature controllers. At higher temperatures the external connectors and seals generally are contained in a water cooled head so that conventional sealing materials can be used and the hazard of burns reduced. The reference handbook edited by Janz (72) gives references to a number of cell designs as does a chapter by Corbett and Duke in the series of volumes edited by Jonassen and Weissberger (73). A recently published laboratory manual (74) describes various types of high pressure and high temperature apparatus with special emphasis on their design, control, and calibration, along with extensive bibliographies. Zambonin (75) has described a general purpose cell which uses a rotating platinum electrode and is designed for operation in fused salts.

High pressure cells mainly are used to contain low boiling solvents like liquid NH_3 or liquid SO_2 at ambient temperature or to study solvents with moderate boiling points at elevated temperatures. Hills (76) has provided a review of polarography at high pressures, describing in general outline the apparatus and some of the results that have been obtained by high temperature, high pressure polarography. The dropping mercury electrode can be readily operated at several thousand atmospheres and one of the interesting results is that the hydrogen overpotential on mercury declines about 1 V over a range of 200°C. Thus above 350°C the reduction of H^+ appears to be fast and reversible on mercury. Considerable detail about the handling of low boiling solvents also is provided in the series of volumes edited by Lagowski (77).

D. CELLS FOR CONDUCTIMETRY

Reliable and precise measurements of electrolytic conductance require attention to the design of cells, electrodes, and measuring circuitry. In Chapter 5 a short discussion is included on the measurement of solution conductance. Extraction of an ohmic resistance from A-C bridge measurements is not a trivial task, particularly in solutions with high resistance (such as organic solvents) or very low resistance (molten salts). Expositions of the principles are provided in two monographs (78,79) that emphasize aqueous solutions. Shedlovsky and Shedlovsky (80) recently have provided a concise review of the principles and experimental methods, with emphasis on aqueous solutions, and Loveland (81) has presented a review of conductimetry and high frequency oscillometry that emphasizes analytical applications.

Of perhaps greater interest to electrochemists who work in organic sol-

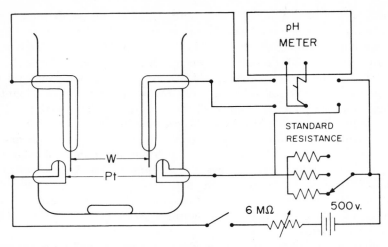

Figure 3-19. Four-electrode cell system for D-C conductometric titrations. The tungsten electrodes (W) act as sensors to the ion migration that results from the small current between the platinum electrodes (Pt).

vents is a review which surveys measurements of the conductance of electrolytes in a number of solvents (82); a brief discussion of experimental methods and a more extensive discussion of conductance data with tabulations of conductance parameters is included. For conductivity measurements in organic solvents, cells patterned after the design by Kraus (83) are widely used because the same batch of solvent (usually laboriously purified) can be used for a number of measurements. Solutions of different concentrations are prepared in the cell, either by serial dilution of a concentrated solution or by the addition of preweighed amounts of salt. The latter are dispensed from cups that are contained in a small rotating turntable which fits on top of the cell (84). A cell patterned after the Kraus design is available commercially (Beckman Instruments, Fullerton, Calif.).

Troublesome frequency dependence in difficult systems has prompted some workers to use D-C conductance methods. Such methods usually employ four electrodes, two to carry the current and two to measure the potential drop without carrying current. The method has been used to measure the conductance in molten alkali nitrates (85); silver electrodes in thin-walled compartments are used to sense the iR drop through the solution.

A four-electrode cell also has been employed to measure resistance in conductimetric titrations (86) (see Figure 3-19). Here the aim is to achieve

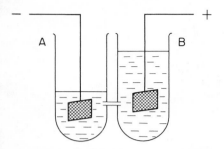

Figure 3-20. Capillary-tube D-C conductance cell with platinum mesh electrodes (6 cm²
each). Half-cell compartment volumes, 100 ml; capillary, 2 cm×0.16 mm ID. As shown, flow
from compartment *B* controls conductance.

simplicity of measurement; an A-C bridge is not required and the potential
drop can be measured with a pH meter or high impedance voltmeter. The
two center electrodes in the cell of Figure 3-19 do not carry appreciable
current, so the resistance measurement is not disturbed by polarization. The
two outer electrodes have net faradaic currents at the electrode-solution
interface, but the currents are so small that the composition of the solution is
not significantly altered.

Another stable and sensitive conductivity cell has been described which
measures the D-C current that passes through a small volume of solution
confined in a small capillary tube (87). The very simple apparatus is
illustrated in Figure 3-20. When filled with 0.1 *F* KCl the cell exhibits an
Ohm's law behavior with an applied voltage from 5 to 100 V (approximately 5 to 100 mA of current). The cell also can be used for conductimetric
titrations, with the titration carried out in vessel *B*. Because some liquid
flows through the capillary during a titration, a correction factor is used to
obtain the most accurate results (although corrections usually are less than
0.2%).

E.   MICRO CELLS

Cells that employ thin layers of solution (thin-layer electrodes) have special
virtues that have been detailed in Chapter 2. In addition, the dipping
thin-layer cell is especially useful for the measurement of chronopotentiometric *n* values (88). It is constructed from the components of a
micrometer such that the cell thickness can be varied to permit a plot of the
transition time versus solution thickness. This allows the determination of *n*,
the number of electrons involved in the electrochemical reaction. An ac-

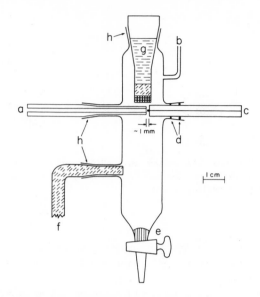

Figure 3-21. Flow cell with horizontal DME for continuous monitoring. (*a*) Sample introduction tube; (*b*) solution outlet; (*c*) DME capillary; (*d*) O-ring seals; (*e*) waste mercury; (*f*) agar salt bridge to counter electrode; (*g*) SCE reference electrode with agar plug and fine glass frit; (*h*) ground glass joints.

curacy of $\pm 4\%$ has been demonstrated for a number of compounds in different organic solvents.

In most electrochemical measurements solutions are made up to an arbitrary volume which usually is at least 1 cm^3. However a few microcells have been described for work with solution volumes that are well below 1 cm^3. The coulometric determination of silver ion in cell volumes as small as 20 μl (formed by a thin copper sheet and a cavity of beeswax) has been discussed (89). For work with the hanging mercury drop electrode, a cell has been described (90) for use with volumes as small as 0.3 ml. An interesting cell for the micro combustion of small samples has been described (91). After combustion, electrolyte is added and the solution volume is measured by a calibrated sidearm, then displaced by mercury into a compartment that contains a dropping mercury electrode.

F. FLOW AND CIRCULATION CELLS

Voltammetric techniques can be used to monitor a flowing stream continuously; for example, the effluent from an ion exchange column. Of course, the material whose detection is sought must be electroactive. An effective cell design has been described (Figure 3-21) which employs a

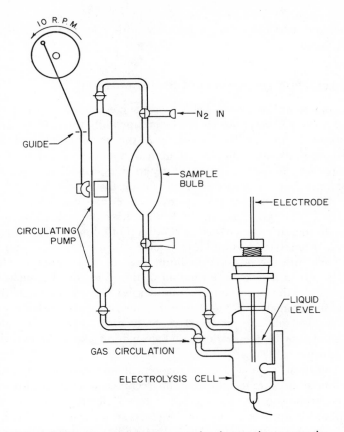

Figure 3-22. Gas-recirculating system for electroactive gas samples.

vertical orifice (Smoler) dropping mercury electrode (92). The cell possesses a low holdup time, a small volume, and a low flushout time, all of which insure rapid response. It also can be operated for long periods of time without dismantling or cleaning. Because of the low noise of the signal (about 2 nA), high sensitivity is achieved, even in the presence of high background current. Changes of concentration as small as $3 \times 10^{-6}\ F$ HCl in the presence of oxygen-saturated 1 F KCl can be detected with good accuracy (100 nA current signal on a 32-μA background current).

Figure 3-22 illustrates a cell that is useful for the circulation of an electroactive gas through a solution (93). Such a system is particularly desirable when the supply of gas is so limited that it cannot be continuously bubbled through the solution. The gas is pumped by means of a loosely fitting Teflon-coated piston which is driven with a reciprocating motion by an external magnet.

G.   CELLS FOR SPECTROELECTROCHEMISTRY

Spectroscopic techniques have been used in conjunction with electrochemistry in a variety of ways, which can be grouped into three areas; the direct optical study of the electrode interface, the measurement of photon-stimulated currents, and the use of optical and electron spin resonance spectroscopy for the *in situ* study of reactive intermediates (particularly radical anions) that are produced electrochemically.

Direct optical studies of the electrode interface have been reviewed recently (94). The principal methods that have been used include: ellipsometry to measure the change of condition of elliptically polarized light as it passes from one medium into another through a thin film and is reflected back into the original medium; specular reflectance or electoreflectance (95,96) to detect the formation of thin films on the metal surface and to determine the electronic structure of solid surfaces; X-ray diffraction; and laser interferometry (97) to determine concentration gradients near the electrode surface that are produced by passage of current. The references that are cited provide the reader with information about the experimental methods and the cells.

There has been a great deal of theoretical and experimental work on photon-stimulated currents. Most of this centers around the questions of whether electrons can be photoejected from an electrode surface (98) and whether photochemical excitation of the reactant or the electrode can reduce the barrier to electron transfer (98–100). In addition there has been some study of reactive intermediates that are produced near an electrode by flash photolysis (101).

The third area of interest has been the observation by optical and electron spin resonance spectroscopy of intermediates that are produced electrochemically. Electron spin resonance is a useful technique for identifying species which have unpaired electrons and recent reviews have documented the power of ESR for unravelling complicated reaction pathways (102–104). A number of cells have been described for use with this technique which fall into two categories—the flow cell in which the reactive intermediate is generated externally and flows into the cavity (105); and the *in situ* generation system where electrodes are placed inside the resonant cavity of the spectrometer (106).

Many optical techniques make use of transparent electrodes such as

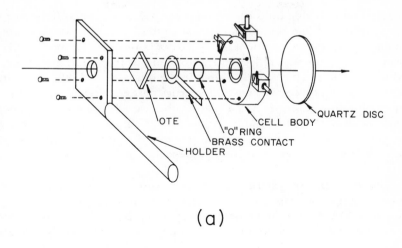

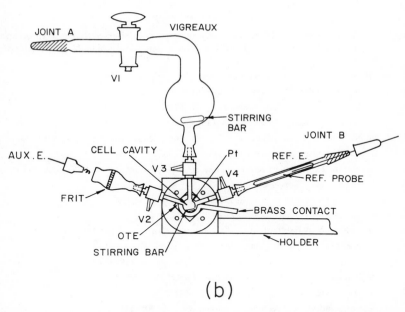

Figure 3-23. Cell system for spectroelectrochemistry by use of optically transparent electrodes (OTE).

conducting tin oxide films coated on glass (107), thin metal films coated on glass (108), and fine gold-mesh electrodes (109). The first two types are used mainly in the visible region of the spectrum [although thin metal grids have been used for internal reflection spectroscopy in the infrared region (110)], while the gold-mesh electrode is used in the ultraviolet, visible, and infrared. The optical and electrochemical properties of these electrodes are described more fully in Chapter 2. A cell that has been used with the transparent tin oxide coated electrode is shown in Figure 3-23. In combination with a rapid scanning spectrophotometer this cell provides the means to obtain spectra of intermediates that are involved in biological electron transfer sequences (107).

The principle of external generation also has been used for optical measurements of transient intermediates. Flow cells (111,112) and cells which provide for the circulation of the solution past an electrode and into a spectrophotometric cell in a closed loop have been described (113,114).

References

1. W. B. Schaap and P. S. McKinney, "Recent Studies in High-resistance Polarography," in *Polarography 1964*, Vol. 1, G. J. Hills, ed., Interscience Publishers, New York, 1966, pp. 197–214.

2. W. B. Schaap and P. S. McKinney, *Anal. Chem.*, **36**, 1251 (1964).

3. J. E. Mumby and S. P. Perone, *Chem. Instrum.*, **3**, 191 (1971).

4. D. Ilkovic, *Coll. Czech. Chem. Commun.*, **8**, 13 (1936).

5. L. Nemec, *J. Electroanal. Chem.*, **8**, 166 (1964).

6. I. M. Kolthoff, J. C. Marshall, and S. L. Gupta, *J. Electroanal. Chem.*, **3**, 209 (1962).

7. J. Devay, *Acta Chim. Acad. Sci. Hung.*, **35**, 255 (1963).

8. W. L. Belew, D. J. Fisher, M. T. Kelley, and J. A. Dean, *Chem. Instrum.*, **2**, 297 (1970).

9. R. Bezman, *Anal. Chem.*, **44**, 1781 (1972).

10. R. V. Sedletskii and B. E. Limin, *Elektrokhimiya*, **8**, 22 (1972).

11. J. E. Harrar and I. Shain, *Anal. Chem.*, **38**, 1148 (1966).

12. B. D. Cahan, A. Nagy, and M. A. Genshaw, *J. Electrochem. Soc.*, **119**, 64 (1972).

13. G. L. Booman and W. B. Holbrook, *Anal. Chem.*, **37**, 795 (1965).

14. J. E. Harrar and C. L. Pomernacki, *Anal. Chem.*, **45**, 57 (1973).

15. G. L. Booman and W. B. Holbrook, *Anal. Chem.*, **35**, 1793 (1963).

16. C. Kasper, *Trans. Electrochem. Soc.*, **78**, 131 (1940).

17. T. Kimura and G. R. Freeman, *J. Chem. Educ.*, **50**, A85 (1973).

18. N. W. Alcock, *Anal. Chem.*, **40**, 1397 (1968).

19. I. Morcos and E. Yeager, *Electrochim. Acta*, **15**, 953 (1970).

20. J. H. Yoe and H. J. Koch, *Trace Analysis*, John Wiley and Sons, Inc., New York, 1957.

21. G. G. Eichholz, A. E. Nagel, and R. B. Hughes, *Anal. Chem.*, **37**, 863 (1965).

22. R. D. DeMars and I. Shain, *Anal. Chem.*, **29**, 1825 (1957).

23. R. A. Durst and B. T. Duhart, *Anal. Chem.*, **42**, 1002 (1970).

24. R. V. Coyne and J. A. Collins, *Anal. Chem.*, **44**, 1093 (1972).

25. D. E. Robertson, *Anal. Chem.*, **40**, 1067 (1968).

26. J. Chatt, *Pure Appl. Chem.*, **24**, 425 (1970).

27. F. I. Scott, Jr. and R. E. Rutherford, *Amer. Lab.*, 43 (August, 1970); Electron Technology, 626 Schuyler Ave., Kearny, N.J. 07032.

28. C. A. Seitz, W. M. Bodine, and C. L. Klingman, *J. Chromatographic Sci.*, **9**, 29 (1971).

29. L. Meites, *Polarographic Techniques*, 2nd ed., Interscience Publishers, New York, 1965, p. 87.

30. J. J. Lingane, *Electroanalytical Chemistry*, 2nd ed., Interscience Publishers, Inc., New York, 1958, p. 244.

31. G. W. Ewing and J. E. Nelson, *J. Chem. Educ.*, **46**, 292 (1969).

32. J. S. Hetman, in *Polarography 1964*, Vol. 1, G. J. Hills, ed., Interscience Publishers, Inc., New York, 1966, p. 389.

33. Ref. 29, p. 89.

34. D. J. C. Yates, *J. Chem. Educ.*, **44**, 699 (1967).

35. D. F. Shriver, *The Manipulation of Air-Sensitive Compounds*, McGraw-Hill Book Company, New York, 1969, pp. 230–231.

36. Vacuum/Atmospheres Company, 4652 W. Rosecrans Ave., Hawthorne, Calif. 90250.

37. W. Bahmet and P. A. Hersch, *Anal. Chem.*, **43**, 803 (1971); P. A. Hersch, *Amer. Lab.*, 29 (August, 1973).

38. I. D. Eubanks and F. J. Abbott, *Anal. Chem.*, **41**, 1708 (1969).

39. F. A. Keidel, *Anal. Chem.*, **31**, 2043 (1959).

40. Ref. 35, p. 145.

41. Ref. 35, p. 154.

42. A. J. Bard and K. S. V. Santhanam, "Application of Controlled Potential Coulometry to the Study of Electrode Reactions," in *Electroanalytical Chemistry*, Vol. 4, A. J. Bard, ed., Marcel Dekker, Inc., New York, 1970, p. 304.

43. A. J. Bard, *Pure Appl. Chem.*, **25**, 379 (1971).

44. J. L. Mills, R. Nelson, S. G. Shore, and L. B. Anderson, *Anal. Chem.*, **43**, 157 (1971).

45. C. D. Schmulbach and T. V. Oomen, *Anal. Chem.*, **45**, 820 (1973).

46. R. F. Amlie and T. A. Berger, *J. Electroanal. Chem.*, **36**, 427 (1972).

47. L. Meites and H. Hofsass, *Anal. Chem.*, **31**, 119 (1959).

48. I. Piljac, B. Grabaric, and I. Filipovic, *J. Electroanal. Chem.*, **42**, 433 (1973).

49. G. Marinenko and J. K. Taylor, *Anal. Chem.*, **40**, 1645 (1968).

50. E. L. Eckfeldt and E. W. Shaffer, Jr., *Anal. Chem.*, **37**, 1534 (1965).

51. J. Lindberg, *J. Electroanal. Chem.*, **40**, 265 (1972).

52. P. P. L. Ho and M. M. Marsh, *Anal. Chem.*, **35**, 618 (1963).

53. S. W. Feldberg and C. E. Bricker, *Anal. Chem.*, **31**, 1852 (1959).

54. Ref. 30, p. 381.

55. Ref. 30, p. 375.

56. A. Bewick, M. Fleischmann, and M. Liler, *Electrochim. Acta*, **1**, 83 (1959).

57. D. T. Pence and G. L. Booman, *Anal. Chem.*, **38**, 1112 (1966).

58. E. R. Brown, D. E. Smith, and G. L. Booman, *Anal. Chem.*, **40**, 1411 (1968).

59. J. E. Harrar and I. Shain, *Anal. Chem.*, **38**, 1148 (1966).

60. A. J. Bard, *Anal. Chem.*, **35**, 1125 (1963).

61. R. G. Clem, *Anal. Chem.*, **43**, 1853 (1971).

62. H. Lund and P. Iversen, "Practical Problems in Electrolysis," in *Organic Electrochemistry*, M. M. Baizer, ed., Marcel Dekker, Inc., New York, 1973, Chap. IV.

63. L. Eberson and H. Schäfer, *Fortschr. Chem. Forsch.*, **21**, 1 (1971).

64. J. Chang, R. F. Large, and G. Popp, "Electrochemical Synthesis," in *Techniques of Chemistry*, Vol. 1, *Physical Methods of Chemistry*, A. Weissberger and B. W. Rossiter, eds., Part IIB, *Electrochemical Methods*, Wiley-Interscience, New York, 1971, Chap. X.

65. I. M. Kolthoff and J. J. Lingane, *Polarography*, Vol. 1, 2nd ed., Interscience Publishers, New York, 1952, p. 29.

66. R. C. Buchta and D. H. Evans, *Anal. Chem.*, **40**, 2181 (1968).

67. R. P. Van Duyne and C. N. Reilley, *Anal. Chem.*, **44**, 142; 153; 158 (1972).

68. K. Bechgaard and V. D. Parker, *J. Am. Chem. Soc.*, **94**, 4749 (1972).

69. L. L. Miller and E. A. Mayeda, *J. Am. Chem. Soc.*, **92**, 5818 (1970).

70. R. J. Wilson, L. F. Warren, and M. F. Hawthorne, *J. Am. Chem. Soc.*, **91**, 758 (1969).

71. B. B. Graves, *Anal. Chem.*, **44**, 993 (1972).

72. G. J. Janz, ed., *Molten Salts Handbook*, Academic Press, New York, 1967.

73. J. D. Corbett and F. R. Duke, "Fused Salt Techniques," in *Technique of Inorganic Chemistry*, Vol. 1, H. B. Jonassen and A. Weissberger, eds., Interscience Publishers, 1963, Chap. 3.

74. G. C. Ulmer, ed., *Research Techniques for High Pressure and High Temperature*, Springer-Verlag, New York, 1971.

75. P. G. Zambonin, *Anal. Chem.*, **41**, 868 (1969).

76. G. J. Hills and P. J. Ovenden, "Electrochemistry at High Pressure," in *Advances in Electrochemistry and Electrochemical Engineering*, Vol. 4, P. Delahay and C. W. Tobias, eds., Interscience Publishers, New York, 1966, p. 185.

77. J. J. Lagowski, ed., *The Chemistry of Non-Aqueous Solvents*, Academic Press, New York, Vol. 1, 1966, Chap. 7; Vol. 2, 1967, Chap. 6; Vol. 3, 1970, Chap. 2.

78. H. S. Harned and B. B. Owen, *The Physical Chemistry of Electrolytic Solutions*, 3rd ed., Reinhold Publishing Corp., New York, 1958, Chap. 6.

79. R. A. Robinson and R. H. Stokes, *Electrolyte Solutions*, Butterworths Scientific Publications, London, 1955.

80. T. Shedlovsky and L. Shedlovsky, "Conductometry," in *Techniques of Chemistry*, Vol. 1 *Physical Methods of Chemistry*, A. Weissberger and B. W. Rossiter, eds., Part IIA, *Electrochemical Methods*, Wiley-Interscience, New York, 1971, Chap. III.

81. J. W. Loveland, "Conductometry and Oscillometry," in *Treatise on Analytical Chemistry*, Vol. 4, Interscience Publishers, Part I, 1963, Chap. 51, p. 2569.

82. B. Kratochvil and H. L. Yeager, *Fortschr. Chem. Forsch.*, **27**, 1 (1972).

83. H. M. Daggett, E. J. Bair, and C. A. Kraus, *J. Am. Chem. Soc.*, **73**, 799 (1951).

84. R. L. Kay, B. J. Hales, and G. P. Cunningham, *J. Phys. Chem.*, **71**, 3925 (1967).

85. L. A. King and F. R. Duke, *J. Electrochem. Soc.*, **111**, 712 (1964).

86. R. P. Taylor and N. H. Furman, *Anal. Chem.*, **24**, 1931 (1952).

87. O. Hello, *Anal. Chem.*, **44**, 646 (1972).

88. J. E. McClure and D. L. Maricle, *Anal. Chem.*, **39**, 236 (1967).

89. S. S. Lord, Jr., R. C. O'Neill, and L. B. Rogers, *Anal. Chem.*, **24**, 209 (1952).

90. W. L. Underkofler and I. Shain, *Anal. Chem.*, **33**, 1966 (1961).

91. R. M. Parkhurst, *Anal. Chem.*, **33**, 320 (1961).

92. E. Scarano, M. G. Bonicelli, and M. Forina, *Anal. Chem.*, **42**, 1470 (1970).

93. P. E. Toren, *Anal. Chem.*, **35**, 120 (1963).

94. B. E. Conway, "Special Techniques in the Study of Electrode Processes and Electrochemical Adsorption," in *Techniques of Electrochemistry*, Vol. 1, E. Yeager and A. J. Salkind, eds., Wiley-Interscience, New York, 1972, Chap. 5.

95. T. Takamura, Y. Sato, and K. Takamura, *J. Electroanal. Chem.*, **41**, 31 (1973).

96. D. Laser and M. Ariel, *J. Electroanal. Chem.*, **35**, 405 (1972).

97. R. N. O'Brien and F. P. Dieken, *J. Electroanal. Chem.*, **42**, 25; 37 (1973).

98. H. Berg, H. Schweiss, E. Stutter, and K. Weller, *J. Electroanal. Chem.*, **15**, 415 (1967).

99. G. C. Barker, *Electrochim. Acta*, **13**, 1221 (1968).

100. N. Martinus, D. M. Rayner, and C. A. Vincent, *Electrochim. Acta*, **18**, 409 (1973).

101. G. L. Kirschner and S. P. Perone, *Anal. Chem.*, **44**, 443 (1972).

102. K. W. Bowers, "ESR of Radical Ions," in *Advances in Magnetic Resonance*, Vol. 1, J. S. Waugh, ed., Academic Press, New York, 1965, pp. 317–396.

103. R. N. Adams, *J. Electroanal. Chem.*, **8**, 151 (1964).

104. Ya. P. Stradyn' and R. A. Gavar, "Electrochemical Generation of Free Radical Ions and Use of Electron Paramagnetic Resonance for their Investigation," in *Progress in Electrochemistry of Organic Compounds*, Vol. 1, A. N. Frumkin and A. B. Ershler, eds., Plenum Press, New York, 1971, pp. 1–42.

105. P. H. Rieger, J. Bernal, W. H. Reinmuth, and G. K. Fraenkel, *J. Am. Chem. Soc.*, **85**, 683 (1963).

106. I. B. Goldberg and A. J. Bard, *J. Phys. Chem.*, **75**, 3281 (1971).

107. F. M. Hawkridge and T. Kuwana, *Anal. Chem.*, **45**, 1021 (1973).

108. W. Von Benken and T. Kuwana, *Anal. Chem.*, **42**, 1114 (1970).

109. M. Petek, T. E. Neal, and R. W. Murray, *Anal. Chem.*, **43**, 1069 (1971).

110. D. Laser and M. Ariel, *J. Electroanal. Chem.*, **41**, 381 (1973).

111. R. E. Sioda and W. Kemula, *J. Electroanal. Chem.*, **31**, 113 (1971).

112. J. Q. Chambers and R. N. Adams, *Mol. Phys.*, **9**, 413 (1965).

113. J. Janata and H. B. Mark, Jr., *Anal. Chem.*, **39**, 1896 (1967).

114. A.-M. Gary, E. Piemont, M. Roynette, and J. P. Schwing, *Anal. Chem.*, **44**, 198 (1972).

Four

SOLVENTS AND ELECTROLYTES

I. Introduction

The choice of solvent in an electrochemical investigation usually is dictated by circumstances. For example, an electrochemical technique frequently is used to study a solvent-solute system that has been studied already by other techniques. The focus is on the particular system and the information that can be gleaned from the electrochemical study. However, if there is a choice of the solvent to be used, some rational criteria can be used to choose the optimum one.

A. THE PHYSICAL CHEMICAL PROPERTIES OF SOLVENTS AND THEIR RELEVANCE TO ELECTROCHEMISTRY

The solvent properties of electrochemical importance include: protic character (acid-base properties): anodic and cathodic voltage limits (related to redox properties and protic character); mutual solubility of the solute and solvent; and physical-chemical properties of the solvent (dielectric constant and polarity, donor or solvating properties, liquid range, viscosity, and spectroscopic properties). Practical factors also enter into the choice and include the availability and cost of the solvent, ease of purification, toxicity, and general ease of handling.

1. Protic Character

The protic character of the solvent is an important consideration because electrochemical intermediates (particularly radical anions) frequently react rapidly with protons. The classification of solvents into protic or aprotic solvents is somewhat arbitrary. A simple classification (1) is that protic solvents (such as hydrogen fluoride, water, methanol, formamide, and ammonia) are strong hydrogen-bond donors, exchange protons rapidly, and include solvents with hydrogen bound to more electronegative atoms (such as fluorine, oxygen, and nitrogen). Aprotic solvents generally have hydrogen bound only to carbon and are at best poor hydrogen-bond donors; they are weakly acidic and proton exchange occurs slowly, even with D_2O. Common aprotic solvents include aliphatic and aromatic hydrocarbons, chlorinated hydrocarbons, ethers, acetone, nitromethane, nitrobenzene, acetonitrile, benzonitrile, dimethylformamide, N-methyl-2-pyrrolidone, propylene carbonate, dimethylsulfoxide, sulfolane, and hexamethylphosphoramide.

A classification of solvents can be developed which is based on the stability of the radial anion produced by reduction of aromatic hydrocarbons, such as naphthalene and anthracene. The solvent reactions of such anions have been widely studied (2) and have generally been found to go by a sequence of reactions in either a protic solvent or in the presence of a proton donor in an aprotic solvent (3).

$$R + e^- \rightleftharpoons R^- \cdot \qquad (4\text{-}1)$$

$$R^- \cdot + HX \rightarrow RH \cdot + X^- \qquad (4\text{-}2)$$

$$RH \cdot + e^- \rightarrow RH^- \qquad (4\text{-}3a)$$

or

$$RH \cdot + R^- \cdot \rightarrow RH^- + R \qquad (4\text{-}3b)$$

$$RH^- + HX \rightarrow RH_2 + X^- \qquad (4\text{-}4)$$

The reversible one-electron transfer to form an anion radical, $R^- \cdot$, is followed by an irreversible chemical protonation to form $RH \cdot$ which is subsequently reduced itself [the reduction potential of the species $RH \cdot$ has been shown to be more positive (3) than that of the parent, R] and then undergoes another irreversible protonation reaction. In a protic solvent, the reactions proceed rapidly to the final product, RH_2. In a rigorously purified aprotic solvent, the intermediate anion radical, $R^- \cdot$, has an appreciable lifetime and only slowly reacts, principally with adventitious impurities in the solvent. Thus the stability of aromatic anion radicals can be taken as a measure of the protic character of a solvent.

More recently, it has been suggested that the mechanism of reduction of aromatic hydrocarbons proceeds via reaction 4-3b rather than 4-3a (4), but this will have no effect on a classification that is based on the stability of the radical anion produced in reaction 4-1.

Both kinetic and thermodynamic factors are important in determining protic character. Although the equilibrium concentration of solvated protons in a protic solvent like water or ethanol may be very small, there is a low activation energy for dissociation or exchange of protons, and these labile protons can rapidly react with any species having an appreciable proton affinity. Aprotic solvents usually have a lower equilibrium concentration of solvated protons (perhaps by as much as a factor of 10^{10}) and the activation energy for dissociation and exchange is substantially higher (5,6), which results in low proton availability in the solvent.

In mechanistic studies, aprotic solvents are preferred because electrochemical intermediates (particularly anion radicals) often have sufficient stability to enable their identification by spectroscopic techniques such as electron spin resonance or ultraviolet absorption spectroscopy. Follow up chemical reactions frequently proceed slowly enough to allow rate measurements.

Aprotic solvents may be the preferred solvents in organic polarographic analysis. This is said in spite of the fact that Zuman (7) has stated that "wherever possible aqueous solutions are preferred." Woodson and Smith (8) have reported that D-C, fundamental harmonic A-C, or second harmonic A-C polarographic responses are obtained for 19 of 24 pharmaceutically important compounds in acetonitrile. Barbiturates, salicylates, corticosteroids, alkaloids, sulfa drugs, and estrogens are included. Many of the compounds give ideal, one-electron reversible (diffusion controlled) waves, whereas the corresponding aqueous solution response frequently is irreversible or nonexistent. Similar conclusions may be justified for other dipolar aprotic solvents. The advantages of aprotic solvents over water include: generally greater solubilities of organic compounds; a wider range of D-C potentials for the observation of both oxidation and reduction processes; minimization of troublesome adsorption effects; and more numerous reversible electrode reactions which make the use of highly sensitive A-C polarographic methods feasible. A balanced view must be taken however. Not all of the advantages cited for aprotic solvents may be realizable in any one system. Frequently an analytical sample is presented in an aqueous solution and the removal of water and redissolution of the sample in an organic solvent may be troublesome. Where there is a choice of solvents, the behavior of the compound should be investigated in one or more dipolar aprotic solvents.

2. Voltage Limits

The voltage limits of a solvent define the "window" of accessible electronic energy levels available for electron-transfer processes. Although the solvent may have intrinsic limits based on its oxidation-reduction properties, the practical working limits depend also on the nature of the working electrode material and the composition of the supporting electrolyte. In practical terms, the voltage limits are a system property. Table 4-1 summarizes the practical working limits for a number of solvent, electrode-supporting electrolyte systems, expressed *versus* the aqueous calomel or other reference electrode. Because of the uncertainty in junction potentials (see Chapter 2), there is a considerable uncertainty in the estimated values *versus* the aqueous calomel reference electrode. The choice of limits also is somewhat arbitrary and dependent on current density; therefore the limits have an uncertainty of about $\pm 0.2 \, V$. These values generally are for rigorously purified solvents; the presence of proton donors will seriously diminish the negative (cathodic) limits, particularly on platinum and other solid-metal electrodes with low overvoltages for hydrogen ion reduction. The anodic and cathodic limits also are affected by the composition of the supporting electrolyte and the temperature of the electrolysis-cell system. One study (9) indicates that the voltage limits are greater at low temperatures.

Several studies of chemical oxidation-reduction reactions in nonaqueous solvents have been made (10). Not all dipolar aprotic solvents exhibit good stability toward oxidation or reduction. While some solvents extend the range toward reducing conditions, others withstand oxidizing conditions. For example, dimethylformamide can be used to study reactions involving powerful reductants such as Cr(II), but is readily oxidized by mild oxidizing agents. Conversely, nitrobenzene is resistant to oxidation but quite susceptible to reduction. Although these purely chemical reactions usually parallel their electrochemical analogs, there is not always an exact correspondence between chemical and electrochemical electron-transfer reactions.

3. Solvent Polarity

The Dipole Moment and Dielectric Constant. The net dipole moment of a molecule is given by the vector sum of the bond moments and is a function of the charge separation and geometry of the molecule. Because of the geometry factor and the possibility that bond moments may partially or exactly cancel, the dipole moment is probably a less useful measure of the ability of a solvent to promote dissociation of an ionic solute than is the dielectric constant, ϵ. The orientation of solvent dipoles in an electrostatic field tends to partially cancel the electrostatic field, and in a point-charge electrostatic model, the force between positive and negative point charges is

Table 4-1 Voltage Range on Mercury and Platinum in Selected Solvents and Supporting Electrolytes[a]

Supporting Electrolyte	Acetonitrile	Propylene Carbonate	Dimethylformamide	Dimethylsulfoxide
TEAP[b]				
Hg	0.6 to −2.8	0.5 to −2.5	0.5 to −3.0	0.25 to −2.8
Pt		1.7 to −1.9	1.6 to −2.1	0.7 to −1.85
TBAI[c]				
Hg	−0.6 to −2.8		−0.4 to −3.0	−0.4 to −2.85
Et$_4$NBF$_4$				
Hg	—to −2.7		—to −2.7	
Pt	+2.3 to—	+2.4 to—		
NaClO$_4$				
Hg	1.8 to −1.5		0.5 to −2.0	0.25 to −1.90
Pt			1.6 to −1.6	0.7 to −1.85

[a] Measured potentials are in volts, versus aqueous SCE.
[b] TEAP = tetraethylammonium perchlorate.
[c] TBAI = tetra-n-butylammonium iodide.

inversely proportional to the bulk dielectric constant of the medium. A large dielectric constant therefore promotes the dissociation of an ionic solute which in turn lowers the solution resistance. Whenever possible, a solvent with a substantial dielectric constant should be used in electrochemical work to minimize the solution resistance. This will minimize ohmic losses and diminish the problem of potential-control errors (see Chapter 3).

The dipole moments and dielectric constants of aprotic solvents range from near-zero values for the hydrocarbons to moderate values for solvents like dimethylformamide ($\epsilon = 36.7$) and acetonitrile ($\epsilon = 36$). At dielectric constants much below 15, substantial ion association begins to take place. In accord with the classification of Kratochvil (10) a value of 25 for the dielectric constant is used as the dividing line between nonpolar and dipolar solvents. The choice is arbitrary because there is no sharp change of solute dissociation behavior in this region and the dielectric constants of the various solvents fall on a continuum. Others have suggested values from 15 to 30 for the dividing line.

The Gutmann Donor Number.   Donor or *ionizing* solvents promote the ionization of covalent compounds to form intimate (or contact) ion-pairs. This property is largely a function of the coordinating power, or Lewis-base strength, of the solvent and is quantitatively expressed by the donor number. The *dissociating* power of a solvent, which is its ability to promote dissociation of an ionic solute or a contact ion-pair, is largely a function of the dielectric constant.

Although solvents may be classified as "donor solvents" (Lewis bases) and "acceptor solvents" (Lewis acids), most of the more widely used nonaqueous solvents are donor solvents. Some acceptor solvents, such as SO_2, BrF_3, $AsCl_3$, or the liquid hydrogen halides, have proved to be useful in coordination chemistry (11,12). Ionization is promoted in a donor solvent by solvation of cations and in an acceptor solvent by solvation of anions. For example, arsenic(III) iodide is ionized in a donor solvent, D, according to the reaction

$$2D + AsI_3 \rightarrow D_2AsI_2^+ I^-$$ (4-5)

while triphenylchloromethane is ionized in an acceptor solvent (such as liquid hydrogen chloride) according to the reaction

$$Ph_3CCl + HCl \rightarrow Ph_3C^+ HCl_2^-$$ (4-6)

The donor properties and protic character of a solvent are linked to a certain extent because protic solvents can solvate anions by strong hydrogen bonding as well as by ion-dipole interactions. The ease of solvation increases with decreasing size and increasing electronegativity of the anion; SCN^- $< I^- < N_3^- < Br^- < Cl^- \ll F^- \approx OH^-$ (13). Anions are much less strongly solvated by dipolar aprotic solvents, because of the absence of hydrogen bonding. In this case the large polarizable anions are more easily solvated than the small, more electronegative anions and the solvation series above is reversed. Cations are less strongly solvated in protic solvents than anions; the reverse is true for dipolar aprotic solvents (13).

To measure the donor power of a solvent, Gutmann (14) has proposed that one determines the enthalpy of coordination of the donor solvent toward a standard reference acceptor, $SbCl_5$, in a reference solvent of low donor power such as 1,2-dichloroethane. The quantity $-\Delta H_{D\text{-}SbCl_5}$ is easily determined at high dilution and may be considered a semi-quantitative measure of the coordinating properties of a donor solvent toward an acceptor molecule. It has been termed the donor number, DN_{SbCl_5}. Table 4-2 summarizes the donor numbers for a number of solvents.

Although a good *dissociating* solvent will have a high dielectric constant, it may not necessarily behave as a good ionizing solvent. Water is both an ionizing solvent ($DN = 18$) and a dissociating solvent ($\epsilon = 81$), while nitromethane has poor ionizing properties ($DN = 2.7$) but is a moderate dissociating solvent ($\epsilon = 36$). On the other hand, tributylphosphate is an ionizing solvent ($DN = 24$) but its dissociating properties are extremely poor ($\epsilon = 6.8$).

Solvents with a large donor number tend to associate with hydrogen donors such as water or chloroform. This may explain the finding (15) that the stability of the anthracene and naphthalene anion radicals in the presence of small amounts of added water is greater in dimethylformamide (DMF) and dimethylsulfoxide (DMSO) than in acetonitrile. This stability order parallels the donor numbers of the solvents and reflects the greater tendency of DMF and DMSO to associate with water. The latter causes the protons on water to be less available to protonate the anion radical. Because organic solvents almost always contain traces of water in the range 10^{-3} to 10^{-2} M unless rigorously dried, solvents with large donor numbers are preferred where the maximum stability of an anion radical is desired.

Lewis Acid-Base Character and Stability of Anion and Cation Radicals. Most of the so-called dipolar aprotic solvents have appreciable Lewis-base character with donor numbers greater than 10 and autoprotolysis constants smaller than 10^{-20} ($pK > 20$) (16). They solvate cations better than anions, and radical anions often have appreciable stability in the rigorously purified solvents.

Radical cations that are produced by electrochemical oxidation are not stable in solvents with appreciable base character. This results because such radicals are subject to attack by available nucleophiles, and solvents that contain donor electron pairs are good nucleophiles. Cation radicals are most stable in solvents that are good Lewis acids and show negligible basic properties. Some of the solvent systems that have been employed to stabilize electrochemically produced cation radicals include nitromethane and nitrobenzene (17), dichloromethane (18), trifluoroacetic acid-dichloromethane (1 : 9) (19), nitromethane-AlCl$_3$ (20), and AlCl$_3$-NaCl (1 : 1) (21). Organic chemists should be familiar with the stabilization of carbonium ions by "superacid" media (22). These media usually contain fluorosulfuric acid, or mixtures of fluorosulfuric acid with antimony pentachloride and sulfur

Table 4-2   Donor Number, DN$_{SbCl_5}$, of Some
Organic Solvents

Solvent	DN$_{SbCl_5}$
1,2-Dichloroethane	—
Nitromethane	2.7
Nitrobenzene	4.4
Acetic anhydride	10.5
Benzonitrile	11.9
Acetonitrile	14.1
Sulfolane	14.8
Propylene carbonate	15.1
Ethylene sulfite	15.3
Propionitrile	16.1
Ethylene carbonate	16.4
n-Butyronitrile	16.6
Acetone	17.0
Ethyl acetate	17.1
Water	18.0
Diethyl ether	19.2
Tetrahydrofuran	20.0
Trimethylphosphate	23.0
Tributylphosphate	23.7
Dimethylformamide	26.6
N,N-Dimethylacetamide	27.8
Tetramethylurea	29.6
Dimethylsulfoxide	29.8
N,N-Diethylformamide	30.9
N,N-Diethylacetamide	32.2
Pyridine	33.1
Hexamethylphosphoramide	38.8

dioxide, and are potent solvents for the production and stabilization of organic cations.

Other Solvent Polarity Scales. In addition to the method proposed by Gutmann, a number of other methods have been proposed to define operationally solvent donor/acceptor power or solvent polarity. These are based on various physical phenomena and include shifts in infrared stretching frequencies of —O—D and —C≡O bonds (23), proton NMR shift of $CHCl_3$ in the solvent relative to that in cyclohexane (24), retention characteristics in gas chromatography (25), and absorption frequency shifts of solvatochromic dyes (26,27).

A number of empirical solvent-polarity scales are based on solvatochromic dyes (28). One of these, the Kosower Z parameter, is the transition energy (in kcal per mole) of the charge-transfer absorption band of the 1-ethyl-4-carbomethoxy-pyridinium iodide ion-pair. Solvent molecules are oriented about the ion-pair in its ground state in such a way as to minimize the electrostatic free energy of the dipole of the dye molecule. Excitation of the ion-pair produces a species whose major dipole component is in the plane of the aromatic ring rather than perpendicular to it. Because the solvent molecules cannot adjust to the new orientation of the dipole in the time of the electronic transition, the solvent stabilization of the excited state is much less than that of the ground state. Hence the transition energy becomes a sensitive measure of the electrostatic part of the solvation of an ion-pair or dipole and the parameter Z is a measure of the ability of the medium to stabilize ion-pairs relative to less polar electronically excited states produced by charge transfer.

Several of the polarity scales have been used to correlate and to interpret solvent effects on the rates of solvolysis and other bimolecular nucleophilic displacement reactions (1). However, there is not yet a sufficient body of electrochemical data to establish their usefulness for interpretation of solvent effects.

4. Liquid Range and Vapor Pressure

The liquid range and vapor pressure of a solvent may dictate special experimental requirements. Low boiling solvents such as liquid NH_3 or SO_2 must be used at low temperature or at higher pressures at room temperature. Curiously, the earliest work was carried out with these difficult to handle solvents, and the techniques have been described (29,30). Solvents such as acetonitrile or dichloromethane have substantial vapor pressures at room temperature and the practice of removing dissolved oxygen by bubbling an inert gas through the solution can lead to a loss of solvent, cooling of the solution by evaporation, and a safety hazard due to the toxic vapors. For such solvents, presaturation of the inert gas with solvent and adequate venting of the toxic vapors are necessary.

For most thermodynamic studies data must be obtained over a wide range of temperature. This requires that the solvent have a broad liquid range. Some recent work also has pointed out the advantages of operating at low temperatures (9). Hence solvents with a low melting point or glass transition point are desirable; butyronitrile has been shown to be useful down to $-130°C$.

5. Viscosity

In many applications a low viscosity is desirable so that mass transport by diffusion or convection will be facilitated. However, in some instances a solvent of higher viscosity will extend the time range for mass transport by pure diffusional control to periods as long as 40 to 50 sec, which can be advantageous to electroanalytical techniques such as chronopotentiometry (31). At low temperatures the solvent may not appear to crystallize, but may form a rigid glass whose viscosity is so high that mass transport practically ceases; the experimentalist must be alert to this possibility.

The dissolution of an ionic solute sometimes greatly increases solution viscosity. A notable example is the dissolution of quaternary ammonium salts in hexamethylphosphoramide, where effective solution resistance is approximately four times that of a solution in acetonitrile or dimethyl-formamide of the same concentration (32).

6. Solvent Miscibility

Godfrey (33) has reported an empirical miscibility scale based upon the mutual miscibility of pairs of 31 standard solvents. Each standard solvent is assigned a value from 1 (glycerol—low lipophilicity) to 31 (petrolatum—high lipophilicity). A liquid is assigned a miscibility number, M, by determining its miscibility with the standard solvents. The miscibility numbers of a large number of compounds are shown in Table 4-3. The rules for predicting miscibility are: (a) if the M numbers of two compounds differ by 15 or less they probably are miscible in all proportions; (b) an M-number difference of 16 is likely to have borderline miscibility; (c) a difference of 17 or more generally corresponds to immiscibility.

In a substantial number of cases (about 17% of those studied) immiscibility is encountered with solvents at both ends of the lipophilicity scale. This behavior is accounted for by assigning a dual M-number. The first M-number is always less than 16 and defines the boundary of miscibility with solvents of high lipophilicity (approximately $M+15$). Converse statements apply to the second M-number. Two liquids that both have dual M-numbers are usually miscible with each other. Solvents with M-number 16 are likely to be miscible with nearly all the solvents in Table 4-3; hence this central class comprises the "universal" solvents.

7. Spectroscopic Properties

The techniques of optical spectroscopy (ultraviolet, visible, and infrared spectrometry) are often used to examine the reactants or products of an electrode reaction (see Chapters 2 and 3). Obviously the solvent (and supporting electrolyte) must be transparent at the wavelength region of interest; all of the commonly used dipolar aprotic solvents are transparent in the visible region. However, those solvents which contain aromatic or conjugated unsaturation, or doubly bonded oxygen atoms (amides, carboxylic acids, esters, ketones), exhibit a fairly rapid cut-off in the ultraviolet region (see Table 4-4). Among the dipolar aprotic solvents only acetonitrile and ethylene carbonate (and by inference, propylene carbonate) have good transparency in the ultraviolet. For this reason acetonitrile has been used most often in spectroelectrochemistry, particularly in the study of aromatic radical anions. Acetonitrile must be purified rigorously to remove ultraviolet absorbing impurities if a UV cut-off below 200 nm is to be realized. Procedures to do this are described in Section IV of this chapter.

A number of companies supply reagent or spectroquality solvents which have been purified to remove UV absorbing impurities. Some of them, particularly dimethylsulfoxide, may be suitable for general electrochemical use as purchased. However, small quantities of electroactive impurities (particularly water) often are present in spectroquality solvents. Therefore, a particular batch of solvent always should be tested by measurement of the residual current with an appropriate supporting electrolyte and a platinum, gold, or carbon electrode (to test the anodic limits) and a dropping mercury electrode (to test the cathodic limits). The voltage window or domain of electroactivity is a sensitive measure of the adequacy of the purification procedures.

Most organic solvents will show a number of absorption bands in the near and fundamental infrared regions. For infrared spectroelectrochemical work the spectra of the more common solvents are useful and are available from several sources (34,35).

B.   CLASSIFICATION OF SOLVENTS

Solvent classification helps to identify properties useful in solvent selection for individual applications; for example, the study of acid-base reactions, oxidation-reduction reactions, inorganic coordination chemistry, organic nucleophilic displacement reactions, and electrochemistry. Unfortunately, a single classification scheme suited to all areas of nonaqueous solvent study

Table 4-3. Miscibility Numbers of Organic Liquids

23,	Acatal	27	1-Bromododecane
14	Acetic acid	21	Bromoethane
12, 19	Acetic anhydride	24	1-Bromohexane
8	Acetol	24	1-Bromo-3-methylbutane
10	Acetol acetate	26	1-Bromooctane
9, 17	Acetol formate	26	2-Bromooctane
15, 17	Acetone	29	1-Bromotetradecane
11, 17	Acetonitrile	6	1,2-Butanediol
15, 18	Acetophenone	4	1,3-Butanediol
11	N-Acetylmorpholine	3	1,4-Butanediol
14, 18	Acrylonitrile	12, 17	2,3-Butanedione
8, 19	Adipontrile	15	1-Butanol
14	Allyl alcohol	16	2-Butanol
22	Allyl ether	3	2-Butene-1,4-diol
13	2-Allyloxyethanol	15	2-Butene-1-ol
2	2-Aminoethanol	16	2-Butoxyethanol
5	Aminoethylethanolamine	15, 17	2-iso-Butoxyethanol
2	2-(2-Aminoethoxy)ethanol	15	2-(2-Butoxyethoxy)ethanol
12	1-(2-Aminoethyl)piperazine	22	Butyl acetate
6	1-Amino-2-propanol	21	iso-Butyl acetate
25	sec-Amylbenzene	22	sec-Butyl acetate
12	Aniline	15	iso-Butyl-alcohol
20	Anisole	16	t-Butyl alcohol
15, 19	Benzaldehyde	23	iso-Butyl iso-butyrate
21	Benzene	26	Butyl ether
15, 19	Benzonitrile	19	Butyl formate
13	Benzyl alcohol	23	Butyl methacrylate
15, 21	Benzyl benzoate	28	Butyl oleate
29	Bicyclohexyl	26	Butyl sulfide
23	Bis(2-butoxyethyl) ether	15	Butyraldoxime
20	Bis(2-chloroethyl) ether	16	Butyric acid
20	Bis(2-chloroisopropyl) ether	21	Butyric anhydride
18	Bis(2-ethoxyethyl) ether	10	Butyrolactone
5	Bis(2-hydroxyethyl) thiodipropionate	14, 19	Butyronitrile
6	Bis(2-hydroxypropyl) maleate	26	Carbon disulfide
15, 17	Bis(2-methoxyethyl) ether	24	Carbon tetrachloride
11, 19	Bis(2-methoxyethyl) phthalate	25	Castor oil
21	Bromobenzene	21	Chlorobenzene
23	1-Bromobutane	23	1-Chlorobutane
25	Bromocyclohexane	27	1-Chlorodecane
27	1-Bromodecane	11	2-Chloroethanol
		19	Chloroform

Table 4.3 (*Continued*)

22	1-Chloronaphthalene
15, 20	β-Chlorophenetole
16	o-Chlorophenol
23	2-Chloropropane
4	3-Chloro-1,2-propanediol
14	1-Chloro-2-propanol
20	α-Chlorotoluene
29	Coconut oil
14	p-Cresol
11, 18	4-Cyano-2,2-dimethylbutyral- dehyde
28	Cyclohexane
16	Cyclohexanecarboxylic acid
16	Cyclohexanol
17	Cyclohexanone
26	Cyclohexene
29	Cyclooctane
27	Cyclooctene
25	p-Cymene
29	Decalin
29	Decane
18	1-Decanol
29	1-Decene
14	Diacetone alcohol
21	Diallyl adipate
22	1,2-Dibromobutane
21	1,4-Dibromobutane
20	1,2-Dibromoethane
19	Dibromomethane
21	1,2-Dibromopropane
25	1,2-Dibutoxyethane
17	N,N-Dibutylacetamide
23	Di-iso-butyl ketone
22	Dibutyl maleate
22	Dibutyl phthalate
13	Dichloroacetic acid
21	o-Dichlorobenzene
20	1,4-Dichlorobutane
20	1,1-Dichloroethane
20	1,2-Dicholorethane
20	cis-Dichloroethylene
21	trans-Dichloroethylene
20	Dichloromethane
20	1,2-Dichloropropane
20	1,3-Dichloropropane
12	1,3-Dichloro-2-propanol

26	Dicyclopentadiene
26	Didecyl phthalate
1	Diethanolamine
26	Diethoxydimethylsilane
14	N,N-Diethylacetamide
19	Diethyl adipate
21	Diethyl carbonate
5	Diethylene glycol
12, 19	Diethylene glycol diacetate
9	Diethylenetriamine
18	Diethyl ketone
14, 20	Diethyl oxalate
13, 20	Diethyl phthalate
12, 21	Diethyl sulfate
17	2,5-Dihydrofuran
17	1,2-Dimethoxyethane
13	N,N-Dimethylacetamide
10	N,N-Dimethylacetoacetamide
14	2-Dimethylaminoethanol
14, 19	Dimethyl carbonate
12	Dimethylformamide
12, 19	Dimethyl maleate
11, 19	Dimethyl malonate
12, 19	Dimethyl phthalate
16	1,4-Dimethylpiperazine
16	2,5-Dimethylpyrazine
22	Dimethyl sebacate
12, 17	2,4-Dimethylsulfolane
9	Dimethylsulfoxide
24	Dioctyl phthalate
17	p-Dioxane
15, 19	p-Dioxene
26	Dipentene
23	Diphenylmethane
25	Di-iso-propylbenzene
11	Dipropylene glycol
23	Di-iso-propyl ketone
12, 17	Dipropyl sulfone
29	Dodecane
18	1-Dodecanol
29	1-Dodecene
14, 19	Epichlorohydrin
15, 19	Epoxyethylbenzene
5	Ethanesulfonic acid
14	Ethanol
14	2-Ethoxyethanol

Table 4.3   (*Continued*)

13	2-(2-Ethoxyethoxy)ethanol
14, 18	2-(2-Ethoxyethoxy)ethyl acetate
15, 19	2-Ethoxyethyl acetate
19	Ethyl acetate
13, 19	Ethyl acetoacetate
24	Ethylbenzene
21	Ethyl benzoate
17	2-Ethylbutanol
22	Ethyl butyrate
6, 17	Ethylene carbonate
9	Ethylenediamine
2	Ethylene glycol
9, 17	Ethylene glycol bis(methoxy-acetate)
12, 19	Ethylene glycol diacetate
8, 17	Ethylene glycol diformate
10, 19	Ethylene monothiocarbonate
23	Ethyl ether
9	Ethylformamide
15, 19	Ethyl formate
14, 17	2-Ethyl-1,3-hexanediol
17	2-Ethylhexanol
23	Ethyl hexoate
14	Ethyl lactate
16	*N*-Ethylmorpholine
23	Ethyl orthoformate
21	Ethyl propionate
13	2-Ethylthioethanol
21	Ethyl trichloroacetate
20	Fluorobenzene
21	1-Fluoronaphthalene
3	Formamide
5	Formic acid
10	*N*-Formylmorpholine
20	Furan
11, 17	Furfural
11	Furfuryl alcohol
1	Glycerol
3	Glycerol carbonate
13, 19	Glycidyl phenyl ether
29	Heptane
17	1-Heptanol
22	3-Heptanone
23	4-Heptanone
28	1-Heptene
26	Hexachlorobutadiene

30	Hexadecane
29	1-Hexadecene
15	Hexamethylphosphoramide
29	Hexane
5	2,5-Hexanediol
12, 17	2,5-Hexanedione
2	1,2,6-Hexanetriol
17	Hexanoic acid
17	1-Hexanol
27	1-Hexene
5	Hydracrylonitrile
8	1-(2-Hydroxyethoxy)-2-propanol
2	2-Hydroxyethyl carbamate
1	2-Hydroxyethylformamide
12	2-Hydroxyethyl methacrylate
3	2-Hydroxypropyl carbamate
14, 17	Hydroxypropyl methacrylate
22	Iodobenzene
22	Iodoethane
21	Iodomethane
18	Isophorone
25	Isoprene
30	Kerosene
9	2-Mercaptoethanol
24	Mesitylene
18	Mesityl oxide
15, 19	Methacrylonitrile
4	Methanesulfonic acid
12	Methanol
8	Methoxyacetic acid
11, 19	Methoxyacetonitrile
14	3-Methoxybutanol
13	2-Methoxyethanol
12	2-(2-Methoxyethoxy)ethanol
14, 17	2-Methoxyethyl acetate
15	2-Methoxyethyl methoxyacetate
15	1-[(2-Methoxy-1-methyl)ethoxy]-2-propanol
5	3-Methoxy-1,2-propanediol
15	1-Methoxy-2-propanol
11, 17	3-Methoxypropionitrile
15	3-Methoxypropylamine
10	3-Methoxypropylformamide
15, 17	Methyl acetate
19	Methylal
11	2-Methylaminoethanol

Table 4.3 (*Continued*)

19	Methyl iso-amyl ketone	6	3,3'-Oxydipropionitrile
27	2-Methyl-1-butene	15, 19	Paraldehyde
26	2-Methyl-2-butene	7	PEG-200
19	Methyl-iso-butyl ketone	7	PEG-300
13, 19	Methyl chloroacetate	8	PEG-600
8, 17	Methyl cyanoacetate	25	1,3-Pentadiene
29	Methylcyclohexane	7	Pentaethylene glycol
27	1-Methylcyclohexene	9	Pentaethylenehexamine
28	Methylcyclopentane	9	Pentafluoroethanol
17	Methyl ethyl ketone	3	1,5-Pentanediol
14, 19	Methyl formate	12, 18	2,4-Pentanedione
8	2,2'-Methyliminodiethanol	17	1-Pentanol
20	Methyl methacrylate	23	Pentyl acetate
13	Methyl methoxyacetate	16	*t*-Pentyl alcohol
16	*N*-Methylmorpholine	26	Pentyl ether
22	1-Methylnaphthalene	31	Petrolatum (liquid)
26	Methyl oleate	20	Phenetole
7	5-Methyloxazolidinone	12	2-Phenoxyethanol
29	2-Methylpentane	13, 17	1-Phenoxy-2-propanol
29	3-Methylpentane	12, 19	Phenylacetonitrile
14	2-Methyl-2,4-pentanediol	10	*N*-Phenylethanolamine
17	4-Methyl-2-pentanol	22	Phenyl ether
28	4-Methyl-1-pentene	16	2-Picoline
27	cis-4-Methyl-2-pentene	14	PPG-400
13	1-Methyl-2-pyrrolidone	14, 23	PPG-1000
26	Methyl stearate	11	Propanediamine
23	α-Methylstyrene	3	1,3-Propanediol
10, 17	3-Methylsulfolane	4	1,2-Propanediol
29	Mineral spirits	7, 19	Propanesulfone
14	Morpholine	15	1-Propanol
14, 20	Nitrobenzene	15	2-Propanol
13, 20	Nitroethane	19	*iso*-Propenyl acetate
10, 19	Nitromethane	15	Propionic acid
15, 20	2-Nitropropane	13, 17	Propionitrile
17	1-Nonanol	19	Propyl acetate
17	Nonylphenol	19	*iso*-Propyl acetate
30	1-Octadecene	24	*iso*-Propylbenzene
27	1,7-Octadiene	9, 17	Propylene carbonate
29	Octane	17	Propylene oxide
26	1-Octanethiol	26	*iso*-Propyl ether
17	1-Octanol	16	Pyridine
17	2-Octanol	10	2-Pyrrolidone
22	2-Octanone	22	Styrene
28	1-Octene	9, 17	Sulfolane
28	trans-2-Octene (cis isomer 27)	13, 19	1,1,2,2-Tetrabromoethane

Table 4.3 (*Continued*)

19	1,1,2,2-Tetrachloroethane	21	Tricresyl phosphate
25	Tetrachloroethylene	2	Triethanolamine
30	Tetradecane	26	Triethylamine
29	1-Tetradecene	25	Triethylbenzene
7	Tetraethylene glycol	6	Triethylene glycol
9	Tetraethylenepentamine	14	Triethylene glycol monobutyl
23	Tetraethyl orthosilicate		ether
17	Tetrahydrofuran	13	Triethylene glycol monomethyl
13	Tetrahydrofurfuryl alcohol		ether
21	Tetrahydrothiophene	9	Triethylenetetramine
24	Tetralin	14	Triethyl phosphate
16	*N,N,N′,N′*-Tetramethylethylene-	29	Trilsobutylene
	diamine	16	Trimethyl borate
29	Tetramethylsilane	12, 17	Trimethylboroxin
15	Tetramethylurea	12	Trimethyl nitrilotripropionate
29	Tetrapropylene	29	Triisobutylene
4	2,2′-Thiodiethanol	27	2,4,4-Trimethyl-1-pentene
8	1,1′-Thiodi-2-propanol	27	2,4,4-Trimethyl-2-pentene
6, 19	3,3′-Thiodipropionitrile	10	Trimethyl phosphate
20	Thiophene	26	Tripropylamine
23	Toluene	12	Tripropylene glycol
11, 19	Triacetin	20	Vinyl acetate
28	Tributylamine	22	Vinyl butyrate
18	Tributyl phosphate	26	4-Vinylcyclohexene
24	1,2,4-Trichlorobenzene	26	Vinylidenenorbornene
22	1,1,1-Trichloroethane	29	VM&P naphtha
19	1,1,2-Trichloroethane	23	*m*-Xylene
20	Trichloroethylene	23	*o*-Xylene
20	1,2,3-Trichloropropane	24	*p*-Xylene
27	1,1,1-Trichloro-2,2,2-trifluoro-		
	ethane		

has not yet been devised; the criteria that are useful in one area often are not appropriate for another.

Bates (36) has discussed the scheme that was originally proposed by Bronsted for acid-base systems. This divides solvents into four classes: protogenic (proton donating); protophilic (proton accepting); amphiprotic (both proton donating and accepting); and aprotic (neither proton donating or accepting). Bronsted further divided each of these classes into two subdivisions on the basis of dielectric constant, taking $\epsilon = 30$ as the dividing line.

Kratochvil (10) has broadened this classification by replacing the Bronsted acid (proton donor) with a Lewis acid (an electron acceptor) and the Bronsted base with a Lewis base (an electron donor). (A Bronsted acid is

Table 4-4   Ultraviolet Cutoff of
Spectroquality Organic Solvents[a]

Solvent	UV Cutoff (nm)
Acetone	330
Acetonitrile	190
Benzene	280
Benzonitrile	299
n-Butanol	205
Isobutanol	210
Chloroform	245
1,2-Dichloroethane	225
Dichloromethane	233
N,N-Dimethylacetamide	270
N,N-Dimethylformamide	270
p-Dioxane	215
Ethyl acetate	255
Ethylene carbonate (cyclic ester)	215
Methanol	205
2-Methoxyethanol	210
1-Methyl-2-pyrrolidone	261
Dimethylsulfoxide	262
Nitromethane	380
Pyridine	305

[a]Absorbance = 1.0 in 1-cm cell versus water.

a Lewis acid but not necessarily vice-versa.) Solvent-proton interactions are therefore included as one subdivision of this classification, but many solvation reactions of cations with solvents also will be included as reactions of Lewis acid-base systems. This approach still does not solve the problem of fitting specific solvation interactions into the classification scheme. For example, acetonitrile behaves as a good Lewis base toward silver (I) ion, but a poor one toward hydrogen ion. The broader scheme also does not specifically take into account hydrogen-bonding effects in hydroxylic and other solvents, which affect both the dielectric constant and the Lewis acid-base properties of a solvent. Sucrose, for example, is soluble in water ($\epsilon = 78$) because of solvent-solute hydrogen bonding; by contrast, the solubility of sugars in propylene carbonate ($\epsilon = 69$) is slight (10).

There also is difficulty in accomodating fused-salt systems in a classification scheme primarily designed for organic solvents because the dielectric constants are not comparable and the measures of solvent polarity appropriate to organic solvents are not generally useful in fused salts. Therefore, fused salt systems are not discussed nor included in the general classification scheme.

In Table 4-5 is presented a solvent classification based on an extension of the basic scheme of Bronsted, along with representative solvents of each class. In this table an arbitrary value of 25 for the dielectric constant is used as the dividing line between dipolar and nonpolar solvents, which is in accord with the suggestion of Kratochvil (10). In classes 5, 6, and 8 a further subdivision into two groups (based on the ability of the solvent to form a hydrogen bond by donation of a hydrogen atom) is useful. Solvents which are hydrogen-bond donors have a greater tendency to associate and to solvate anions, relative to aprotic solvents (non-hydrogen-bond donors).

II. Role of the Solvent-Supporting Electrolyte System in Electrochemistry

There are several ways in which the solvent-supporting electrolyte system can influence mass transfer, the electrode reaction (electron transfer), and the chemical reactions which are coupled to the electron transfer. The diffusion of an electroactive species will be affected not only by the viscosity of the medium, but also by the strength of the solute-solvent interactions which determine the size of the solvation sphere. The solvent also plays a crucial role in proton mobility; water and other protic solvents produce a much higher proton mobility because of fast solvent proton exchange, a phenomenon which does not exist in aprotic organic solvents.

The medium also has important effects on the structure of the electrical double layer, a crucial factor in electrochemistry because electron transfer takes place in or near the double layer. In this connection three factors are important (37). First, most of the polar organic solvents tend to be strongly oriented in the double layer at the electrode-solution interface. This strongly affects the double layer capacitance, which is not related in a simple way to the dielectric constant of the bulk solvent. Second, adsorption phenomena, which frequently affect the course of an electrochemical reaction, are less pronounced in organic solvents than in water. This results because the energy involved in replacing a polar orgainc solvent molecule by a molecule of electroactive solute generally is higher in organic solvents than in water. Finally, ions of the supporting electrolyte may be *specifically adsorbed* in the double layer, which creates a monomolecular layer of adsorbed ions (the so-called inner Helmholtz layer) that is associated with a characteristic

Table 4-5 A Solvent Classification Scheme for Nonaqueous Solvents

Solvent Class	Dielectric Constant	Brønsted and/or Lewis Acid	Brønsted and/or Lewis Base	Examples
1. Amphiprotic	>25	+	+	Water, methanol, 1,2-ethanediol
2. Amphiprotic	<25	+	+	Ethanol, isopropanol
3. Protic	>25	+	−	Hydrogen cyanide, sulfuric acid
4. Protic	<25	+	−	Acetic acid, hydrogen halides (except HF)
5. Lewis base properties a. Aprotic	>25	−	+	Acetonitrile, dimethylformamide, propylene carbonate, dimethylsulfoxide
b. Hydrogen bond donor				Formamide, 2-pyrrolidone
6. Lewis base properties a. Aprotic	<25	−	+	Pyridine, acetone, tetramethylurea, acetic anhydride
b. Hydrogen bond donor				Ethylenediamine
7. Negligible Lewis acid or base properties	>25	−	−	Nitromethane, nitrobenzene
8. Negligible Lewis acid or base properties a. Aprotic	<25	−	−	Hexane, dichloromethane
b. Hydrogen bond donor				Chloroform

potential, φ_1. At a greater distance from the electrode surface there is presumed to be a second layer (the outer Helmholtz layer) of ions which are not specifically adsorbed, and which have their normal solvation shells with an associated potential, φ_2. This model of the double layer structure has a characteristic feature—the potential does not change in a monotonic fashion in going from the bulk potential of the metal to the bulk potential of the solution. This discontinuous potential gradient can have profound effects on the rates of electron transfer (37).

Most electron transfers that involve organic compounds have rates which tend to lie in the upper range of detection by present electrochemical techniques (38). In the absence of adsorption or fast follow-up chemical reactions, the effect of the medium often can be isolated by measurement of the variation of half-wave potentials for one-electron, reversible systems. For a reduction reaction

$$R + e^- \rightarrow R^- \tag{4-7}$$

the reduction potential (half-wave potential) relative to some reference electrode can be expressed as

$$E_{1/2} = EA + \Delta G_{solv} - K \tag{4-8}$$

where EA is the electron affinity of the neutral molecule in the gaseous state, ΔG_{solv} the difference in free energy of solvation of R and R^- ($\Delta G_R - \Delta G_{R^-}$), and K a constant (39). Implicit in Equation 4-8 is the assumption that there is no ion association between R^- and the cation of the supporting electrolyte.

The half-wave potentials (corrected for changes in liquid-junction potential) for the one-electron reduction of aromatic hydrocarbons generally become more positive (the reduction is easier) as the dielectric constant of the solvent increases (40). This is in accord with the direction of the variation in solvation energy of the radical anions that is predicted by the simple Born theory

$$\Delta G_{R^-} = -\frac{Ne^2}{2r}\left(1 - \frac{1}{\epsilon}\right) \tag{4-9}$$

In deriving Equation 4-9 the ion is assumed to be a sphere of radius r in a continuous medium of dielectric constant ϵ. The changes in ΔG_{solv} that are produced by variations of the solvation energy of the neutral hydrocarbon (ΔG_R) are not expected to be large because there are no strong charge-solvent dipole interactions. Therefore changes in ΔG_{solv} are dominated by changes in the solvation energy of the radical anion.

In the absence of secondary chemical reactions, or other complications, the transfer of a second electron is more difficult because of the repulsion; conversely, the removal of a second electron from a radical cation is more difficult because of the coulombic attraction. However, charge delocalization or structural effects can modify this behavior to the point that reversible two-electron transfers exist in which the addition of successive electrons is not distinguishable (41).

From the point of view of the overall reaction, secondary chemical reactions play a decisive role and are strongly solvent dependent. A radical anion that is produced in a reduction step can react by several pathways. If it is sufficiently basic it may abstract a proton from the solvent or from the small amounts of water that are usually present; even the solute itself may act as a proton donor (42). Neutral radicals can be produced by expulsion of a negative leaving group (such as a halide ion); these may react with one another to form dimers or with the solvent. Several examples for each type of reaction have been cited by Saveant in his discussion of the role of the solvent in organic electrochemistry (38).

The use of an aprotic solvent tends to simplify the possible chemical reactions relative to those in protic solvents. This offers interesting possibilities for the study of primary anion radicals that are produced by simple electron transfer. Radical cations are more difficult to stabilize, but the addition of small amounts of finely divided alumina or trifluoroacetic anhydride to acetonitrile (to scavenge the last traces of water) prolongs their lifetime sufficiently that they may be studied by electrochemical and spectroscopic techniques (43).

III. The Role of the Supporting Electrolyte

A. INTRODUCTION

The use of an indifferent or "inert" supporting electrolyte is indispensable in electrochemistry and affects the solvent medium in several ways: (a) it regulates cell resistance and mass transport by electrical migration; (b) it may control or "buffer" the level of hydrogen ion activity in solution; (c) it may associate with the electroactive solute, as in the complexing of metal ions by certain ligands; (d) it may form ion-pair or micellar aggregates with the electroactive species; (e) it largely determines the structure of the double layer; and (f) it may impose positive or negative voltage limits because of its redox properties.

B. CONTROL OF CELL RESISTANCE

All of the pure organic solvents that are discussed essentially are nonconductors. Without the addition of some electrolyte their resistance is so great that the voltages required to pass even milliampere currents are impracti-

Table 4-6 Solubilities of Tetraalkylammonium Salt Electrolytes and Specific Resistances of the Solutions at 25°C[a]

	Acetonitrile		1,2-dimethoxyethane		Tetrahydrofuran		Dimethylformamide	
	Solubility g/100 ml of Solution (concn, F)	Specific resistance Ω cm (concn, F)	Solubility g/100 ml of Solution (concn, F)	Specific resistance Ω cm (concn, F)	Solubility g/100 ml of solution (concn, F)	Specific resistance Ω cm (concn, F)	Solubility g/100 ml of solution (concn, F)	Specific resistance Ω cm (concn, F)
Et_4NClO_4	26(1.13)	26(0.60)	(<0.01)		(<0.01)		23(1.00)	52(0.60)
$n\text{-}Bu_4NClO_4$	70(2.05)	37(0.60)	31(1.10)	312(1.0)	50(1.48)	368(1.0)	79(2.29)	77(0.60)
Et_4NBF_4	37(1.69)	18(1.0)	(<0.01)		(<0.01)		27(1.24)	38(1.0)
$n\text{-}Bu_4NBF_4$	71(2.21)	31(1.0)	53(1.70)	228(1.0)	65(2.02)	373(1.0)	75(2.34)	69(1.0)
Et_4NBr	7.8(0.37)		(<0.01)		(<0.01)		4.1(0.19)	
$n\text{-}Bu_4NBr$	66(1.99)	48(0.60)	(<0.1)		4.8(0.14)		52(1.57)	106(0.60)

[a]From Ref. 32.

188

Table 4-7 Limiting Ionic Conductivities of Ions in Selected Solvents[a]

Ion	Water[b]	Acetone	Acetonitrile	Propylene Carbonate	Nitromethane	Nitrobenzene	Dimethyl-formamide	Dimethyl-sulfoxide	Sulfolane (40°C)
							λ_i^0		
H^+	349.8	90	100		63		35	16	
Li^+	38.7	72.8	69.3	7.3	55	16.3	25.0	11.4	4.3
Na^+	50.1	78.4	76.9		58	17.8	29.9	13.5	3.6
K^+	73.5	80.6	83.6	12.0	60		30.8	13.9	4.0
Me_4N^+	44.9	97.7	94.5		54.5	17.1	38.9	18.6[c]	4.3
Et_4N^+	32.7	89.0	84.8	13.3	47.6	16.4	35.6	17.1	4.0
$n\text{-}Bu_4N^+$	19.5	67.3	64.1	9.4	34[c]	11.9	25.9	11.6	2.8
Cl^-	76.4	105.2	98.4	20.2	62.5	22.2	55.1	24.2	9.3
Br^-	78.1	115.9	100.7	19.3	62.9	21.6	53.6	24.3	8.9
I^-	76.8	113.0	102.4	18.8	62	20.4	52.3	24.0	7.2
NO_3^-	71.5	120.1	106.4		64	22.6	57.3	27.0	
BF_4^-			108.5						
PF_6^-			104.2						6.0
ClO_4^-	67.3	115.3	103.7	18.8	64	20.9	52.4	24.7	6.7

[a] All values at 25°C unless otherwise indicated; data from Ref. 44 unless otherwise indicated.

[b] R. A. Robinson and R. H. Stokes, *Electrolyte Solutions*, Butterworths Scientific Publications, London, 1955, p. 452.

[c] G. J. Janz and R. P. T. Tomkins, eds., *Nonaqueous Electrolytes Handbook*, Academic Press, New York, 1972.

Table 4-8   Association Constants for Some 1:1 Electrolytes in Acetonitrile[a]

	Cl^-	Br^-	NO_3^-	I^-	BF_4^-	ClO_4^-	PF_6^-
Me_4N^+	56	46		19		7	5
Et_4N^+			5	5		0	
$n\text{-}Bu_4N^+$		2		3			0
Li^+						4	
Na^+				0		10	
K^+				0		14	

From Ref. 44.

[a]Anions arranged in order of increasing crystallographic radius.

cally large. Therefore, the primary function of the supporting electrolyte is to provide a conducting medium. The solvent-supporting electrolyte combination should be chosen to give resistance values that are as small as possible. This will minimize uncompensated iR drop, which leads to potential-control error (see Chapters 2 and 5), and will minimize ohmic heating of the solution in preparative electrochemistry (where the currents are large).

The solubility and solution resistance of some commonly used supporting electrolytes in selected aprotic solvents are listed in Table 4-6. Table 4-7 contains conductance data for a number of other salts and solvents, while Table 4-8 indicates the extent of ion association in acetonitrile, a solvent of moderate dielectric constant. As the dielectric constant decreases, ion association increases, and solution conductivity is not proportional to the concentration of electrolyte (44). Extensive compilations of solution conductances in organic solvents are available in the review by Kratochvil and Yeager (44) and in the handbook edited by Janz and Tomkins, which covers the literature through 1971 (45).

When an electric field is imposed on an electrolyte solution the ions will tend to migrate—the cations moving toward the cathode (negative electrode) and the anions toward the anode. This migration of ions constitutes the flow of current in the cell, and each kind of ion carries a fraction of the current proportional to its mobility and concentration. For a particular value of the current, the addition of an inert electrolyte will reduce the solution resistance which in turn will decrease the electric field according to Ohm's law, $E = iR$. Therefore, the mass transport of an ionic electroactive species that is caused by migration in an electrical field can be reduced to a negligible level by "swamping" the solution with inert electrolyte. Then most of the current will be carried by the ions of the supporting electrolyte.

In polarography and voltammetry the customary practice is to make the supporting electrolyte concentration at least 50 to 100 times the concentration of the electroactive species, so that electrical migration will be suppressed. This is particularly important where quantitative agreement is sought with the equations that are used to calculate limiting or peak currents. These are based on the assumption that mass transport is controlled by pure diffusion.

C. CONTROL OF SOLUTION ACIDITY

Many inorganic and organic redox reactions involve protons, for example

$$+ 2H^+ + 2e^- \rightarrow \qquad\qquad (4\text{-}10)$$

The effect of proton donors on the reduction of aromatic hydrocarbons is discussed in Section IA of this chapter. The importance of potential-pH relations to the understanding and ultilization of the redox behavior of these systems has long been recognized; extensive potentiometric data have been obtained, particularly for inorganic systems. However, few organic redox systems behave reversibly in aqueous solutions and potentiometric data have been obtained for only a limited number of organic compounds; much of this work is summarized in the monograph by Clark (46).

With the discovery of polarography much of the activity in potentiometric studies of organic compounds tapered off. Instead, attention was focused on the measurement of the polarographic half-wave potentials, $E_{1/2}$, which provide similar free energy information. In many cases, half-wave potentials are the only data readily obtained by electrochemical measurements. Polarographic and voltammetric techniques have provided most of the information on the involvement of hydrogen ion in electrochemical processes.

The role of the proton in electrode reactions is too complex to describe fully in this chapter, but Refs. 47 through 49 provide the reader with a full account and access to the large body of literature in this area. For a reversible reduction (such as that for quinone which is represented in Equation 4-10) of the general form

$$O + mH^+ + ne^- \rightarrow RH_m \qquad\qquad (4\text{-}11)$$

the half-wave potential varies with pH according to

$$\frac{dE_{1/2}}{d(\text{pH})} = -\frac{2.3RT}{nF}m \qquad (4\text{-}12)$$

where m is easily evaluated once n is known (50).

The reduction of quinone consumes hydrogen ions, and if the aqueous solution is not well-buffered the hydrogen ions are supplied by water to produce an excess of OH^- at the surface of the dropping mercury electrode. This will increase the pH at the electrode surface and shift the cathodic wave to more negative values. The reverse is true for the oxidation of hydroquinone, in which H^+ will be liberated at the electrode surface. The result is that in an unbuffered solution the composite wave will be split into two waves, while in a well-buffered solution only a single composite wave is observed for an equimolar mixture of quinone and hydroquinone (quinhydrone). This example illustrates the fact that the buffering capacity of the system must be sufficient to react with the H^+ or OH^- ions that are liberated at the electrode surface. Although this might seem simple enough to accomplish, there are several complicating factors. First, the components of some buffer systems (like phosphate or citrate) may interact strongly with some species, particularly in biological systems that contain nucleotides or other metabolic intermediates. An answer to this problem is the use of one of several new or little used hydrogen ion buffers that are compatible with common biological media (51). Bates and coworkers (52) have assigned hydrogen ion activities to one of these buffers, tris(hydroxymethyl) methylglycine, in the physiological range of pH 7.2 to 8.5.

The second complication is more disturbing. Calculations indicate that the pH at the electrode surface for an electrode at -1.5 V versus SCE may be as much as 2 pH units lower than in the bulk of the solution (53,54). This results from the alteration of the distribution of charged species in the double layer due to the potential drop in the diffuse part of the double layer. The resolution of this problem must await further theoretical and experimental work.

If the electron-transfer step in an electrode reaction is preceded by a chemical reaction which involves proton transfer, the polarographic current often will be a complex function of the concentration of the electroactive species, the hydrogen ion concentration, and the rate constants for proton and electron transfer. Currents controlled by the rate of a chemical reaction are called kinetic currents and often are observed in the reduction of electroactive acids (e.g., pyruvic acid), in which the protonated form of the acid is more easily reduced than the anion. A polarogram of pyruvic acid in

Table 4-9   Comparison of pK_a Values of Several Acids in Different Solvents

Acid	Dimethyl-sulfoxide	Dimethyl-formamide	Acetonitrile	Sulfolane	Water
$HClO_4$	Strong	Strong	Strong	2.7	Strong
HSO_3CF_3	Strong	—	—	3.4	Strong
Picric	−0.3	1.2,(1.4)	8.9,(11.0)	—	0.38
HSO_3CH_3	1.76	—	6.1	—	—
HBr	1.0	1.8	5.5	7.4	−9
HCl	2.01	3.4	8.9	12.7	−7
H_2SO_4	1.4	3.0	7.25		−4.0
$HOOCCF_3$	3.45	—			−0.3

unbuffered solution exhibits two waves whose relative wave heights depend on the concentration of pyruvic acid and the solution pH (55).

Elving (49) and Zuman (47) have reviewed the effects of solution acidity on the polarography of organic compounds, principally in aqueous solution. A thorough discussion of kinetic and catalytic currents that involve hydrogen ions has been presented (48), and the irreversible polarographic and voltammetric curves that involve protons in unbuffered and poorly buffered solutions recently have been discussed (55).

Although there is a wealth of data in the literature on acid-base behavior in aprotic solvents (56,57), there are few examples of the use of buffers for polarography and voltammetry in aprotic solvents. This has occurred because most investigators have sought to keep all potential proton donors out of the system and thereby stabilize anion radicals. Although the picric acid-picrate system has been used as a buffer in a number of studies in aprotic solvents, its use in voltammetric work is limited because of the ease of reduction of picric acid.

Most neutral acids are much weaker acids in aprotic solvents than in water. This is due largely to their smaller dielectric constants, which increases the energy required for charge separation in the dissociation process. Table 4-9 summarizes the pK_a values for several acids in five different solvents (58). There is a tendency in solvents with low dielectric constants, which cannot stabilize the anion of weak Bronsted acids (HA) by hydrogen bonding, for the anion (A^-) to hydrogen bond with the undissociated acid to yield the species AHA^- (called a "homoconjugate") (59). This adds a further complication to the measurement of dissociation constants in these solvents.

D. COMPLEX FORMATION

A number of neutral or anionic donor ligands will form complexes with metal ions. These ligands include cyanide, thiocyanate, halides, amines, polyhydroxy carboxylates (such as tartrate), and polyamino carboxylates (such as EDTA). In general, the metal ion-ligand complex will be reduced at more negative potentials than the "free" metal ion (which is actually an aquo or solvent complex ion). This property can be used to eliminate interferences between overlapping polarographic waves of two metal ions by selectively complexing one of them. As an example, the determination of Tl(I) in the presence of Cu(II), Cd(II), Sn(II), Pb(II), Bi(II), and Sb(III) can be accomplished by a combination of EDTA complexation and judicious choice of pH (60). The use of complexing agents to eliminate interferences is sometimes called "masking" and a number of applications in electroanalytical chemistry have been summarized (61).

The use of complexing agents in organic solvents is not widespread because metal ions more often are determined in aqueous solution; this is an area where further research may prove fruitful.

E. ION-PAIRING AND DOUBLE-LAYER EFFECTS

The reduction of metal ions and aromatic compounds in organic solvents is affected strongly by the nature of the supporting electrolyte. In particular the reductions of alkali metal ions in hexamethylphosphoramide (HMPA) are influenced significantly by the type of cation in the supporting electrolyte (62). The addition of small amounts of tetraethylammonium ion interferes with the reduction of Na^+, Li^+, and K^+, while reduction is affected less by the presence of tetrabutylammonium ion. From conductivity studies the size of the solvated cations in HMPA appears to increase in the order: $Et_4N^+ \ll Bu_4N^+ \sim K^+ \sim Na^+ < Li^+$. At extremely negative potentials the smaller solvated ions are attracted to the electrode surface such that Et_4N^+ ion can compete effectively with the considerably larger *solvated* Li^+ ion, while the smaller *solvated* Rb^+ and Cs^+ ions are less affected. The distance of closest approach of the supporting electrolyte cations apparently increases with increasing size of the supporting electrolyte cation (the quaternary ammonium ions). As this distance increases the potential of φ_1 at the inner Helmholtz plane becomes more negative, which accelerates the desolvation and electron transfer at the inner Helmholtz plane. This leads to greater reversibility for the reduction of the alkali metal cations in the presence of tetrabutylammonium ions.

In the electroreduction of aromatic hydrocarbons, nitro compounds, and quinones in aprotic solvents, the first step is the transfer of an electron from the electrode to form a radical anion. Once the radical anion is formed, electron repulsion will decrease the facility with which a second electron transfer occurs. But solvation and ion-pairing diminish the effect of electron repulsion and tend to shift the reduction potential for the addition of the second electron to more positive values. In both cases the extent of the shift will be dependent on the nature of the radical anion, the supporting electrolyte cation, and the solvent. If the radical anion is sufficiently polar, and if the electron transfer is fast, ion-pairing of the radical anion with the cations of the supporting electrolyte will shift the first reduction potential to more positive values also.

There is strong evidence (63) that at least two types of ion pairs exist. All cations in solution tend to be surrounded by solvent molecules. A small ion generates in its vicinity a more powerful electrostatic field than a large ion of the same charge; thus smaller ions tend to produce more rigid solvation shells than large ions.

The association of a cation that is surrounded by a tight solvation shell with an anion proceeds smoothly until the solvent shell comes into contact with the anion. At this stage either the structure of the ion pair, separated by solvent molecules, is preserved (Figure 4-1a) or the solvation shell is squeezed out in a process that leads to a contact pair. This implies that at least two types of ion pair may coexist in solution, each having its own physical and chemical properties; such two-step associations have been revealed by various relaxation experiments. However, ions which weakly interact with the solvent and do not surround themselves with tight solvation shells form contact pairs only. This situation is encountered in poorly solvated liquids and for bulky ions. Those cations that interact strongly with solvent molecules tend to form solvent-separated pairs, especially when combined with large anions.

The available solvation data have been classified according to the nature of the anion (highly polar with localized charge or low charge density with delocalized charge) and the solvating power of the solvent (64). Nitrobenzenes and quinones give rise to polar anions with localized charges which interact sufficiently strongly with cations to cause desolvation and the formation of contact ion pairs in solvents of low or high dielectric constant. Consequently the magnitude of the ion-pairing interaction (positive shift of the reduction potential) becomes larger as the crystallographic radius of the unsolvated cation decreases.

In the case of anthracene, which forms large anions with delocalized charge, the ion-pairing effect is related to the size of the *solvated* cation; the interaction increases as the solvated cation becomes smaller.

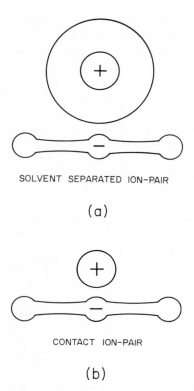

SOLVENT SEPARATED ION-PAIR

(a)

CONTACT ION-PAIR

(b)

Figure 4-1.   Two types of ion pairs that form in solutions.

F.   MICELLAR AGGREGATES

Surface active agents that contain a polar head and a long hydrophobic alkyl "tail" can be used to solubilize organic compounds in water. Although these micellar aggregates may approach colloidal dimensions electron transfer from the surface of the electrode to the organic compound in the micellar aggregate still is possible, and more or less normal polarograms can be obtained (65).

IV. Electrochemical Properties of Water and Selected Organic Solvents

A.   WATER

A laboratory engaged in careful electrochemical work with aqueous solutions or in trace analysis will need facilities for the preparation and storage of highly purified water. Water commonly is contaminated with metals in both dissolved cationic form and in the form of colloidal or particulate matter that is not ionized appreciably (66). Frequently it also is contaminated by bacteria and by organic impurities that cannot be removed by ordinary or oxidative distillation because of the steam volatility of the impurities (67).

The purification of water has been treated exhaustively in the literature, and recent reviews have summarized general conclusions and recommended techniques for the ultrapurification and analysis of water (66,68). There is a general consensus among investigators that: (a) water purified in quartz or plastic apparatus contains a lower level of cationic impurities than water purified in borosilicate or metal apparatus; (b) storage of purified water for periods exceeding 30 days, even in plastic or Teflon containers, will result in an increase in cationic impurities; and (c) resistivity measurements may be used as a survey technique but cannot be relied upon for an unequivocal indication of water quality because they do not indicate nonionic impurities.

1. Removal of Cationic Impurities from Water

Careful analysis of water purified by various methods (see Table 4-10) indicates that the water which is obtained by passing ordinary distilled water through a small monobed deionizer (contained in polyethylene) and a submicron filter is equal or superior (with respect to cations) to water obtained by distillation in conventional quartz stills, and is distinctly superior to the product from systems constructed of metal (66). From the data available in the literature, simple distillation clearly does not produce high purity water. In practice, two effects cause contamination of the distillate. Entrainment is the major factor which prevents the perfect separation of a volatile substance from nonvolatile solids during distillation. Rising bubbles of vapor break through the surface of the liquid with considerable force and throw a fog of droplets (of colloidal dimensions) into the vapor space above the liquid surface. These droplets are carried into the condenser to an extent that depends upon the still design and its operating conditions. A second effect is the flow of a film of water (that wets all internal surfaces of the still) toward the condenser, which is under the combined influence of capillarity and the vapor stream.

Subboiling distillation circumvents both of these problems and is favored by workers at the National Bureau of Standards as a simple and effective technique for removing metallic or cationic impurities. In subboiling distillation, infrared heaters vaporize liquid from the surface without boiling. The vapor is condensed on a tapered cold finger and the distillate is collected in a suitable container. A commercial quartz subboiling still like that shown in Figure 4-2 is available from Quartz Products Corp., Plainfield, N.J.; all-Teflon subboiling stills have been described (69).

Table 4-10 Metal Content of Water Purified by Various Methods

Element	Subboiling Quartz Still, Doubly Distilled Water Feed, Kuehner (1973)[a]	Doubly Distilled in Quartz, Distilled Water Feed, Hughes (1971)	Deionized, Monobed, Polyethylene Apparatus, Distilled Water Feed, Hughes (1971)	Distilled, Two-Stage Commercial Metal Still, City Water Feed, Hughes (1971)	Recirculating Ultrapure System, Metal, Hughes (1971)	Single Pass Deionization of Tap Water Thiers (1957)	Deionization, Carbon Absorption, Deionization, Membrane Filtration Zief and Barnard (1973)[b]
Na	0.06	x	x	1.0	4		1
K	0.09	x	x	x	x	x	
Mg	0.09	0.05	0.01	8	2	2	0.5
Ca	0.08	0.07	0.03	50	10	0.2	1
Sr	0.002	x	x	x	x		
Ba	0.01	x	x	x	x		
B	x	—	—	0.01	0.5	+	3
Al	x	0.5	0.1	10	10		0.1
Tl	0.01	x	x	x	x		
Si	x	5.0	1.0	50	10	+	0.5
Sn	0.02	—	—	5	10	+	0.1
Pb	0.008	—	—	50	10	0.02	0.1
Te	0.004	x	x	x	x		

Ti	x	—	+	—		0.1
Cr	0.02	—	—	0.5	0.02	0.1
Mn	x	—	0.01	0.1	0.02	0.05
Fe	0.05	—	0.1	10	0.02	0.2
Ni	0.02	—	1.0	0.05	0.002	0.1
Cu	0.01	—	50	10	+	0.2
Ag	0.002	—	1.0	0.7	+	0.01
Zn	0.04	—	10	3	0.06	0.1
Cd	0.005	x	x	x		0.1

Parts per billion by weight, ng/g; from Ref. 66 unless otherwise indicated.

—, Sought, not detected; x, not sought; +, detected, but not determined quantitatively.

[a]Kuehner et al., *Anal. Chem.*, **44**, 2050 (1973).

[b]Zief and Barnard, *Chem. Tech.*, 440 (1973).

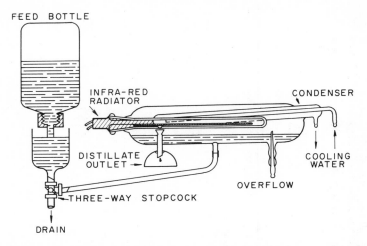

FEED BOTTLE

INFRA-RED RADIATOR

CONDENSER

DISTILLATE OUTLET

COOLING WATER

OVERFLOW

THREE-WAY STOPCOCK

DRAIN

Figure 4-2.   Quartz subboiling still with infrared heating and a cold finger condenser.

Subboiling distillation also is an effective technique for the purification of nitric, hydrochloric, hydrofluoric, perchloric, and sulfuric acids. Typical analyses indicate a 100-fold reduction in the level of metallic impurities (69). While this method is extremely efficient for the removal of metal ions, it offers little purification from impurities with high vapor pressures, such as organic matter and many of the anions.

2. Removal of Organic Impurities

Conway and coworkers (67) have reported that in recent years, both in North America and in Europe, preparation of pure water that is free from organic, surface-active contaminants is impossible by means of distillation, even from alkaline $KMnO_4$, although previously such a procedure was known to be quite adequate. Organic contaminants that now are present commonly in many domestic and industrial water supplies are steam volatile, hence not removed by distillation. The criteria for detecting the presence of such contamination include: (a) the inability to obtain the correct surface tension of pure water (the values generally are too low); (b) the presence of electroactivity at the platinum anode at potentials where reactions are known not to occur for pure solutions; (c) the surface-blocking of electrosorption of H and OH species at platinum due to adsorption of

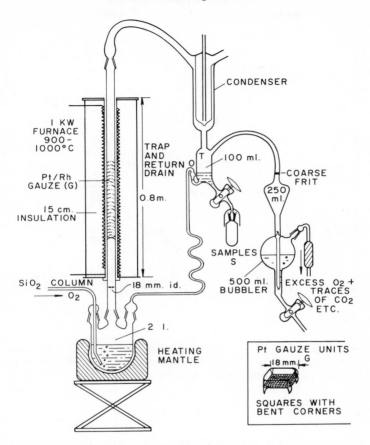

Figure 4-3. Distillation system with a catalytic combustion tube for the removal of organic impurities in water vapor.

organic contaminants; (*d*) the failure to obtain linear log *i* versus *E* relations at the mercury electrode for oxygen-free, preelectrolyzed solutions; and (*e*) the indication of macromolecular contaminants by means of light scattering (usually due to the presence of bacteria).

By passing the water vapor through a column that is packed with platinum gauze and heated to 750 to 800°C in a stream of oxygen the organic materials can be catalytically combusted and entirely removed. The ap-

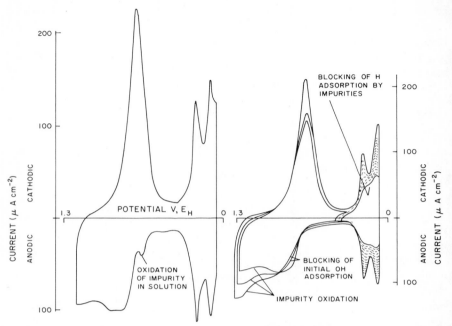

Figure 4-4. Cyclic voltammograms for a platinum electrode. Left curve for redistilled tap water with evidence of an organic impurity. Right curves for distilled water that contains oxidizable organic impurities which chemisorb on the electrode.

paratus shown in Figure 4-3 is designed to recirculate the water continually through the heated zone so that the water in the reservoir is maintained free from organic impurities. The left curve of Figure 4-4 illustrates a current-potential profile for a platinum electrode in redistilled tap water; an oxidation peak that is characteristic of an organic impurity is indicated. The set of curves for the right half of Figure 4-4 indicates the decrease in height of the H adsorption and desorption peaks that results from blocking of the platinum surface by adsorbed organic impurities. The behavior on platinum in a clean aqueous solution is illustrated by Figure 2-23 of Chapter 2. The ability to reproduce every nuance of this current-potential profile can be taken as a primary criterion of the purity of aqueous solutions.

The contribution of the supporting electrolyte to the impurity burden cannot be ignored. Generally the impurities which interfere are heavy metals which may be removed by prolonged electrolysis over a mercury cathode (70). Commercial apparatus is available (Environmental Science Associates, Inc. and Princeton Applied Research Corp.) for removal of heavy metal impurities by this method. Preelectrolysis procedures that use graphite tapes also have been suggested (71) as has the circulation of the solution through beds of charcoal (72).

B. NONAQUEOUS SOLVENTS

Many organic compounds are not soluble in water and the investigator who desires to study their electrochemistry must resort to organic solvents. The solvents most often used are the so-called dipolar aprotic solvents that belong to Class 5a in the classification scheme of Table 4-5. These are solvents with moderately large dielectric constants and low proton availability. As pointed out in Section I this aprotic character tends to simplify the electrochemical reactions; often the primary product is a stable radical cation or anion that is produced by removal or addition of an electron.

Although a number of solvents have been used by different workers, only a few enjoy continued favor. In Table 4-11 the physical properties of more than 50 solvents are listed (not all of them are aprotic). In the following paragraphs some of the properties and purification methods for four solvents are discussed: acetonitrile; propylene carbonate (PC); dimethylformamide (DMF); and dimethylsulfoxide (DMSO). These are the most widely used solvents and probably fulfill the requirements of about 80% of the electrochemical uses to which organic solvents are put. The reader who desires to have more details for other solvents in Table 4-11 should consult references (73) and (74) for physical and electrochemical properties and purification methods.

1. Acetonitrile

Acetonitrile is resistant to both oxidation and reduction, is transparent in the region 200 to 2000 nm, and is an excellent solvent for many polar organic compounds and some inorganic salts. Its dielectric constant of 37 permits reasonably high conductivities, although there is evidence of some association (see Table 4-8). It is not as strong a base as dimethylformamide and dimethylsulfoxide, and therefore does not solvate alkali metal cations as strongly. However acetonitrile forms stable complexes with Ag(I) and Cu(I) ions.

Although somewhat difficult to purify, acetonitrile is stable on storage after purification. It is toxic, with a maximum recommended limit of 40 ppm (75), and the vapor pressure is large enough for this to be a hazard. Its high volatility makes the removal of solvent by evaporation easy (e.g., work

Table 4-11 Physical Properties of Selected Organic Solvents

Solvent	MW	BP (°C)	FP (°C)	Vapor Pressure (mm Hg)	Density (g/cm^3)	Dielectric Constant	Dipole Moment	Viscosity (cP)	Specific Conductance (Ω^{-1}/cm)
Water	18.02	100	0	23.8	0.9970	78.4	1.87	0.89	5.49×10^{-8}
Alcohols									
Methanol	32.04	64.70	−97.68	125.03	0.7866	32.70	2.87_{20}	0.54	1.5×10^{-9}
Ethanol	46.07	78.29	−114.1	59.77	0.7850	24.55	1.66_{20}	1.08	1.35×10^{-9}
2-Propanol	60.10	82.26	−88.0	45.16	0.7813	19.92	1.66_{30}	1.77_{30}	5.8×10^{-8}
1-Butanol	74.124	117.66	−88.62	6.18	0.8060	17.51	1.75	2.27_{30}	9.12×10^{-9}
1,2-Ethanediol	62.07	197.3	−13.	0.12	1.1100	37.7	2.28_{20}	13.55_{30}	1.16×10^{-6}
1,2-Propanediol	76.10	187.6	−60	0.13	1.0328	32.0_{20}	2.25	56.0_{20}	
2-Methoxyethanol	76.10	124.6	−85.1	9.7	0.9602	16.93	2.04	1.60	1.09×10^{-6}
Ethers									
1,2-Dimethoxyethane, (glyme)	90.12	93.0	−58	75.2	0.8621	7.20	1.71	0.46	
Bis(2-methoxyethyl) ether, (diglyme)	134.18	159.76d		3.4	0.9440		1.97	0.98	
Tetrahydrofuran, (THF)	72.11	66	−108.5	197	0.8892_{20}	7.58	1.75	0.46	5×10^{-15}
p-Dioxane	88.11	101.32	11.80	37.1	1.0280	2.21	0.45	1.09_{30}	
Tetrahydropyran	86.14	88	−45		0.8772	5.61	1.55	0.76	
Ketones									
Acetone	58.08	56.29	−94.7	181.72_{20}	0.7844	20.70	2.69_{20}	0.30	4.9×10^{-9}
2-Butanone, (MEK)	72.11	79.64	−86.69	90.6	0.7997	18.51_{20}	2.76	0.37_{30}	3.6×10^{-9}
4-Methyl-2-pentanone, (MIBK)	100.16	116.5	−84	20.0	0.7961	13.11_{20}		0.54	5.2×10^{-8}

Acids and Anhydrides									
Acetic	60.05	117.90	16.66	15.43	1.0437	6.15_{20}	1.68_{30}	1.04_{30}	6×10^{-9}
Acetic anhydride	102.09	140.0	−73.1	5.1	1.0691_{30}	20.7_{19}	2.82	0.78_{30}	5×10^{-9}
Trifluoroacetic	114.02	71.78	−15.25	108	1.4785	8.55_{20}	2.28_{100}	0.86	
Esters and Lactones									
Ethyl acetate	88.11	77.11	−83.97	92.	0.8946	6.02	1.88	0.43	$<1 \times 10^{-9}$
γ-Butyrolactone	86.09	204	−43.53	10_{79}	1.1254	39_{20}	4.12	1.7	
Ethylene carbonate	88.06	238.	36.4	$0.02_{36.4}$	1.3208	89.6_{40}	4.87		$<1 \times 10^{-7}$
Propylene carbonate,[a] (PC)	102.09	241.7	−49.2		1.2	64.4		2.5	1×10^{-8}
Chlorinated alkanes									
Dichloromethane	84.93	39.75	−95.14	435.8	1.3168	8.93	1.14	0.39_{30}	4.3×10^{-11}
Chloroform	119.38	61.15	−63.55	194.8	1.4799	4.81_{20}	1.15	0.51_{30}	$<1 \times 10^{-10}$
1,2-Dichloroethane	98.96	83.48	−35.66	83.35_{20}	1.2458	10.36	1.86	0.73_{30}	4×10^{-11}
Nitro compounds									
Nitromethane	61.04	101.20	−28.55	36.66	1.1313	35.87_{30}	3.56_{20}	0.61	5×10^{-9}
Nitroethane	75.07	114.07	−89.52	20.93	1.0446	28.06_{30}	3.60_{20}	0.64	$5 \times 10^{-7}{}_{30}$
Nitrobenzene	123.11	210.80	5.76	0.28	1.1984	34.82	4.03	1.63_{30}	2.05×10^{-10}
Nitriles									
Acetonitrile, (AN)	41.05	81.60	−43.84	81.81	0.7766	37.5_{20}	3.44_{20}	0.33_{30}	6×10^{-10}
Propionitrile	55.08	97.35	−92.78	44.63	0.7768	27.2_{20}	3.57_{20}	0.39_{30}	8.51×10^{-8}
Succinonitrile	80.09	267	57.88	0.008	0.9867_{60}	$56.5_{57.4}$	3.68_{30}	2.59_{60}	5.64×10^{-4}
Butyronitrile	69.11	117.94	−111.9	19.10	0.7865	20.3_{21}	3.57_{20}	0.52_{30}	
Benzonitrile	103.13	191.10	−12.75	$1_{28.2}$	1.0006	25.20	4.05	1.11_{30}	5×10^{-8}

Table 4.11 (*Continued*)

Solvent	MW	BP (°C)	FP (°C)	Vapor Pressure (mm Hg)	Density (g/cm^3)	Dielectric Constant	Dipole Moment	Viscosity (cP)	Specific Conductance (Ω^{-1}/cm)
Amines									
Butylamine	73.14	77.4	−49.1	91.75	0.7346	4.88_{20}	1.37_{20}	0.68	
Ethylenediamine	60.10	117.26	11.3	$13.1_{26.51}$	0.8860_{30}	12.9	1.90	1.54	9×10^{-8}
Triethylamine	101.19	89.5	−114.7	57.07	0.7230	2.42	0.87	0.36	
Pyridine, (PY)	79.10	115.26	−41.55	$20_{24.8}$	0.9782	12.4_{21}	2.37	0.88	4.0×10^{-8}
1,1,3,3-Tetramethyl-guanidine, (TMG)^b		160			0.9136	11.0		1.40	5×10^{-8}
Amides									
Formamide	45.04	210.5d	2.55	$1_{70.5}$	1.1292	111.0_{20}	3.37_{30}	3.30	$<2 \times 10^{-7}$
N-Methylformamide	59.07	~180	−3.8	0.4_{44}	0.9988	182.4	3.86	1.65	8×10^{-7}
N,N-Dimethylformamide, (DMF)	73.10	153.0	−60.43	3.7	0.9440	36.71	3.86	0.80	6×10^{-8}
Acetamide	59.07	221.15	80.00	$1_{65.0}$	$0.9892_{91.1}$	59_{83}	3.44_{30}	$2.182_{91.1}$	$8.8 \times 10^{-7}{}_{83.2}$
N-Methylacetamide	73.10	206	30.55	1.5_{56}	0.9498_{30}	191.3_{32}	$4.39_{20.1}$	3.23_{35}	$2 \times 10^{-7}{}_{40}$
N,N-Dimethylacetamide, (DMAC)	87.12	166.1	−20	1.3	0.9366	37.78	3.72	0.84_{30}	
1,1,3,3-Tetramethylurea, (TMU)	116.16	175.2	−1.2	$10_{61.2}$	0.9687_{20}	23.06	3.47		$<6 \times 10^{-8}$
2-Pyrrolidone	85.11	245	25	10_{122}	1.107		3.55	13.3	
N-Methylpyrrolidone, (1-methyl-2-pyrrolidone)	99.13	202	−24.4	4_{60}	1.0279	32.0	4.09_{30}	1.67	$1 - 2 \times 10^{-8}$

Sulfur compounds									
Dimethylsulfoxide, (DMSO)	78.13	189.0	18.54	0.600	1.0958	46.68	3.9	1.10	2×10^{-9}
Sulfolane	120.17	287.3d	28.45	5.0_{118}	1.2614_{30}	43.3_{30}	4.81	10.29_{30}	$<2 \times 10^{-8}_{30}$
Dimethylsulfite[c]		126	−141		1.2073_{24}	22.5		0.77_{30}	
Ethylene sulfite[d]		171			1.4375_{20}	41.8_{20}	3.68_{20}		$5 \times 10^{-8}_{20}$
Phosphorus compounds									
Tributylphosphate	266.32	289	<-80	0.8_{114}	0.9760	7.96_{30}	3.07	3.39	
Hexamethylphosphor-amide, (HMPA), (HMPT)	179.20	233	7.20	0.07_{30}	1.027_{20}	30_{20}	5.54	3.47_{20}	

The physical properties (except BP, FP) are measured at 25°C unless the temperature is indicated by a subscript. From Ref. 73 unless otherwise indicated.

[a] Ref. 81.
[b] Yao et al., *J. Electrochem. Soc.*, **115**, 999 (1968).
[c] Gutmann and Scherhaufer, *Monatsh. Chem.*, **99**, 1686 (1968).
[d] Popov et al., *J. Phys. Chem.*, **71**, 1756 (1967).

207

up of a reaction mixture for product identification). Both radical cations and anions react with traces of water in acetonitrile and, because it does not hydrogen bond to water like dimethylsulfoxide and dimethylformamide, radical anions generally have a shorter half-life in acetonitrile than in DMSO and DMF (15). When the last traces of water are scavenged by alumina or trifluoroacetic anhydride, stable radical cations can be produced in acetonitrile (43).

Purification procedures have been reviewed recently (76,77); acetonitrile that has been obtained as a byproduct of the Sohio process (synthesis of acrylonitrile from propylene and ammonia) is recommended as the starting material. It ordinarily will contain water, ammonia, methylamine, ethylamine, acetone, propionitrile, methacrylonitrile, and acrylonitrile as impurities, which may be measured by gas chromatography (76,78). The most recent work indicates that Matheson, Coleman, and Bell (MC/B) spectrograde acetonitrile is suitable for spectroelectrochemical use without purification; its only detectable impurities are propionitrile (0.06%) and water (0.01%) (76,78).

A procedure has been described to produce pure dry acetonitrile with good electrochemical and optical properties (76). The method begins with MC/B practical grade acetonitrile: Step 1 reflux over anhydrous aluminum chloride (15 g/liter) for 1 hr prior to rapid distillation; step 2 reflux over alkaline permanganate (10 g $KMnO_4$ and 10 g Li_2CO_3/liter) for 15 min prior to rapid distillation; step 3 reflux over potassium bisulfate (15 g/liter) for 1 hr prior to rapid distillation; step 4 reflux over calcium hydride (2 g/liter) for 1 hr prior to a careful fractionation from a packed column at high reflux ratio; the middle 80% fraction is retained. The overall yield is 60% and the product should have an ultraviolet cut-off at a wavelength below 200 nm (1-cm cell versus water) and a 5-V voltage window (with a platinum electrode to measure the anodic limit and a dropping mercury electrode to measure the cathodic limit).

2. Propylene Carbonate

Propylene carbonate (4-methyldioxolone-2, PC, $C_4H_6O_3$) is a cyclic ester with a low vapor pressure, relatively low toxicity, high dielectric constant ($\epsilon = 69$), and a large liquid range (-49 to $+242°C$) although thermal decomposition begins to take place at 150°C. The solvent can be supercooled by about 25°C. Hydrolysis in the presence of acids or bases is the principal chemical decomposition reaction. It is relatively nonhygroscopic and noncorrosive.

Propylene carbonate is produced in the United States by the Jefferson Chemical Company via a reaction of CO_2 and propylene. It is a clear, colorless (when pure) liquid capable of dissolving a variety of organic and inorganic compounds. It has been recommended as especially suitable for electrochemistry because of its low background currents and resistance to oxidation (79,80). Both radical cations and anions are stable in the solvent. The solvent has been used widely in battery research (81).

As obtained from the manufacturer, the impurities of propylene carbonate which have been identified by gas chromatography (82) include CO_2, water, propylene oxide, allyl alcohol, 1,2-propylene glycol, 1,3-propylene glycol, and lesser amounts of unidentified materials. The solvent may be purified readily by vacuum fractional distillation, with a reflux ratio of 10:1 at 0.5 to 1 torr. At this pressure the column head temperature is 72 to 75°C. Another suggestion is to heat propylene carbonate with sodium carbonate and potassium permanganate (both 10 g/liter) for 2 hr prior to vacuum distillation.

3. Dimethylformamide

Dimethylformamide (DMF) is a colorless, mobile liquid that is miscible with water. Its slight amine odor results from hydrolysis of the DMF by absorbed water to give dimethylamine and formic acid; the purified solvent cannot be stored for more than a day or so at room temperature without some decomposition. DMF, with a dielectric constant of 37, is a good solvent for a wide range of organic compounds and it also will dissolve many inorganic perchlorates, iodides, and lithium chloride. Nitrates are soluble, but they decompose (83). The vapor is toxic and individuals should not be exposed for long periods to a concentration greater than 10 ppm by volume (84).

DMF has been widely used as an electrochemical solvent, especially for the reduction of aromatic hydrocarbons (85). The polarography of a number of metal ions in DMF also has been reviewed (86). In general, the voltage range attained in reductions is comparable to acetonitrile and dimethylsulfoxide, but DMF is less suitable for the study of oxidations. It has been suggested that the cyclic amide, N-methylpyrrolidone, may have most of the favorable properties of DMF, but with less tendency to hydrolyze (87,88). However, it is less available and more expensive.

Although reagent grade DMF may be used directly in some work, particularly if it has been kept free of water, it is commonly purified just before use by drying with molecular sieves and vacuum distillation. One purification procedure (89) is to dry reagent grade DMF with Linde AW-500 Molecular Sieves (which are superior to the more commonly used Linde 4A) for at least 24 hr. Then the DMF is slurried with about 60 grams of P_2O_5/500 ml and distilled under nitrogen at 2.5 to 8 torr, which corresponds to a distillation temperature of 33 to 49°C. The first fraction of 100 ml is discarded, and the second fraction of about 300 ml is collected and stored under nitrogen in a freezer at −20°C. The purified solvent, which should be used within 48 hr after preparation, contains less than 10 ppm of water (as measured by Karl Fischer titration) and less than 4×10^{-6} M acidic or basic impurities. The specific conductance should be 3×10^{-7} mho/cm.

4. Dimethylsulfoxide (DMSO)

Butler has thoroughly reviewed the applications of DMSO in electrochemistry (90). It is a particularly useful solvent because it has a high dielectric constant and is sufficiently resistant to both oxidation and reduction to provide a fairly wide potential range. It is, however, not as resistant as acetonitrile or propylene carbonate to oxidation and these latter two solvents are preferred over DMSO for this purpose.

DMSO is rather polar, with a high donor number, but the claims that it is a strongly associated liquid have been refuted by Amey (91). He has measured the Kirkwood correlation factor for dimethylformamide, dimethylsulfoxide, and nitrobenzene, and has found it to be near 1.0 over a wide temperature range for all three of these solvents. This indicates that their properties are due primarily to large, but nonspecific, molecular dipole-dipole interactions. There is evidence that DMSO strongly associates with water (92), which may account for the fact that radical anions generally are more stable in DMSO than in acetonitrile when both solvents contain traces of water (15).

Pure DMSO is nearly odorless (usually traces of dimethylsulfide are detectable by nose) and is essentially nontoxic. However, it penetrates the skin rapidly, and may carry with it toxic solutes dissolved in the DMSO, so that DMSO solutions should be treated with respect.

The reagent grade solvent generally contains few impurities beyond traces of water and can be used without further purification for many electrochemical purposes. For some applications the presence of traces of water can significantly affect the properties of the solvent. For example, dimethylsulfoxide and dimethylformamide that contain 100 ppm of water dissolve NaCN and NaN_3 to a much greater extent than rigorously purified solvents (93). DMSO can be purified by drying over Molecular Sieves, distilling under vacuum, and then flash vacuum distilling from a small quantity of freshly prepared KNH_2 (under vacuum in a rotary evaporator) and holding the product under vacuum for several hours to remove traces of NH_3. The product should contain less than 10 ppm of water by Karl Fischer titration and less than $5 \times 10^{-6} M$ acidic or basic impurities.

V. Preparation and Purification of Supporting Electrolytes

In both polarographic and preparative electrochemistry in aprotic solvents the custom is to use tetraalkylammonium salts as supporting electrolytes. In such solvent-supporting electrolyte systems electrochemical reductions at a mercury cathode can be performed at -2.5 to -2.9 V versus SCE. The reduction potential ultimately is limited by the reduction of the quaternary ammonium cation to form an amalgam, $(R_4N \cdot)Hg_n$, $n = 12$-13. The tetra-n-butyl salts are more difficult to reduce than the tetraethylammonium salts and are preferred when the maximum cathodic range is needed. On the anodic side the oxidation of mercury occurs at about $+0.4$ V versus SCE in a supporting electrolyte which does not complex or form a precipitate with the Hg(I) or Hg(II) ions that are formed.

By use of platinum, gold, or carbon electrodes a much more positive potential can be attained than on mercury; the anodic limit depends on the negative counter ion. Halide counter ions do not provide a good anodic range because they are readily oxidized; perchlorate or tetrafluoroborate counter ions generally are preferred. The tetrafluoroborate ion gives a slightly greater anodic range than perchlorate ion, which can be oxidized in the presence of tetrafluoroborate ion (94). The exact nature of the oxidation product is somewhat obscure, but recent evidence indicates that the eventual product is ClO_2 (95).

A number of quaternary ammonium perchlorates and tetrafluoroborates have been prepared, and their solubilities and specific resistance values have been measured in several solvents (see Table 4-6) (32). The resistance values are about tenfold higher in ethers like tetrahydrofuran and 1,2-dimethoxyethane relative to solutions in acetonitrile and dimethylformamide. This is due to the greater ion association in the ethers because of their smaller dielectric constants.

Concentrated solutions of supporting electrolytes in hexamethylphosphoramide become very viscous, which causes a twofold increase in resistance over solutions in acetonitrile and dimethylformamide. The increased viscosity may be due to the greater solvation of the cations by the more basic HMPA.

For routine work, the tetrafluoroborate salts are recommended because they are somewhat easier to purify and dry. They also may be safer than the perchlorates, which under conditions of high temperature and high acidity can decompose violently. However, tetraethylammonium perchlorate, which is particularly convenient to prepare because it can be recrystallized from water, has been prepared in many laboratories in batches as large as a kilogram. Obviously care must be taken to see that the initial reaction mixture that contains perchloric acid is never allowed to become overheated or evaporated to dryness.

The following procedures have been used by House and coworkers (32) for the preparation of quaternary ammonium perchlorates and tetrafluoroborates. They can be scaled up to prepare larger quantities when needed.

1. Tetra-n-Butylammonium Tetrafluoroborate

A solution of 8.4 g (25 mmoles) of $(n\text{-Bu})_4\text{NBr}$ in a minimum volume of water (about 18 ml) is treated with 3.6 ml (about 26 mmoles) of aqueous 48 to 50% HBF_4. The resulting mixture is stirred at 25°C for 1 min and the crystalline salt is collected on a filter, washed with water until the washings are neutral, and dried. The crude salt (6.3 g or 79%, mp 155–175°C) is recrystallized three times from ethyl acetate-pentane mixtures to separate 6.0 g (75%) of $(n\text{-Bu})_4\text{NBF}_4$ as white needles; mp (after drying) 162–162.5°C.

2. Tetraethylammonium Tetrafluoroborate

A solution of 5.3 g (25 mmoles) of Et_4NBr in about 8 ml of water is reacted with HBF_4 and concentrated. Next, it is diluted with ethyl ether and filtered to yield 4.6 g (85%) of the crude salt; mp 375–378°C with decomposition. Two recrystallizations from a methanol-petroleum ether (bp 30–60°) mixture yields 3.7 g (69%) of pure Et_4NBF_4 as white needles; mp (after drying) 377–378°C with decomposition.

3. Tetra-n-butylammonium Perchlorate

A saturated aqueous solution of 8.4 g (25 mmoles) of $(n\text{-Bu})_4\text{NBr}$ in 18 ml of water is treated with 2.1 ml (about 26 mmoles) of aqueous 70 to 72% HClO_4. After the resulting insoluble perchlorate salt has been collected, washed with cold water, and dried, the yield is 8.0 g (94%); mp 197–199°C. Two recrystallizations from an ethyl acetate-pentane mixture yield 7.6 g (90%) of pure $(n\text{-Bu})_4\text{NClO}_4$ as white needles which are dried at 100° under reduced pressure; mp 212.5–213.5°C.

4. Tetraethylammonium Perchlorate

The same procedure is used to convert 5.3 g (25 mmoles) of Et_4NBr (in about 8 ml of water) to 4.7 g (81%) of the crude perchlorate salt, which is cooled before filtration. The salt is recrystallized from water to yield 4.3 g (75%) of the pure Et_4NClO_4 as white needles which are dried at 100° under reduced pressure; mp 351–352.5°C with decomposition.

References

1. A. J. Parker, *Chem. Rev.*, **69**, 1 (1969).

2. M. E. Peover, "Electrochemistry of Aromatic Hydrocarbons and Related Substances," in *Electroanalytical Chemistry*, Vol. 2, A. J. Bard, ed., Marcel Dekker, Inc., New York, 1967, pp. 1–51.

3. G. J. Hoijtink, J. van Schooten, E. de Boer, and W. Y. Aalbersberg, *Rec. Trav. Chim.*, **73**, 355 (1954).

4. M. Fujihira, H. Suzuki, and S. Hayano, *J. Electroanal. Chem.*, **33**, 393, (1971).

5. E. F. Caldin, *Fast Reactions in Solution*, John Wiley and Sons, Inc., New York, 1964, pp. 239–251.

6. C. D. Ritchie, "Interactions in Dipolar Aprotic Solvents," in *Solute-Solvent Interactions*, J. F. Coetzee and C. D. Ritchie, eds., Marcel Dekker, Inc., New York, 1969, pp. 246–271.

7. P. Zuman, *Organic Polarographic Analysis*, The Macmillan Co., New York, 1964, p. 61.

8. A. L. Woodson and D. E. Smith, *Anal. Chem.*, **42**, 242 (1970).

9. R. P. Van Duyne and C. N. Reilley, *Anal. Chem.*, **44**, 142 (1972).

10. B. Kratochvil, *Crit. Rev. Anal. Chem.*, **1**, 415, (1971); *Rec. Chem. Prog.*, **27**, 253 (1966).

11. V. Gutmann, *Coordination Chemistry in Non-Aqueous Solutions*, Springer-Verlag, New York, 1968, Chapter 5.

12. V. Gutmann, *Fortschr. Chem. Forsch.*, **27**, 61 (1972).

13. D. Martin, A. Weise, and H. J. Niclas, *Angew. Chem. Int. Edit.*, **6**, 318 (1967)

14. V. Gutmann and E. Wychera, *Inorg. Nucl. Chem. Letters*, **2**, 257 (1966).

15. J. R. Jezorek and H. B. Mark, Jr., *J. Phys. Chem.*, **74**, 1627 (1970)

16. I. M. Kolthoff, "Review of Fundamentals of Polarography in Inert Organic Solvents," in *Polarography 1964*, Vol. 1, G. J. Hills, ed., Interscience Publishers, New York, 1966, p. 2.

17. L. Marcoux, P. Malachesky, and R. N. Adams, *J. Am. Chem. Soc.*, **89**, 5766 (1967).

18. J. Phelps, K. S. V. Santhanam, and A. J. Bard, *J. Am. Chem. Soc.*, **89**, 1752 (1967).

19. K. Bechgaard and V. D. Parker, *J. Am. Chem. Soc.*, **94**, 4749 (1972).

20. D. Bauer and A. Foucault, *J. Electroanal. Chem.*, **39**, 385 (1972).

21. H. L. Jones, L. G. Boxall, and R. A. Osteryoung, *J. Electroanal. Chem.*, **38**, 476 (1972).

22. R. J. Gillespie, *Accounts Chem. Res.*, **1**, 202 (1968).

23. T. Kagiya, Y. Sumida, and T. Inoue, *Bull. Chem. Soc. Japan*, **41**, 767; 773; 779 (1968).

24. R. W. Taft, D. Gurka, L. Joris, P. von R. Schleyer, and J. W. Rakshys, *J. Am. Chem. Soc.*, **91**, 4801 (1969).

25. H. G. Harris and J. M. Prausnitz, *J. Chromatographic Sci.*, **7**, 685 (1969).

26. C. Reichardt, *Angew. Chem. Int. Edit.*, **4**, 29 (1966).

27. C. Reichardt and K. Dimroth, *Fortschr. Chem. Forsch.*, **11**, 1 (1968).

28. M. Mohammad and E. M. Kosower, *J. Phys. Chem.*, **74**, 1151 (1970).

29. A. I. Popov, "Techniques with Nonaqueous Solvents," in *Technique of Inorganic Chemistry*, Vol. 1, H. B. Jonassen and A. Weissberger, eds., Interscience Publishers, New York, 1972, p. 37.

30. J. J. Lagowski, ed., *The Chemistry of Non-Aqueous Solvents*, Vol. 1, Academic Press, New York, 1966, Chap. 7; Vol. 2, 1967, Chap. 6; Vol. 3, 1970, Chap. 2.

31. R. N. Adams, *Electrochemistry at Solid Electrodes*, Marcel Dekker, Inc., New York, 1969, p. 33.

32. H. O. House, E. Feng, and N. P. Peet, *J. Org. Chem.*, **36**, 2371 (1971).

33. N. B. Godfrey, *Chem. Tech.*, 359 (1972).

34. *Spectroquality Solvents—Spectra, Physical Properties, Specifications and Typical Uses*, MC/B Manufacturing Chemists, 2909 Highland Avenue, Norwood, Ohio 45212.

35. *Spectra Collection of Commercial Solvents*, Sadtler Research Laboratories, Inc., 3316 Spring Garden St., Philadelphia, Penn. 19104.

36. R. G. Bates, "Medium Effects and pH in Nonaqueous Solvents," in *Solute-Solvent Interactions*, J. F. Coetzee and C. D. Ritchie, eds., Marcel Dekker, Inc., New York, 1969, pp. 50–53.

37. G. Cauquis, in *Organic Electrochemistry*, M. M. Baizer, ed., Marcel Dekker, Inc., New York, 1973, Chap. 1, Sections IV and VI.

38. J. M. Saveant, *J. Electroanal. Chem.*, **29**, 87 (1971).

39. Ref. 2, p. 40.

40. Ref. 2, pp. 11–15.

41. A. J. Bard and J. Phelps, *J. Electroanal. Chem.*, **25**, App. 2 (1973).

42. H. B. Mark, Jr., *Rec. Chem. Prog.*, **29**, 217 (1968).

43. O. Hammerich and V. D. Parker, *Electrochim. Acta*, **18**, 537 (1973).

44. B. Kratochvil and H. L. Yeager, *Fortsch. Chem. Forsch.*, **27**, 1 (1972).

45. G. J. Janz and R. P. T. Tomkins, eds., *Nonaqueous Electrolytes Handbook*, Academic Press, New York, 1972.

46. W. M. Clark, *Oxidation-Reduction Potentials of Organic Systems*, The Williams and Wilkins Co., Baltimore, 1960.

47. P. Zuman, "Effects of Acidity in the Elucidation of Organic Electrode Processes," in *Progress in Polarography*, Vol. III, P. Zuman and L. Meites, eds., Wiley-Interscience, New York, 1972, pp. 73–156.

48. S. G. Mairanovskii, *Catalytic and Kinetic Waves in Polarography*, Plenum Press, New York, 1968.

49. P. J. Elving, *Pure Appl. Chem.*, **7**, 423 (1963).

50. L. Meites, *Polarographic Techniques*, 2nd ed., Interscience Publishers, New York, 1965, p. 282.

51. N. E. Good, G. D. Winget, W. Winter, T. N. Connolly, S. Izawa, and R. M. Singh, *Biochemistry*, **5**, 467 (1966).

52. R. G. Bates, R. N. Roy, and R. A. Robinson, *Anal. Chem.*, **45**, 1663 (1973).

53. Ref. 48, p. 164.

54. S. G. Mairanovskii, *J. Electroanal. Chem.*, **4**, 161 (1962).

55. R. Guidelli, G. Pezzatini, and M. L. Foresti, *J. Electroanal. Chem.*, **43**, 83; 95 (1973).

56. M. M. Davis, *Acid-Base Behavior in Aprotic Organic Solvents*, National Bureau of Standards Monograph 105, United States Department of Commerce, available from Superintendent of Documents, U. S. Government Printing Office, Washington, D.C. 20402, issued August 1968.

57. C. D. Ritchie, "Interactions in Dipolar Aprotic Solvents," in *Solute-Solvent Interactions*, J. F. Coetzee and C. D. Ritchie, eds., Marcel Dekker, Inc., New York, 1969, Chap. 4.

58. R. L. Benoit and C. Buisson, *Electrochim. Acta*, **18**, 105 (1973).

59. Ref. 56, p. 88.

60. E. Temmerman and F. Verbeek, *J. Electroanal. Chem.*, **19**, 423 (1968).

61. D. D. Perrin, *Masking and Demasking of Chemical Reactions*, Wiley-Interscience, 1970, Chap. 10; Vol. 33 of *Chemical Analysis*, P. J. Elving and I. M. Kolthoff, eds.

62. K. Izutsu, S. Sukura, K. Kuroki, and T. Fujinaga, *J. Electroanal. Chem.*, **32**, App. 11 (1971).

63. M. Szwarc, *Accounts Chem. Res.*, **2**, 87 (1969).

64. L. A. Avaca and A. Bewick, *J. Electroanal. Chem.*, **41**, 405 (1973).

65. P. G. Westmoreland, R. A. Day, Jr., and A. I. Underwood, *Anal. Chem.*, **44**, 737 (1972).

66. R.C. Hughes, P. C. Murau, and G. Gundersen, *Anal. Chem.*, **43**, 691 (1971).

67. B. E. Conway, H. Angerstein-Kozlowska, W. B. A. Sharp, and E. E. Criddle, *Anal. Chem.*, **45**, 1331 (1973).

68. J. W. Mitchell, *Anal. Chem.*, **45**, 492 A (1973).

69. E. C. Kuehner, R. Alvarez, P. J. Paulsen, and T. J. Murphy, *Anal. Chem.*, **44**, 2050 (1973).

70. A. M. Azzam, J. O'M. Bockris, B. E. Conway, and H. Rosenberg, *Trans. Faraday Soc.*, **46**, 918 (1950).

71. M. Rosen, H. H. Bauer, and P. J. Elving, *J. Electrochem. Soc.*, **117**, 878 (1970).

72. D. A. Jenkins and C. J. Weedon, *J. Electroanal. Chem.*, **31**, App. 13 (1971).

73. J. A. Riddick and W. B. Bunger, *Organic Solvents*, 3rd ed., Wiley-Interscience, New York, 1970; Vol. II of *Techniques of Chemistry*, A. Weissberger, ed.

74. C. K. Mann, "Nonaqueous Solvents for Electrochemical Use," in *Electroanalytical Chemistry*, Vol. 3, A. Bard, ed., Marcel Dekker, Inc., New York, 1969, pp. 57–134.

75. Ref. 73, p. 805.

76. M. Walter and L. Ramaley, *Anal. Chem.*, **45**, 165 (1973).

77. J. F. Coetzee, *Pure Appl. Chem.*, **13**, 429 (1966).

78. T. A. Kowalski and P. J. Lingane, *J. Electroanal. Chem.*, **31**, 1 (1971).

79. R. F. Nelson and R. N. Adams, *J. Electroanal. Chem.*, **13**, 184 (1967).

80. D. B. Clark, M. Fleischmann, and D. Pletcher, *J. Electroanal. Chem.*, **42**, 133 (1973).

81. R. Jasinski, "Electrochemistry and Application of Propylene Carbonate," in *Advances in Electrochemistry and Electrochemical Engineering*, Vol. 8, C. W. Tobias, ed., Wiley-Interscience, 1971, p. 253.

82. R. J. Jasinski and S. Kirkland, *Anal. Chem.*, **39**, 1663 (1967).

83. Ref. 74, p. 76.

84. Ref. 73, pp. 839–840.

85. R. Dietz and M. E. Peover, *Discuss. Faraday Soc.*, **45**, 154 (1968).

86. J. B. Headridge, M. Ashraf, and H. L. H. Dodds, *J. Electroanal. Chem.*, **16**, 116 (1968).

87. H. Lund and P. Iversen, "Practical Problems in Electrolysis," in *Organic Electrochemistry*, M. M. Baizer, ed., Marcel Dekker, Inc., 1973, p. 221.

88. M. Breant and J. L. Sue, *J. Electroanal. Chem.*, **40**, 89 (1972).

89. C. D. Ritchie and G. H. Megerle, *J. Am. Chem. Soc.*, **89**, 1447 (1967).

90. J. N. Butler, *J. Electroanal. Chem.*, **14**, 89 (1967).

91. R. L. Amey, *J. Phys. Chem.*, **72**, 3358 (1968).

92. D. H. Rasmussen and A. P. McKenzie, *Nature*, **220**, 1315 (1968).

93. C. D. Ritchie, G. A. Skinner, and V. G. Badding, *J. Am. Chem. Soc.*, **89**, 2063 (1967).

94. M. Fleischmann and D. Pletcher, *Tetrahedron Lett.*, 6255 (1968).

95. G. Bontempelli, F. Magno, and G. A. Mazzochin, *J. Electroanal. Chem.*, **42**, 57 (1973).

Five

INSTRUMENTATION

I. Measurement Instrumentation

A. INTRODUCTION

In this chapter an overview of electrochemical instrumentation is provided with emphasis in two broad areas—measurement and control. These naturally overlap as do the two primary techniques for their implementation—analog and digital instrumentation. A significant fraction of the chapter is devoted to the use of operational amplifiers for measurement and control. Amplifiers will be treated as "black boxes" which can be effectively used without detailed knowledge of their internal circuitry. The reader who would like more information should consult the list of general references (1–4) at the end of the chapter. Even these sources may not provide sufficient information for the reader to construct a particular instrument; for detailed circuit diagrams and explanations one must often go to the original periodical and manufacturers' literature. The applications notes and specification sheets for a particular device generally provide the most detailed source of information that is available; some of these sources are indicated at the appropriate places. A mathematical background sufficient to deal with the representation of impedances by complex numbers in simple A-C circuits is assumed. The general references contain brief sections on the j-operator (1–4) and phasor (4) representation of complex impedances for the reader who needs to refresh his mathematical skills. Short sections on the Laplace transform are contained in Refs. 4 and 5. The general references that have been selected contain a minimum of mathematics and are intended for an audience of nonspecialists. Fuller mathematical treatments are available elsewhere (6,7).

At the end of this chapter there is a short section on signal conditioning, noise, and the extraction of information from noisy signals.

1. Some Historical Notes

One of the first applications of vacuum-tube electronics to electrochemical measurements was the use of a vacuum triode as a null-point detector in measurements of cell potential (8). The use of automatic electronic control in electrochemistry was ushered in by Hickling (9) in 1942 when he described an instrument for the control of potential, which he called a "potentiostat." This instrument employed a three-electrode circuit and had a response time of the order of a second. It was followed during the succeeding 15 years by a number of other potentiostats whose evolution has been traced by Lingane (10). In the late 1950s the utility of the operational amplifier for measurement and control was recognized. Both the terminology and practice largely derived from the field of analog computation and linear-feedback control theory; Booman (11) and De Ford (12) were among the first to describe the application of the general principles to the design of electroanalytical instrumentation. In particular, the detailed description of the De Ford modular instrument (which was widely circulated but unfortunately never published) did much to stimulate this development.

After the initial reports in 1957 there followed a period of intense development of operational-amplifier instrumentation; the symposia of the Analytical Division of the American Chemical Society in 1963 (13) and 1966 (14) reported a variety of new instruments and applications. The appearance of solid-state amplifiers spurred further development by providing lower noise levels and better performance at lower cost; solid-state devices have now totally supplanted vacuum-tube operational amplifiers. High performance integrated-circuit operational amplifiers are now available at a unit cost of less than $10 (one-tenth of the cost only a few years ago).

All of the instruments based on operational amplifiers operate in the *analog signal domain* [i.e., the control and measurement information is represented by a continuously variable signal (voltage, current, charge, etc.)]. The alternative *digital signal domain* processes information by means of discrete voltage or current levels whose sequence in time can be used for control of logic circuitry and to represent digital numbers, usually in the binary number format. The data-domain concept has been applied to the analysis of instrumentation systems (15) and illustrates how the signal flow may change from one data domain to another by means of interdomain conversions. The analog-to-digital converter (ADC) and digital-to-analog converter (DAC) are examples of interdomain converters that are important components of many instrumental systems.

The intensive development of digital computers during the period of the 1960s also has provided inexpensive digital devices of infinite variety which have found many applications in electrochemical instrumentation. These range from the use of logic circuits for timing and control in hybrid analog-digital circuitry to the use of minicomputers for on-line control of experiments, data acquisition, and data analysis.

2. Measurement Instrumentation

All electrical measurements involve some sort of operation which in effect defines the quantity measured. The "operation" gives concrete meaning to a property which might otherwise remain rather abstract, a notion that has been developed by Bridgman because of its utility for analysis of complex problems in physics (16). In Bridgman's terminology there are operations which are predominantly instrumental, as in comparing the length of an object to a meter stick, but there are also mental operations—a viscosity, for example, involves a "mental" analysis of the quantities that enter into the definition of viscosity.

The operations which define "absolute" electrical measurements are prescribed by physical laws—the magnitude of a current is defined by the force that exists between two wires (or two coils) of measurable shape and size that carry the same current. In this case, the operation consists of measuring the force between two sets of coils with a device called a current balance (17). These operations which measure a property according to its basic definition can be undertaken only by laboratories enjoying exceptional facilities [e.g., the National Bureau of Standards (NBS)]. The absolute value of the ohm, the farad, and the henry are accurately linked by impedance measurements at a known frequency and calculation of the self-inductance of a coil or the mutual inductance of two coils. The accuracy of such absolute measurements, which depends on the linear dimensions of the coils and their precise construction, lies between 1 and 3 ppm (18).

These laborious measurements obviously are not practical for everyday use. Hence secondary standards have been constructed such that the accuracy of routine measurements in the laboratory depends on the comparison of a property of the unknown with that of the secondary standard. Secondary resistance standards are coils of manganin wire maintained at the NBS. Galvanic cells (like the Weston standard cell shown in Figure 5-1b) are maintained as secondary voltage standards (18). Carefully constructed capacitors and inductors also are maintained as standards. Some recently developed techniques provide the means to check the stability of the secondary standards for the electric units—measurement of the gyromag-

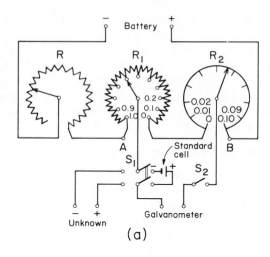

(a)

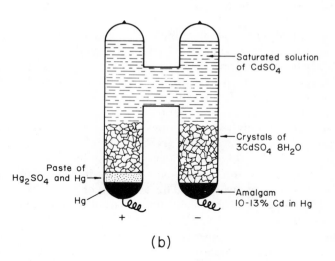

(b)

Figure 5-1. (a) Student potentiometer; (b) Weston standard cell.

netic ratio of the proton for the ampere and the measurement of the ratio h/e by the Josephson effect for the volt (18,19).

3. Comparison Methods

Comparison methods are used to measure voltage, resistance, inductance, and capacitance. These methods in effect compare a secondary standard which is traceable back to NBS measurements with the unknown voltage, resistance, and so on. High quality comparison instruments can be submitted to NBS or commercial testing laboratories for calibration. Or high quality secondary standards may be purchased so that calibration may be done in the home laboratory. These measurements generally are at least an order of magnitude less precise than the measurements of the primary standards (generally in the range 10 to 100 ppm).

B. VOLTAGE MEASUREMENTS

The level of precision that is required for voltage measurements in electrochemical work is summarized in Table 5-1. Most electrochemical cells will have voltages in the range 0 to 2 V. Linear amplifiers and voltage measuring devices with a precision of 0.01% are readily available so that precision measurements of voltage are not especially difficult unless the source impedance is very high. For the latter condition careful attention to shielding the signal source from noise pick-up is necessary. Voltages of less than 10 mV often are amplified with a linear amplifier before measurement.

The potentiometer is the classical instrument for the precise measurement of voltage. The diagram of Figure 5-1a represents a simple "student potentiometer" whose precision depends on the precision of the individual resistance elements of R_1 and the linearity of the slidewire, R_2. In practice, the taps on R_1 and R_2 are adjusted to give a dial reading of 1.018 (numerically equal to the voltage of the Weston standard cell). The series resistance, R, is then adjusted until the current flowing in the voltage divider generates an iR drop between the two tap points that is exactly equal to 1.018 V, as indicated by zero deflection of the galvanometer when switch S_2 is closed. This operation standardizes the slidewire. The measurement of the unknown is made by switching the unknown cell in place of the standard cell and adjusting R_1 and R_2 until a null is reached; the readings on the calibrated taps of R_1 and R_2 then give the unknown voltage. The measurement accuracy depends on the short-term stability of the battery or power-supply voltage, although slow drifts can be compensated by frequent restandardization.

When the internal resistance of the potential source is as large as 10 to 100 MΩ (as in glass electrode cells), the current that flows is too small to cause an observable deflection of most galvanometers. In addition the passage of even 10^{-8} A through a glass electrode will alter its potential by changing the H^+ activity at the glass membrane. This difficulty can be circumvented by replacing the galvanometer with a null detector of high input impedance, such as an electrometer. Although voltage measurements can be made directly with a vibrating reed electrometer to less than $\pm 0.25\%$ error, measurement accuracy is increased by nulling the input voltage with a

Table 5-1   Precision of Voltage Measurements in Electrochemistry

Type of Voltage Measurement	Level of Precision and/or Accuracy
Measurements on electrochemical cells	
(a) Potentiometric titrations	± 5 mV
(b) pH Measurements (± 0.1 pH unit)	± 5 mV
(c) pH Measurements (± 0.01 pH unit) in measurement of blood pH and CO_2 tension	± 0.5 mV
(d) Specific ion electrodes ($\pm 1\%$ accuracy in activity)	$\pm \dfrac{0.2}{n}$ mV
(e) Thermodynamic measurements of activity coefficients, equilibrium constants, etc.	± 0.1–0.01 mV
Voltammetric cells	
(f) Polarographic half-wave potentials	± 1–10 mV
(g) Voltammetric peak potentials on solid electrodes	± 5–20 mV
Miscellaneous	
(h) Current measurements by Ohm's law, $i = E/R$	Commensurate with the precision of the resistor (generally 0.1% or better)
(i) Thermocouple measurements of temperature ($\pm 0.1\,^{\circ}$C)	10–50 μV (dependent on the thermocouple pair)

precision potentiometer (such as the Leeds and Northrup Type K-3) and using the electrometer as a null indicator (20,21). This allows a measuring precision of ± 0.01 millivolt which corresponds to a short term sensitivity of ± 0.0002 pH unit. A stability of about 0.002 pH unit per hour can be achieved—assuming that the temperature of the solution is controlled to $\pm 0.01\,^{\circ}$C and all other conditions are favorable (20). This measurement principle also is used in some pH meters, with a tenfold reduction in precision and cost (22).

For differential measurements where the ion activity of two different solutions is to be compared by use of high-resistance electrodes (or where a glass electrode is used as a reference electrode), a differential input, high impedance amplifier must be used. Rechnitz has described two such differential amplifiers, one with approximately a 2-msec response time (23) and a faster 2-μsec version (24).

At the other end of the scale in precision, an inexpensive potentiometer can be improvised by connecting a ten-turn potentiometer (with a calibrated dial divided into 1000 divisions) in series with a mercury battery and variable resistor, like the circuit of Figure 5-1a. When combined with a sensitive null indicator a sensitivity of ± 1 mV can be obtained; the accuracy is dependent on the linearity of the potentiomenter (25).

1. Linear Amplifiers

Linear amplifiers, which are widely used in instruments such as vacuum-tube voltmeters (26) and direct reading pH meters (27), yield output currents or voltages that are proportional to the input voltage. Most pH meters of recent design use high-input impedance amplifiers with field-effect transistor or varactor bridge input stages. Carefully designed amplifiers can have very low drift and linearities with less than 0.01% error. To realize the full benefit of the precision available, the readout device must be either a potentiometer and null indicator or a digital voltmeter.

Digital voltmeters are now available in a wide range of prices and specifications, from \$50 for a $2\frac{1}{2}$-digit panel meter with a single range to \$2000 for a $5\frac{1}{2}$-digit multirange system with error limits of $\pm 0.003\%$. The cost of digital multimeters, which measure A-C and D-C voltage and current, resistance, and sometimes frequency, has dropped continally as the price of large-scale integrated circuitry has lowered parts counts and assembly costs; multimeters with 0.1% error limits are now available for less than \$300. They are convenient for routine measurements and trouble shooting, and will undoubtedly replace the vacuum-tube voltmeter or multimeter in the future.

2. Voltage Standards

One of the most satisfactory voltage standards is the Weston standard cell, shown in Figure 5-1b. The cell is represented by the schematic

$$Cd(10\% \text{ amalgam})/CdSO_4 \cdot \tfrac{8}{3}H_2O/CdSO_4(\text{satd. soln.})/Hg_2SO_4/Hg \quad (5\text{-}1)$$

It has a voltage of 1.018 V at 25°C, and has good voltage stability but a high temperature coefficient (28). The Weston cell is sensitive to rough handling and should not be tipped more than 45° from the vertical and never turned upside down. The cells are best maintained at constant temperature in a thermostat that is designed to hold a group of cells whose voltages may be intercompared.

The unsaturated Weston cell (saturated at 4°C with $CdSO_4$) is more rugged and has a negligible temperature coefficient. These cells are the ones usually sold commercially for incorporation into instruments as voltage standards. The voltage of the unsaturated Weston cell tends to decrease with time (the average value of a group of cells decreased 0.07 mV/year, and the average decrease of 4% of the cells was 0.28 mV/year (29).

Solid-state devices, like the Zener diode (30), also can provide stable and precise reference voltages, but they must be calibrated against voltage standards like the Weston cell. Weston cells and Zener diodes can be incorporated into voltage calibrators or voltage reference sources, which, by precise amplification, can provide a range of accurately known voltages for the calibration of voltage-measuring equipment.

C.   CURRENT MEASUREMENTS

Many instruments incorporate meters to monitor voltages and currents. The meter movement usually is of the D'Arsonval type (31) in which the interaction of a current-carrying coil with the field of a permanent magnet produces a deflection of a pointer (connected to the coil) that is proportional to the current. In recent designs, the moving coil is fixed to a band of spring-metal that is anchored at both ends so that the restoring force is torsional. This type of movement is much more rugged than the pivots and coil-spring suspension used in the older D'Arsonval movement and represents an increasing percentage of the market. The error of such meters is about ±0.5% of the full scale reading.

Currents may be measured more accurately by measuring the iR drop across a precision resistor through which the current is passed. The voltage drop is measured with a potentiometer or digital voltmeter for highest accuracy. A particularly useful method is to place the resistor in the negative feedback loop of an operational amplifier. The current flowing through the resistor in the feedback loop generates a voltage at the output of the amplifier that is equal to the iR drop across the resistor. The voltage output can be measured with a potentiometer or digital voltmeter. This particular circuit is called a current follower (32) and is incorporated in the potentiostat circuits described later in this chapter.

The measurement of very small currents (down to 10^{-14} A) requires the use of devices which have negligibly small leakage currents at their inputs. The frequency response and sensitivity of various electrometer input devices is summarized in a recent article on high-speed current measurements (33).

D.   BRIDGE MEASUREMENTS OF RESISTANCE, CAPACITANCE, AND INDUCTANCE

Bridge measurements provide the most direct way to compare unknown impedances with known standard impedances. The impedance may be a pure resistance, capacitance, or inductance, or may be represented by some series or parallel combination. The D-C Wheatstone bridge (34) provides a simple and direct way to measure a pure resistance. An A-C supply voltage also can be used if there are no reactive components in the standard or unknown resistances and the conditions for bridge balance will be the same as in the D-C bridge.

1. Measurement of Solution Conductance

Because the measurement of solution conductance (or resistance) is a common procedure that provides useful information to the electrochemist, the principles of a bridge measurement are illustrated by such an application.

The measurement of the ohmic resistance for an electrolyte is more difficult than it is for a metal wire. A variety of chemical and physical processes occur at the electrode/solution interface which must be separated from the voltage drop associated with the migration of ions through the bulk electrolyte.

The resistivity (or specific resistance) of a conductor may be thought of as the resistance of a cube of 1-cm edge, with the current assumed to be uniform and perpendicular to two opposite faces of the cube. The resistance of a conductor of uniform cross-section is given by

$$R = \rho \frac{l}{A} \tag{5-2}$$

where R is the resistance of the conductor in Ω, A the uniform cross-sectional area in cm^2, l the length of the conductor in cm, and ρ the resistivity or specific resistance in Ω-cm. The conductivity or specific conductance is defined as the reciprocal of the specific resistance

$$\kappa = \frac{1}{\rho} \qquad \Omega^{-1}/cm \tag{5-3}$$

Because the conductance of an ionic solution depends on the number of ionic charges, it is convenient to define it in terms of the conductance per unit concentration of ionic constituent,

$$\Lambda = \frac{\kappa}{C} \tag{5-4}$$

where Λ is the equivalent conductance in $cm^2/(\Omega)$ (equivalent) and C is the concentration in equivalents/cm^3. Because the number of equivalents/cm^3 is equal to the normality (N) times 1000 [where N is the molarity times the number of moles of positive (or negative) charge per mole of ionic solute], the equivalent conductance is given by

$$\Lambda = \frac{1000\kappa}{N} \qquad \Omega^{-1} \text{ cm}^2/\text{equivalent} \qquad (5\text{-}5)$$

Thus Λ represents the specific conductance of a hypothetical solution of the electrolyte containing 1 equivalent/cm^3. An experimental value of Λ may be determined by measurement of R, l/A (the cell constant), and N, and use of the relationship

$$\Lambda = \frac{1000(l/A)}{NR} \qquad (5\text{-}6)$$

Most conductance cells do not have uniform cross-sectional area, which requires that the cell constant be determined by calibration with solutions of known conductivity. The cell constant, l/A, may be thought of as the ratio of *effective* length and cross-sectional area of the conducting path. The conductivity standards are based on the careful measurements of Jones and Bradshaw for aqueous KCl solutions; some of their values are tabulated in Table 5-2.

Combined with densities, molecular weights, and transference numbers (fractions of the current carried by the various ionic constituents), the conductivity yields the relative velocities of the ionic constituents under the influence of an electric field. The mobilities [velocity per unit electric field, cm^2/(sec)(V)] depend on the size and charge of the ion, the ionic concentration, temperature, and solvent medium. In dilute aqueous solutions of dissociated electrolytes, ionic mobilities decrease slightly as the concentration increases. The equivalent conductance extrapolated to zero electrolyte con-

Table 5-2 Specific Conductivity of Standard Potassium Chloride Solution, in Ω^{-1}/cm, for Cell Constant Determination

Temperature (°C)	KCl(g) in 1000 g of solution		
	71.1352	7.41913	0.745263
0	0.06518	0.007138	0.0007736
18	0.09784	0.011167	0.0012205
25	0.11134	0.012856	0.0014088

From G. Jones and B. C. Bradshaw, *J. Am. Chem. Soc.*, **55**, 1780 (1933).

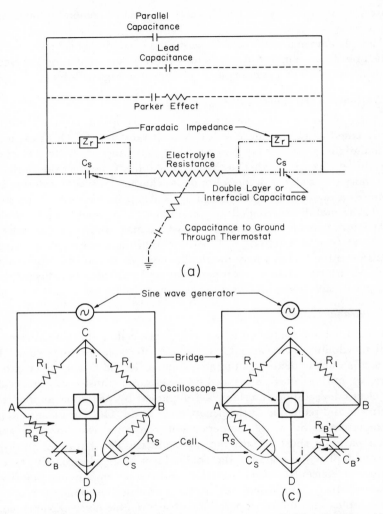

Figure 5-2. (*a*) Electrical equivalent circuit for a conductance cell; (*b*) A-C bridge with the cell impedance balanced by a series R-C combination; (*c*) A-C bridge with the cell impedance balanced by a parallel R-C combination (see Table 5-3).

centration may be expressed as the sum of independent equivalent conductances of the constituent ions

$$\Lambda^\circ = \lambda_+^{\ 0} + \lambda_-^{\ 0} \tag{5-7}$$

The concentration dependence of equivalent conductance is the principal source of our knowledge of ionic interactions (which generally lower mobilities).

The phenomena important in electrolytic conductance have been discussed by Braunstein and Robbins (35) and are represented by the electrical equivalent circuit of a conductance cell shown in Figure 5-2a.

2. Double Layer Capacitance

At each electrode/solution interface there is a substantial capacitance (represented as C_S in Figure 5-2a). A positively charged electrode tends to preferentially attract a layer of negative ions (and a negative electrode a layer of positive ions). Next to this layer of ions at the electrode surface will be a more diffuse layer of ions of opposite charge. This local ordering of ions produces a double-layer which rapidly trails off in the bulk of the solution due to thermal disordering (Brownian motion). The double-layer constitutes a capacitor, capable of storing charge, whose magnitude is of the order of 10 to 100 $\mu F/cm^2$ of electrode surface. Although there is usually a reasonably broad potential region where the double-layer capacitance is approximately constant, it does change with potential and can also be affected by the presence of substances adsorbed at the electrode surface.

3. Faradaic Effects

If a small voltage is applied to a conductance cell, a charging current flows until the double-layer capacitor is charged to its equilibrium value. If the voltage is increased, the charge accumulated in the double-layer capacitor increases until a voltage is reached where electron-transfer takes place across the electrode/solution interface, accompanied by oxidation at the positive electrode, and reduction at the negative electrode. This Faradaic process can be represented by a voltage-dependent resistance (nonlinear resistance) which partially short circuits the double-layer. The Faradaic process also tends to deplete the region of the double-layer of the electroactive species to produce a concentration-polarization and an accompanying polarization-resistance. In the equivalent circuit, the Faradaic impedance is represented as a complex impedance in parallel with the double-layer capacitance.

4. Ohmic Resistance

In the electrolytic solution, the charge carriers which constitute the flow of current are ions (rather than electrons as in a metal wire). The cations migrate toward the cathode and the anions toward the anode under the influence of the electric field. Velocity-dependent retarding forces oppose the accelerating force of the electric field, and result in a constant drift velocity at a given field. As the ions move through the medium, the frictional forces between ions and solvent molecules dissipate energy in the form of heat. To a close approximation the solution resistance is described by Ohm's law over a large voltage range. However, high electric fields (10^4 to 10^5 V/cm) can produce an additional conductance (the Wien effect). The latter results from changes in the interionic attraction that are brought about by the high field strength and the degree of dissociation of weak electrolytes (36).

5. Alternating Current and Frequency Effects

With an A-C rather than a D-C voltage applied to the electrodes, the processes above reverse themselves with the period of the alternating voltage. But each process proceeds at a different rate (with a characteristic relaxation time) so that their relative contributions to energy dissipation vary with frequency. As the frequency is increased concentration-polarization can be reduced or eliminated. By reducing the applied voltage the Faradaic processes can be reduced or eliminated, particularly if the electrode reaction is reversible (fast electron transfer in both directions).

Although increased frequencies (to several thousand hertz) reduce the complications of Faradaic polarization effects, other problems develop as the frequency increases. The most significant are the effects of stray capacitance [between connecting leads to the cell, to ground through the thermostat, or through stray currents created by a less than optimum design geometry of the cell, sometimes called the Parker effect (37)]. Just as there are no perfect gases or ideal solutions, there are no perfect (frequency independent) electrical components—pure resistances, pure capacitances, or pure inductances. Therefore the equivalent circuit representation of the cell shown in Figure 5-2a is an approximation, but one that is usually adequate.

6. A-C Conductance Measurements

An A-C bridge is a direct and accurate way of measuring the A-C conductance of a solution to obtain the solution resistance. The method works well for resistances between 1 and 10 kΩ, and can be extended to 60 kΩ provided several experimental conditions are met. These are that C_S of Figure 5-2a must be fairly large so that its impedance is small with respect to R_S, and that the impedance of the interelectrode capacitance must be large compared to the series combination of C_S and R_S. The conditions usually are met by platinizing the conductance electrodes to obtain a large C_S (by increasing the surface area), by using a source-frequency of at least 1 kHz, and by adjusting the cell constant so that the measured resistances are in the proper range (38). Under these conditions, the cell can be represented by an ohmic resistance in series with the double-layer capacitances at the electrodes. The applied A-C voltage is made small enough so that the Faradaic effects will be negligible.

A-C bridge measurements provide only an absolute impedance and phase shift so that there is no way to distinguish between either a series or parallel R-C combination by use of a pure sinusoidal waveform. Thus the series R-C combination of the cell may be matched by the series combination of Figure 5-2b or the parallel combination of the Wien bridge shown in Figure 5-2c. Either an oscilloscope or phase angle voltmeter (39) are recommended as null indicators because they allow both the phase shift and the imbalance potential to be nulled.

In both of these bridge circuits, the upper arms will be assumed to be pure

Table 5-3 Impedance Balance Conditions for the Two Bridge Circuits of Figure 5-2

Series Bridge (Figure 5-2b)   Parallel Bridge (Figure 5-2c)

$$Z_S = Z_B \qquad\qquad Z_S = Z_{B'}$$
$$R_S = R_B \qquad\qquad R_S \neq R_{B'}$$
$$C_S = C_B \qquad\qquad C_S \neq C_{B'}$$

$$Z_S = R_S + \frac{1}{j\omega C_S}\,^a \qquad\qquad \frac{1}{Z_{B'}} = \frac{1}{R_{B'}} + j\omega C_{B'}\,^a$$

$$Z_B = R_B + \frac{1}{j\omega C_B} \qquad\qquad R_S = \frac{R_{B'}}{1 + (\omega R_{B'} C_{B'})^2}$$

$$R_{B'} = R_S \left[1 + \frac{1}{(\omega R_S C_S)^2}\right]$$

$$C_{B'} = \frac{C_S}{1 + (\omega R_S C_S)^2}$$

$^a j = \sqrt{-1}$; $\omega = 2\pi f$, where f is the frequency.

resistance arms of equal value, but the calculations can be readily extended to cover the case where they are not equal (40). The balance conditions for the two bridge circuits of Figure 5-2b,c (summarized in Table 5-3) show that when the bridge is balanced, both resistances and both capacitances in the lower arms of the series bridge will be equal. This results because equality of the two complex impedances, Z_B and Z_S, requires equality of the real parts and of the imaginary parts, independently. [The reader unfamiliar with the j-operator notation for representing the complex impedance is referred to Refs. 1 to 4.] A similar result would be obtained for balance of a parallel combination by another parallel combination.

In balancing a series R-C combination with a parallel R-C combination

(Figure 5-2c), the resistances and capacitances in the two lower arms will not be equal, and the balancing resistance and capacitance will change with frequency (i.e., show frequency dispersion). The true resistance at the balance point must be calculated with the series R-C bridge equations of Table 5-3. If the value of $\omega C_{B'} R_{B'}$ is sufficiently small (or $\omega R_X C_S$ sufficiently large) R_S will equal $R_{B'}$ and the calculation is not necessary. Fortunately this is usually true for moderately dilute aqueous solutions where typical values might be $R_S = 1000\Omega$, $f = 1000$ Hz, and $C_S = 100$ μF, so that R_S and $R_{B'}$ will differ by only about 3 ppm. However, for measurements in molten salts (with high conductivities), or in cells with low double-layer capacitance the calculation will be necessary.

It might seem more convenient to eliminate the calculation by employing a bridge with a series R-C balancing arm as in Figure 5-2b. However, the parallel combination is used more frequently because the parallel capacitance required to compensate a series capacitance of 100 μF at 1000 Hz with a resistance $R_S = 1000$ Ω is only about 300 pF. Small capacitances can be obtained with higher accuracy and less frequency dependence (and less cost) than large ones.

In the most precise work, the oscillator and detector usually are coupled to the bridge by means of electrostatically shielded transformers to prevent direct coupling of the oscillator signal into the detector. A modified bridge circuit that employs a Wagner ground (41) also may be used to obtain the most precise bridge balance.

Balancing a bridge is a rather slow process, so that the method cannot be used to follow a rapidly changing solution resistance (as encountered in kinetic measurements). A number of circuits (42–44) have been published which allow the resistance of a solution to be recorded continuously. A method which employs the successive application of constant current pulses of equal magnitude, but of opposite sign, has been described which works well in both low resistance and high resistance solutions (up to 100 kΩ) and is well suited to the fast measurement of solution resistances (38).

E.   RECORDING DEVICES

Until quite recently, the servorecorder and oscilloscope were about the only instruments used in electrochemical laboratories for recording and retrieving information from an electrochemical experiment; they probably are still the most widely used. However, minicomputers and fast analog-to-digital converters with some form of punched tape or printed output are widely used, and FM tape recorders and a new data recording device, the transient recorder, have appeared on the scene recently. Table 5-4 summarizes in a qualitative way some of the properties of these recording devices. The

Table 5-4 Comparison of Recording Devices

	Servo-recorder	Thermal Writing Oscillograph	Light Beam Oscillograph	FM Tape Recorder	Oscilloscope and Camera	Storage Oscilloscope	Transient Recorder
Digital output	No	No	No	No	No	No	Yes
Electrical analog output	Yes[a]	No	No	Yes	No	No	Yes
Low cost (<$2000)	Yes	Yes	Yes	No[b]	Yes	Yes	Yes[c]
Easy to use	Yes	Yes	Yes	No	Yes	Yes	Yes
Easy to obtain hard copy	Yes	Yes	Yes	No	Yes	Yes	Yes
Presignal recording	Yes	Yes	Yes	Yes	No	No	Yes
Multiple channel capability	Yes	Yes	Yes	Yes	Yes[d]	Yes[d]	Yes
Wide frequency response (>1000 Hz)	No	No	Yes	Yes	Yes	Yes	Yes

[a]When equipped with a retransmitting potentiometer.
[b]Some less expensive recorders are now coming on the market.
[c]For single channel recorder with 1.6% amplitude resolution.
[d]With dual beam or switched beam oscilloscope.

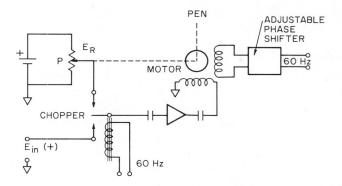

Figure 5-3. Schematic diagram for a potentiometric servo-recorder.

analog-to-digital converter is discussed in greater detail in the section on digital instrumentation.

1. Servorecorders and Oscillographs

Figure 5-3 presents a schematic diagram of a servorecorder. A chopper alternately samples the signal and the reference voltage from the slidewire; the difference between these two potentials is amplified and appears across one winding of the servomotor. The phase of this signal relative to the phase of the second winding, which is connected to the line voltage, is determined by the polarity of the amplified difference-signal. The servomotor rotates clockwise or counterclockwise depending on the phase. The motor is mechanically linked to the tap of the potentiometer, and the direction of rotation of the servomotor is such as to move the tap until E_R becomes equal to E_{in}. The gain and damping of the feedback loop is carefully adjusted so that the pen follows closely the variations in the input signal. Most servore-corders have a limited frequency response (5 Hz or less).

The X-Y recorder has two perpendicular servobalanced channels and is popular for cyclic voltammetry that uses scan times longer than a few seconds. The parameters critical to recorder performance have been re-viewed (45,46) as well as the effect of recorder response time on the shape of the recorded traces for square waves, triangular waves, and gaussian peaks.

The thermal-writing oscillograph uses a heated stylus of low inertia to record a trace on specially prepared paper. Its frequency response is im-proved over the servorecorder and generally extends to at least 50 Hz. The light-pen oscillograph writes with a beam of visible or ultraviolet light on sensitized paper and its frequency response extends to several kilohertz. A fast writing oscillograph can spew out formidable quantities of paper in a short time and the cost of the paper is not negligible.

2. Oscilloscopes

The cathode-ray tube oscilloscope is handy for observing repetitive waveforms present in a circuit and for troubleshooting. When equipped with a camera, it also can be used to record information. This is most commonly done in the "single-shot" mode, in which a nonrepetitive signal is recorded by triggering the sweep of the oscilloscope on a steeply changing part of the signal response, with the sweep trace recorded by the camera.

The storage oscilloscope stores the trace on a persistent phosphor which may be photographed later at the operator's leisure. It is convenient because the experiment can be repeated until all conditions are optimum as indicated by the stored trace. The stored trace is then photographed at considerable saving in film over the use of a conventional oscilloscope and camera.

The X-Y or dual-beam oscilloscopes are useful instruments for the observation of phase relations; the dual-beam scope allows the simultaneous viewing of both the input and output signals in an experiment.

The frequency response of modern oscilloscopes extends to 10 MHz or greater, which is more than adequate for any fast electrochemical work. The taking of data points from a small oscilloscope photograph is tedious, and the use of cartographic devices to ease this burden has been discussed (47).

3. FM Tape Recorders

Although the FM tape recorder does not allow the user to view the signals as they are recorded, this instrument does allow playback (electrical regeneration) of the recorded signals with the proper interface. The tape recorder is often interfaced with a computer to process the data. The frequency response of the tape recorder is generally from D-C to 20 kHz, but exotic (and expensive) machines are available with responses to 1 MHz that use tape speeds of 60 to 120 in./sec.

4. The Transient Recorder

A versatile new instrument for recording and retrieving information is the transient recorder (Biomation, Inc.); a block diagram is shown in Figure 5-4. It is essentially a fast analog-to-digital converter combined with a storage memory. The contents of the memory can be obtained in digital form, displayed in analog form on a cathode ray tube, or recorded on a servorecorder (on a slowed-down time scale) in analog form. The transients that are recorded by this device can range from 20 sec to a few microseconds in duration. The instrument has a number of advantages over the oscilloscope

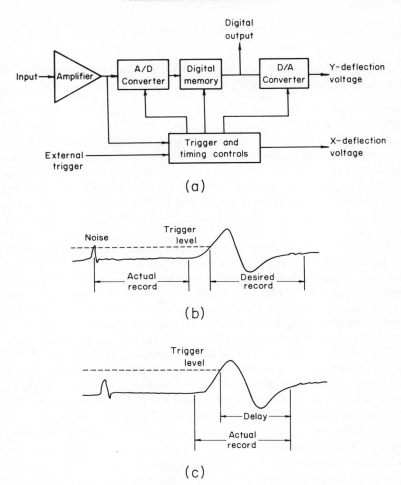

Figure 5-4. Transient recorder. (*a*) Block diagram; (*b*) trace with low level triggering; (*c*) trace with a higher trigger level.

in convenience of readout. Because it can record information continuously, storing new data and erasing old data as it goes along, the baseline can be obtained easily prior to the signal of interest. This presignal recording capability is useful in experiments where baseline information is needed. It also allows the trigger level to be set at a higher value, which provides greater immunity from false triggering.

II. Control Instrumentation

A. INTRODUCTION TO THE USE OF OPERATIONAL AMPLIFIERS IN
MEASUREMENT AND CONTROL

Electrochemical instrumentation based on operational amplifiers was first described in the late 1950s. The versatility of this approach was immediately apparent and since then well over a hundred publications have appeared which discuss operational amplifier circuits; almost every electrochemical laboratory (in the United States at least) contains homemade or commercial instrumentation that uses operational amplifiers. An instrumentation philosophy centered around these devices has now become a part of the teaching and practice of instrumental analysis, and has found its way into the textbooks.

Each of the general references (1–5) contain a chapter on operational amplifiers and their applications. In addition references (48–53) are recommended as more detailed sources of information. Table 5-5 contains a list of manufacturers of operational amplifiers. Most of these companies also manufacture power supplies for operational amplifiers (op amps), multipliers, analog-to-digital and digital-to-analog converters, active filters, and other devices that are useful in instrumentation. This list is not exhaustive and the *Electronics Buyers Guide* (published annually by McGraw-Hill) should be consulted for a more extensive list. Detailed specification sheets and valuable applications literature are available from most of the manufacturers. In particular the applications handbook published by Philbrick Researches, Inc., (now Teledyne Philbrick) is highly recommended.

Originally the term "operational amplifier" was used in the computing field to describe amplifiers that performed various mathematical operations for analog computation. The application of negative feedback around a high gain D-C amplifier produces a circuit with a precise gain characteristic which depends only on the external feedback. By the proper selection of feedback components, operational amplifier circuits can be used to add, subtract, integrate, differentiate, and perform other mathematical operations.

As practical operational amplifier techniques became more widely known, they were used in many control and instrumentation applications. The following sections describe some of the characteristics of operational amplifiers and some simple applications in electrochemical instrumentation and give references to many more which have been described in the literature.

Table 5-5 A Selected List of Manufacturers of Operational Amplifiers[a,b]

Operational Amplifiers, Active Filters, A-D and D-A Converters, Analog Multipliers, Logarithmic Amplifiers, and so on

Analog Devices, Inc., 221 Fifth St., Cambridge, Mass. 02142.
Burr-Brown Research Corp., International Airport Industrial Park, Tucson, Ariz. 85734.
Optical Electronics, Inc., P. O. Box 11140, Tucson, Ariz. 85734.
Teledyne Philbrick, Allied Drive at Route 128, Dedham, Mass. 02026.
Zeltex, Inc., 1000 Chalomar Road, Concord, Calif. 94520.

Linear Integrated Circuits

Fairchild Semiconductor, 313 Fairchild Drive, Mountain View, Calif. 94040.
Motorola Semiconductor Products, Inc., Technical Information Center, P. O. Box 20912, Phoenix, Ariz. 85036.
RCA Electronic Components, 415 S. Fifth St., Harrison, N. J. 07029.
Texas Instruments, Inc., Components Group, P. O. Box 5012, M. S. 84, Dallas, Tex. 75222.

---

a A more extensive list can be found in the *Electronics Buyers' Guide, '73*, McGraw-Hill Book Company, New York.
b A survey of published circuits for electroanalytical instrumentation indicates that the companies listed are the principal sources of discrete component and integrated circuit operational amplifiers.

1. Symbols and Terminology

Although there is not complete standardization of the symbols used to represent operational amplifiers, the symbols used here are commonly encountered in the manufacturers' literature.

In the most common types of differential amplifiers, the amplifier is represented by a triangle with two input terminals and one output terminal. The potentials of the input and output terminals are usually, but not always, referred to ground. The ground is symbolized by $\downarrow$, as illustrated in Figure 5-5. The power supply and input bias connections usually are omitted, and where it is understood that the input and output are referred to ground, the ground line common to both frequently is omitted also. The symbol then

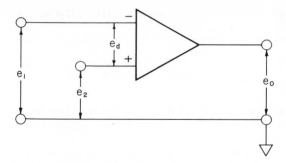

Figure 5-5. Differential amplifier.

becomes that shown in Figure 5-6. A positive input applied to the noninverting ($+$) terminal gives a positive output, and a positive input applied to the inverting ($-$) terminal gives a negative output.

The gain, g, of the amplifier is defined as

$$e_o = -ge_d \tag{5-8}$$

where e_o is the output voltage (open loop) and e_d is the voltage at the inverting ($-$) terminal measured with respect to the voltage at the noninverting ($+$) terminal (i.e., the differential voltage $e_1 - e_2$ of Figure 5-5).

2. Common-Mode Error

The output of a differential amplifier is proportional to the difference between the $-$ and $+$ inputs. Ideally, if the same signal (common-mode voltage) is applied to both inputs, there should be no output signal. The ability of an amplifier to reject a common-mode signal is specified by the common-mode rejection ratio (CMRR). This is defined as

$$\text{CMRR} = \frac{\text{gain of amplifier to difference signal}}{\text{gain of amplifier to common-mode signal}} \tag{5-9}$$

A high common-mode rejection ratio is a desirable characteristic for an amplifier.

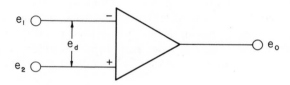

Figure 5-6. Differential amplifier with input biasing and common ground line omitted.

Table 5-6 Operational Amplifier Specifications

	Teledyne Philbrick Model 1026 FET Input	Burr-Brown Model 3308/12C FET Input	Ideal
Output voltage	±10 V	±10 V	Large
Output current	±5 mA	±5 mA	Large
Open-loop D-C gain	100,000	200,000	Infinite
Full power output frequency response	100 kHz	100 kHz	High frequency
Input voltage range			
D-C common mode limit	±10 V	±10 V	Large
Common mode rejection ratio	10,000	10,000	Infinite
Input impedance			
Differential	$10^{11}\ \Omega$	$10^{11}\ \Omega$	Infinite
Common mode	$10^{11}\ \Omega$	$10^{11}\ \Omega$	Infinite
Input offset voltage			
Initial offset at 25°C	Adj. to zero	Adj. to zero	Zero
Versus temperature	±50 μ/°C	±50 μ/°C	Zero
Input bias current			
Initial bias at 25°C	−50 pA	−50 pA	Zero
Versus temperature	Doubles each +10°C	Doubles each +10°C	Zero
Noise (referred to input)			
Voltage, 0.01–10 Hz	5 μV p-p	3 μV p-p	Zero
0.1–10 kHz	2.2 μV rms	5 μV rms	Zero
Current, 0.01–10 Hz	0.2 pA p-p	0.5 pA p-p	Zero
0.1–10 kHz	2 pA rms	1 pA rms	Zero
Power requirements	±10–18 V at ±10 mA	±15 V at ±10 mA	
Cost (1973)	$12	$12	

3. Common-Mode Limit

When an amplifier is connected in a differential amplifier or follower configuration both inputs will be at the same potential and the common-mode voltage limit of the amplifier (as specified by the manufacturer) must be observed. In addition most manufacturers of solid-state amplifiers will specify a maximum differential input voltage which may not be exceeded. These values are typically about 10 V for solid-state amplifiers.

Table 5-6 specifies the characteristics of an ideal operational amplifier and of two typical differential amplifiers of modest performance, both solid-state. The important differences between vacuum-tube and solid-state amplifiers are cost, size, weight, voltage drift, and noise level; all favor the solid-state amplifiers. Most solid-state amplifiers have relatively low voltage outputs (typically ± 10 V) compared to the 50 to 100 V outputs of vacuum-tube amplifiers. However the increased voltage output requires a larger and more costly power supply (typically ± 300 V) than the usual ± 15-V power supplies used with solid-state amplifiers. Recently, high voltage output solid-state amplifiers have become available, so that solid-state amplifiers are used almost exclusively in recently designed operational amplifier instrumentation. This will be even more true in the future because inexpensive high performance integrated-circuit amplifiers are now widely available. Finally, solid-state amplifiers are available with a wide range of characteristics and can be specially designed for high input impedance (10^{12} Ω), low voltage drift (± 1 $\mu V/day$), or with good high frequency response.

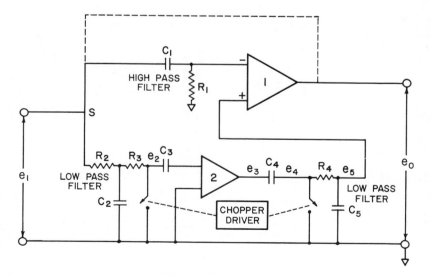

Figure 5-7. Chopper stabilized differential amplifier.

4. Chopper Stabilization of Drift

For many purposes, the drift of several millivolts per day that occurs with vacuum-tube operational amplifiers is not tolerable. Some applications may require ultralow voltage drifts (± 1 μV/day), and even solid-state differential amplifiers with low drift characteristics may not be adequate. One of the most effective ways to reduce the drift of an operational amplifier and increase its overall gain is to employ a chopper amplifier in tandem with the differential amplifier as shown in Figure 5-7.

The chopper amplifier is an inverting amplifier with a D-C gain of 1000 or greater. It "sees" the error at point s (the summing point) of the amplifier with respect to ground, amplifies the error signal by a factor of 1000, inverting it in the process, and applies it to the $+$ (noninverting) terminal of the differential amplifier. The output of the differential amplifier acts to reduce the error at the summing point to zero by means of an external feedback loop (represented by the dotted line). The operation of a chopper amplifier is illustrated by Figure 5-8.

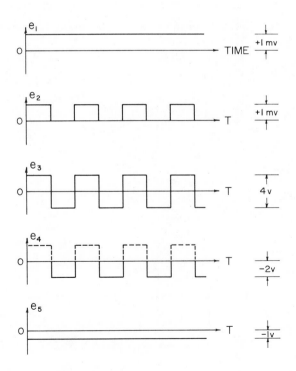

Figure 5-8. Operation of a chopper stabilized amplifier (refer to circuit of Figure 5-7).

A blocking capacitor, C_b, commonly is inserted to reduce the current drawn from the signal source to very low levels—typically below 10^{-9} A. High frequency signals pass through the blocking capacitor so that high frequencies are not limited by the slower response of the chopper amplifier.

The chopper may be a mechanical-reed switch driven by an A-C drive coil at the 60-Hz line frequency or a photochopper, which consists of a photoresistor coupled to a light source modulated at any arbitrary frequency.

The symbol for the chopper stabilized amplifier shown in Figure 5-7 usually is abbreviated to that shown in Figure 5-9a or b. Again, the reference point common to input and output has been omitted. Because the + (noninverting) input is used for drift stabilization, the amplifier may no longer be used in the differential configuration. Such an amplifier is commonly called a "single-ended input" amplifier with input and output both referred to ground.

5. Two Important Principles Useful in Simple Circuit Analysis

Inherent in the properties of the ideal amplifier are the principles that:

1. The amplifier will draw a negligible amount of current at either input.
2. The − and + inputs will be at the same potential, provided there is a closed feedback loop.

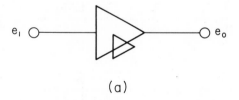

(a)

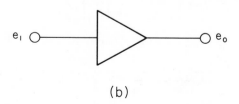

(b)

Figure 5-9. Symbols for a chopper stabilized amplifier.

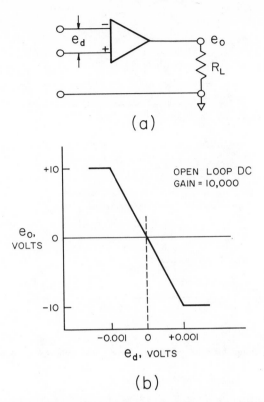

Figure 5-10. (*a*) Open-loop amplifier; (*b*) out-put voltage as a function of input voltage (voltage crossing detector).

These two principles place inherent restraints on circuit design and provide the basis to begin a circuit analysis. Real amplifiers approach the ideal closely enough so that the closed-loop gain relations (derived below) are directly applicable to real circuits—to within a few tenths of one percent in most cases. Several useful operational amplifier configurations are discussed here in detail.

6. Open-Loop Operation

In the open-loop amplifier shown in Figure 5-10*a*, a differential input voltage of only ± 1 mV will drive the amplifier to saturation as shown in Figure 5-10*b*. The open-loop amplifier, therefore, can function as a voltage-crossing detector or switch. If one input monitors a selected reference voltage and the other follows a signal from the system, the output of the operational

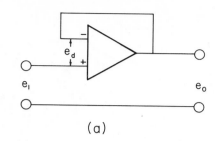

(a)

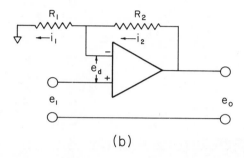

(b)

Figure 5-11. (*a*) Voltage follower amplifier ($e_o = e_1$); (*b*) follower amplifier with gain [$e_o = e_1(1 + R_2/R_1)$].

amplifier will switch across its full output voltage range as the two voltages cross. Such an element can be used to detect the endpoint of a potentiometric titration, for example. Unless the operational amplifier is designed to function well in the open-loop mode, it may be prone to "latch-up" and introduce a significant lag before the amplifier switches states.

Within the output limits of this particular amplifier, the maximum potential difference between the − and + terminals is about 1 mV. As the gain approaches infinity, the difference approaches zero.

7. Closed Feedback-Loop Configurations

By far the greatest utility is obtained from operational amplifiers when negative feedback is employed. The following circuits illustrate useful applications.

The Voltage Follower. The operation of the voltage follower is illustrated by Figure 5-11*a*. The output of the amplifier is connected directly to the inverting input. For this circuit, the differential voltage, e_d, is defined as

$$e_d = e_o - e_1 \qquad (5\text{-}10)$$

The output voltage, e_o, is given by rearrangement

$$e_o = e_1 + e_d \qquad (5\text{-}11)$$

and substitution of the amplifier gain in place of e_d from Equation 5-8 gives

$$e_o = e_1 - \frac{e_o}{g} \qquad (5\text{-}12)$$

If the gain, g, is very large, then $e_o = e_1$. Note that the closed-loop gain of the follower is precisely unity (if g is large enough) and that the signal is not inverted. Unity-gain followers are used as electrical buffers or impedance matching amplifiers to prevent the loading of a signal source by a circuit or device which draws an appreciable amount of current. The input impedance of the follower is typically greater than $100\,\text{M}\Omega$ and the output impedance is low. The follower can furnish voltage or current to the load (up to the voltage and current limits of the amplifier) without loading the source.

*The Noninverting Amplifier or Follower with Gain..** The voltage at the negative input of this amplifier is $e_{in} + e_d$ and, because the current in R_1 equals the current in R_2 (assuming negligible amplifier input current)

$$i_1 = i_2 \qquad (5\text{-}13)$$

where

$$i_1 = \frac{e_1 + e_d}{R_1} \qquad (5\text{-}14)$$

and

$$i_2 = \frac{e_o - (e_d + e_1)}{R_2} \qquad (5\text{-}15)$$

Substitution of Equation 5-8 for e_d in Equations 5-14 and 5-15 and recognition that g is very large yields (from Equation 5-13)

$$\frac{e_1}{R_1} = \frac{e_o - e_1}{R_2} \qquad (5\text{-}16)$$

*Refer to Figure 5-11*b*.

or

$$e_o = e_1\left(1 + \frac{R_2}{R_1}\right) \tag{5-17}$$

The closed loop gain is $[1 + (R_2/R_1)]$. This circuit is widely used in control instrumentation where noninverting gain is required; a gain of 100 can be achieved. This amplifier also has a high input impedance.

*The Inverting Amplifier.** The inverting amplifier configuration is used most often in control instrumentation. To achieve this the differential input amplifier is made an inverting amplifier by grounding the positive input and applying the input signal through input resistance R_1 to the inverting input, as shown in Figure 5-12a. The amplifier will maintain the negative input at the same potential as the positive input, ground potential in this case, by

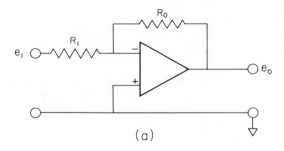

(a)

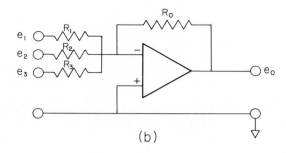

(b)

Figure 5-12. (a) Voltage inverter amplifier ($e_0 = -e_1 R_0/R_1$); (b) summing inverter amplifier

$$[e_0 = -(e_1 R_0/R_1 + e_2 R_0/R_2 + e_3 R_0/R_3)].$$

*Refer to Figure 5-12.

virtue of the feedback loop that contains R_o. Because the amplifier input current is negligible, equal currents will flow in R_1 and R_o such that

$$\frac{e_1 - e_d}{R_1} = \frac{e_d - e_o}{R_o} \tag{5-18}$$

Because $e_d = -e_o/g$ (Equation 5-8) and the amplifier gain, g, is large, this gives

$$e_o = -e_1 \frac{R_o}{R_1} \tag{5-19}$$

The closed loop gain is precisely determined by the ratio R_o/R_1. The input impedance is essentially equal to R_1 because the input voltage sees ground potential at the summing point, s. Point s is not physically grounded but is at ground potential and is sometimes called a *virtual ground*.

When choosinginput and feedback resistors, the characteristics of the amplifier must be kept in mind. For example, if the amplifier has a ± 10-V ± 5-mA output capability, resistance values of less than 2000 Ω cannot be used in the feedback loop; otherwise the amplifier would draw more than 5 mA at its rated voltage output of 10 V. At the other extreme, the use of 10 MΩ might mean that the amplifier input current would not be negligible compared to the currents flowing in the input and feedback resistors; this would create a current-offset error.

The input offset-voltage of the amplifier can usually be biased out so that it will be less than 0.1 mV. This means that to achieve errors of less than $\pm 1\%$ the input voltage must be greater than 10 mV.

A number of inputs can be connected to the summing point of an inverting amplifier as shown in Figure 5-12b. In this case the sum of the current inputs will equal the feedback current and

$$e_o = -\left(e_1 \frac{R_o}{R_1} + e_2 \frac{R_o}{R_2} + e_3 \frac{R_o}{R_3} + \cdots \right) \tag{5-20}$$

If $R_o = R_1 = R_2 = R_3$, the output voltage e_o is the algebraic sum of the input voltages and the configuration is called an adding inverter. This configuration is used in the circuit designed to control the potential of an electrode that is described in Section IIB.

8. Current Output

Up to this point the output of the inverting amplifier has been considered in terms of voltage, but the amplifier also can be thought of as a current supplying device. This is accomplished by placing the load in the feedback loop as shown in Figure 5-13.

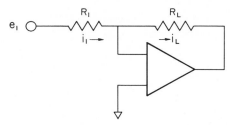

Figure 5-13. Inverter amplifier with current output through load resistance, R_L ($e_L = -\frac{e_1}{R_1}$).

The current flowing through the load resistance, R_L, is independent of the load resistance, within the current and voltage output capabilities of the amplifier

$$i_L = -\frac{e_1}{R_1}$$ (5-21)

If the load, for example, is a current drawing device such as a meter, the meter read-out or meter current is independent of the resistance of the meter and depends only on the input voltage and resistance.

9. The Integrator and Differentiator

Figure 5-14a illustrates the configuration of a potential-time integrator. At any instant of time, the charge stored on the feedback capacitor is given by

$$q = Ce_o$$ (5-22)

where q is the charge in coulombs, C the capacitance in farads, and e_o the output voltage in volts; the summing point, s, is maintained at ground potential by virtue of the closed feedback loop that contains the capacitor. If the amplifier input current is negligible, the feedback current will be equal to the input current and

$$i = \frac{dq}{dt} = C\frac{de_o}{dt} = -\frac{e_1}{R}$$ (5-23)

Rearrangement of the right-hand terms and integration gives

$$e_o = -\frac{1}{RC}\int_0^t e_1\,dt$$ (5-24)

The integrator fulfills two important functions in electrochemical instru-

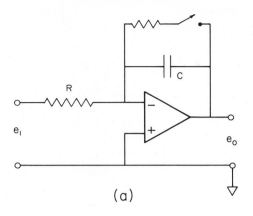

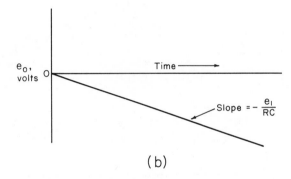

Figure 5-14. (*a*) Circuit for potential-time integrator; (*b*) output for a constant voltage input.

mentation. It can be used to integrate a voltage signal proportional to the current which flows through the cell in controlled potential coulometry; the time-integral of the current is proportional to the total number of coulombs of charge which have passed through the cell. It also is used to generate a linear voltage-sweep or a voltage-ramp for D-C polarography or cyclic voltammetry. In this application, the input voltage is fixed, with the slope of the ramp determined by the time constant, R-C. Thus substitution into Equation 5-24 gives

$$e_o = - \frac{e_1}{RC} t \qquad (5\text{-}25)$$

with t the time of the voltage-sweep.

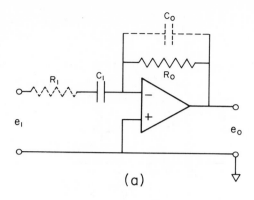

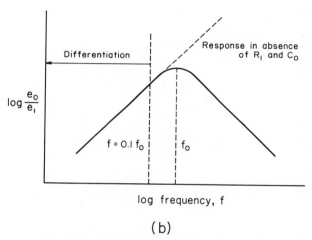

Figure 5-15. (a) Circuit for voltage differentiator ($e_0 = -R_0C_1\dfrac{de_1}{dt}$); (b) output as a function of frequency. Choose R_1 and C_0 so that

$$R_1C_1 = R_0C_0 = 1/2\pi f_0.$$

Care must be taken in the choice of components to achieve an integrator with low voltage drift. To obtain large time constants, large capacitors (1-10 μF) must be used in combination with large values of resistance (up to 10 MΩ). This requires that the input current of the amplifier be extremely small (preferably less than 10 pA). Low drift field-effect-transistor (FET) input or varactor bridge operational amplifiers are best suited to this application. The capacitor must have a high quality dielectric, such as polystyrene, and high resistance across the input terminals. All input wiring must have high quality insulation (Teflon), and the capacitor and its input circuitry should be hermetically sealed. If these precautions are not taken, the normal atmosphere in a chemical laboratory will rapidly deteriorate the quality of the insulation; this causes a noticeable increase in the drift of the integrator.

In the differentiator configuration, shown in Figure 5-15a, the capacitor and resistor are switched. Because the input current and the current in the feedback resistor are equal this yields

$$i = \frac{dq}{dt} = -C_1 \frac{de_1}{dt} = \frac{e_0}{R_0} \tag{5-26}$$

or

$$e_0 = -R_0 C_1 \frac{de_1}{dt} \tag{5-27}$$

Integration is inherently a smoothing process, while differentiation is just the opposite. The ideal differentiator (as illustrated) is not practical in its simple form because any noise is differentiated to produce even greater noise. A practical differentiator is constructed by limiting its frequency response by insertion of a resistor in series with the input capacitor and a capacitor in parallel with the feedback resistor. The output of the practical differentiator as a function of frequency is shown in Figure 5-15b. It differentiates accurately at low frequencies, but the gain decreases at high frequencies as the network begins to act as an integrator.

10. Reactive Elements

The resistances R_1 and R_o in Figure 5-12a may be replaced by impedances Z_1 and Z_o to represent reactive elements. The most general approach is to write Equation 5-19 in operational form

output voltage transform = network transform × input excitation transform

$$\tag{5-28}$$

or

$$e_o(p) = -\frac{Z_o(p)}{Z_1(p)} e_1(p) \tag{5-29}$$

where $e_1(p)$ represents the Laplace transform of the input excitation voltage; the other quantities also represent the Laplace transform of the respective

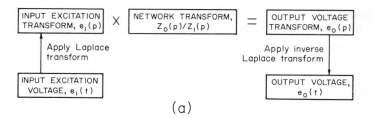

(a)

SOME COMMON TRANSFER IMPEDANCES

Network   Transfer Impedance, Z(p)

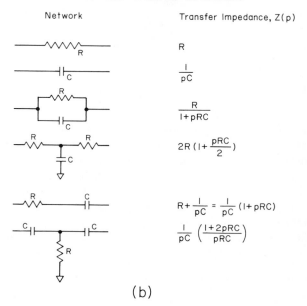

(b)

Figure 5-16. (a) Reactive element transform relationships; (b) common transfer impedances.

functions. The Laplace transform is defined by

$$\bar{F}\left[F(t)\right] = \int_0^\infty e^{-pt} F(t)\, dt \qquad (5\text{-}30)$$

where $F(t)$ represents any function such that the integration defined by Equation 5-30 may be performed. Laplace transform methods are particularly suited to problems where the output voltage and input excitation voltage are functions of time and the input and output impedances consist of passive networks of resistors and/or capacitors. In the time domain, the output and input may be related by a linear differential equation. In the Laplace domain, the input and output voltage transforms and the network transforms will be represented by polynomials in the Laplace variable, p. The overall relationships are illustrated schematically in Figure 5-16a.

As an example of the general approach, consider the problem of the output voltage obtained when a step input is applied to the integrator configuration of Figure 5-14a. If the voltage on all network capacitors is zero at time $t = 0$, then the Laplace variable p represents the operator d/dt (54), that is

$$pe_o = \frac{de_o}{dt} \tag{5-31}$$

where e_o represents the output voltage.

For this system the input and feedback currents are equal so that

$$\frac{-e_o}{R} = Cpe_o \tag{5-32}$$

or

$$e_o = \frac{-1}{pRC} e_1, \tag{5-33}$$

where p is treated as a variable in the Laplace domain.

The quantity $1/pRC$ represents the *network transform* or *transfer impedance*, $Z_o(p)/Z_1(p)$, where $Z_o(p) = 1/pC$ and $Z_1(p) = R$. Figure 5-16b contains the transfer impedances or networks transforms for a number of networks taken from reference (55).

Now consider how the output will vary when a step input (e_1 V) is applied at time $t = 0$, as shown in Figure 5-14b. The Laplace transform of the input excitation voltage is given by evaluating the right-hand side of Equation 5-30 for $F(t) = e_1$. The result is

$$\bar{F}[e_1] = \frac{e_1}{p}. \tag{5-34}$$

The output-voltage transform is given by the product

$$e_o(p) = \left(\frac{e_1}{p}\right)\left(-\frac{1}{pRC}\right) = -\frac{e_1}{p^2 RC} \tag{5-35}$$

Table 5-7   LaPlace Transforms

Function	Transform
1. $F(t)$	$\int_0^\infty e^{-pt}F(t)dt = \overline{F}(p)$
2. Unit impulse	1
3. $f(t) = 0 \quad t < 0$ step $\quad f(t) = A \quad t > 0$	$\dfrac{A}{p}$
4. t	$\dfrac{1}{p^2}$
5. e^{at}	$\dfrac{1}{p-a}$
6. $\dfrac{1}{a}\sin at$	$\dfrac{1}{p^2+a^2}$
7. $\cos at$	$\dfrac{p}{p^2+a^2}$
8. $\dfrac{d}{dt}[F(t)]$	$p\overline{F}(p) - F(+0)$
9. $\int_0^t F(b)\,db$	$\dfrac{1}{p}\overline{F}(p)$

Applying the inverse Laplace transform* to Equation 5-35 gives

$$e_o(t) = \left(-\frac{e_1}{RC}\right)t, \qquad (5\text{-}36)$$

the result obtained previously.

*The Laplace transform of $F(t) = t$ may be shown to be $1/p^2$,

$$\overline{F}[t] = \int_0^\infty e^{-pt}t\,dt = \frac{1}{p^2},$$

by use of integration by parts. Consequently the inverse transform of $1/p^2$ is t. Tables of Laplace transforms and inverse transforms are available (see Table 5-7 and Ref. 55 to 57).

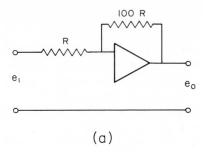

(a)

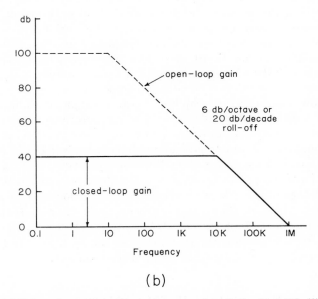

(b)

Figure 5-17. (a) Voltage inverter amplifier with an open-loop D-C gain of 100; (b) Bode plot for amplifier.

11. Frequency Response and Stability

At D-C and low frequency the open-loop gain of an operational amplifier may be regarded as a number, g, which is defined by Equation 5-8 and is sometimes expressed in decibels (dB)

$$\text{open-loop gain in dB} = 20 \log g \qquad (5\text{-}37)$$

To be stable with a variety of feedback elements the gain of the amplifier must decrease (gain "roll-off") with frequency (48). The frequency response of the amplifier often is plotted as log *gain* versus log *frequency* (called a Bode plot), and is illustrated in Figure 5-17*b* (the dotted line represents the open-loop gain of the amplifier).

The closed-loop gain of the inverter shown in Figure 5-17*a* is plotted as the solid line of Figure 5-17*b*. In order for the circuit to be stable the slope of the open-loop gain curve where it intersects the closed-loop gain curve must not exceed 40 dB per decade (12 dB/octave). To allow sufficient margin of stability many amplifiers have a roll-off of about 20 dB/decade (6 dB/octave), like that shown in Figure 5-17*b*.

Note that gain must be sacrificed to obtain a greater frequency response range. For the circuit of Figure 5-17b the cutoff frequency of a 100-dB amplifier is 10 Hz (actually the gain is down 3 dB at this point because the curve shown is only an approximation to the real curve which shows curvature at the break point). The high-frequency cutoff of a 40 dB (gain of 100) amplifier extends to 10 kHz. The frequency response range of an amplifier is called its bandwidth, and the bandwidth of an amplifier often can be adjusted by external or internal compensation.

Instability arises in an amplifier because of phase shifts around the feedback loop. The Bode diagram, although constructed from only gain-frequency information, provides phase-angle information as well. The normal operation of an operational amplifier is inverting, and phase angles are expressed relative to this normal inversion or 180° phase lag. Thus only the additional phase shift is reflected in the Bode plots and is related to the rate of change of the gain with frequency. On the log-log plot a zero slope corresponds to a zero phase shift, a slope of -20 dB/decade (as in Figure 5-17*b*) represents a 90° lag, -40 dB/decade a 180° lag, and so on. Thus a -40 dB/decade slope at the intersection of the open-loop and closed-loop gain curve verges on positive feedback, hence instability.

B.   POTENTIAL CONTROL INSTRUMENTATION—
THE POTENTIOSTAT

A potentiostat is a controller circuit that maintains the potential between the working electrode and the reference electrode equal to some signal-generator potential, which may be a constant voltage or a time-varying signal. In its fundamental operation the controller reacts to the difference between these two potentials through a negative feedback circuit—containing the counter electrode—in such a way as to reduce the difference to zero. In addition, this operation must be performed without drawing significant current through the reference electrode.

This control operation is commonly performed by an operational amplifier, usually of the differential type, which has high open-loop gain and is designed to remain stable with large amounts of negative feedback from output to input.

An optimum design should meet several standards: (a) the circuit should provide accurate potential control with no iR drops through the signal generator or current measuring resistor that lead to potential-control errors; (b) a design allowing a stabilized (single-ended) amplifier is preferred over one requiring a differential amplifier, because of the lower drift of the

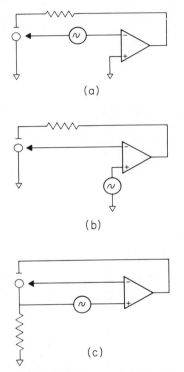

(a)

(b)

(c)

Figure 5-18. Circuit diagrams for single-amplifier potentiostats.

stabilized amplifier; (c) because it is most convenient to refer all signal inputs and outputs to a common ground, a circuit design that allows the signal generator and the current measuring device (load resistor) each to have one terminal grounded is desirable.

In a single-amplifier potentiostat it is not possible to meet each one of these standards. Schwarz and Shain (58) have made a classification that shows that 18 generalized configurations are possible using a single ampli-

Table 5-8 **Characteristics of Potentiostats**

| | Figure Number | | | | |
	5-18a	5-18b	5-18c	5-20a	5-20b
(a) Is the circuit free of potential control error caused by iR drop in the signal generator?	Yes	Yes	Yes	No[a]	No[a]
(b) Is the circuit free of potential control error caused by iR drop in the load resistor?	Yes	Yes	Yes	Yes	Yes
(c) Can stabilized (single-ended) amplifiers be used?	Yes	No	No	Yes[b]	Yes[b]
(d) Can one end of the signal generator be grounded?	No	Yes	No	Yes	Yes
(e) Can the current measuring device have one input grounded?	No	No	Yes	Yes	Yes

[a] Yes, if the internal resistance of the signal source is negligible compared to the resistance of the input resistors connected to the input of the control amplifier.

[b] Except for follower amplifiers, which require a differential input.

258

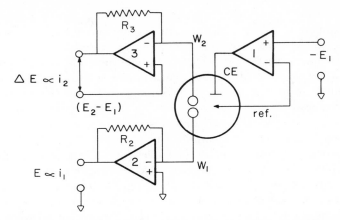

Figure 5-19. Potentiostat with independent control of the potential for two working electrodes, W_1 and W_2, relative to a common reference electrode.

fier, a signal generator, and a current-measuring load resistor. Two basic grounding schemes are possible with nine different configurations in each. Three of these single-amplifier potentiostat circuits are shown in Figure 5-18, all of them using the "type A" grounding scheme described by Schwarz and Shain.

It is not difficult to understand qualitatively how the potentiostat operates if two basic principles are kept in mind: (*a*) the amplifier input currents are assumed to be negligibly small; and (*b*) if there is a closed feedback loop from the output to the inverting input, the amplifier will maintain the inverting input at the same potential as the noninverting (+) input.

In the circuit of Figure 5-18*a*, the inverting input is a virtual ground because the + input is grounded. By means of the feedback loop through the counter electrode, the amplifier forces current through the working electrode until $E_{cell} = E_{sg}$ where E_{cell} is the potential of the working electrode measured with respect to the reference electrode and E_{sg} is the potential of the signal generator (referred to the reference electrode). The current flowing through the cell also flows through the resistor in the feedback loop. Because neither end of the resistor is grounded, the read-out device must have a floating (i.e., differential) input. Analysis of the mode of operation of the circuits of Figure 5-18*bc* follows a similar line and is left as an exercise for the reader. Their characteristics are qualitatively summarized in Table 5-8.

The circuit of Figure 5-19 is designed to independently control the potential of two working electrodes, W_1 and W_2, with respect to a common reference electrode. It has been used in twin-electrode thin layer cells (59, 60) and can be used to control the ring and disc potentials of a ring-disc electrode (see Chapter 2).

The operation of this circuit is outlined to illustrate further the principles of operational amplifiers and their use in electrochemical intrrumentation. Amplifier 2 has its noninverting ($+$) input grounded, so that the amplifier will maintain its inverting ($-$) input at ground potential (virtual ground) by means of the closed feedback loop that contains resistor R_2. Therefore W_1 is maintained at ground potential (assuming negligible resistance for the connecting leads and the electrode itself). Cathodic current (taken as positive in the conventional sense) that flows from the counter electrode, CE, and through the working electrode will also flow through R_2. (Amplifier input currents are assumed to be negligible.) This will generate at the output of amplifier 2 a voltage, $E = -i_1 R_2$, that is proportional to the current flowing through W_1. This configuration is called a *current follower* and is widely used in controlled potential circuits because it maintains the working electrode at ground potential yet still provides a single-ended voltage output proportional to current.

Because W_1 is maintained at ground potential, the circuit is seen to be equivalent to that of Figure 5-18b, with potential control of W_1 enforced by amplifier 1 through the feedback loop that contains the counter electrode and the reference electrode. The cell comprised of W_1 and the reference electrode is maintained at the potential given by

$$E_{W_1} - E_{\text{ref}} = E_{\text{cell},1} = E_1 \tag{5-38}.$$

E_1 may be thought of as the control potential of working electrode W_1. For example, to maintain W_1 at -0.8 V versus the reference electrode, a potential of $+0.8$ V ($-E_1$) is applied to the noninverting ($+$) input of amplifier 1, referred to ground.

Working electrode W_2 is maintained at the potential $E_2 - E_1$ (measured with respect to ground) by virtue of the feedback loop around amplifier 3 that contains R_3. Starting at working electrode W_1 (at virtual ground) the potential drops around the loop comprised by W_1, the reference electrode, W_2, and $E_2 - E_1$ can be summed. This gives

$$-E_{\text{cell},1} + E_{\text{cell},2} - (E_2 - E_1) = 0 \tag{5-39}$$

or in view of Equation 5-38

$$-E_1 + E_{\text{cell},2} - (E_2 - E_1) = 0 \tag{5-40}$$

therefore

$$E_{\text{cell},2} = E_2 \tag{5-41}$$

To maintain simultaneously W_1 at -0.8 V versus the reference electrode and W_2 at $+0.3$ V versus the reference electrode, a voltage of $+0.8$ V $(-E_1)$ is applied to the noninverting $(+)$ input of amplifier 1 and a potential of $+1.1$ V $(E_2 - E_1)$ is applied to the inverting $(-)$ input of amplifier 3, both inputs referred to ground.

The current that flows through working electrode W_2 also flows through R_3 to generate a potential drop across R_3. The output voltage at amplifier 3, measured with respect to ground, is equal to $-i_2R_3 + (E_2 - E_1)$ V. Thus a differential voltage measurement, as indicated in Figure 5-19, is necessary to obtain an output voltage proportional to the current flowing through W_2. This requires a read-out device (such as a recorder) with a floating input, because neither point may be grounded. Note that the current flowing through the counter electrode is the algebraic sum of the currents that flow through W_1 and W_2, where cathodic currents are taken as positive and anodic currents as negative.

Figure 5-20a represents a potentiostat configuration that has been widely used for over a decade. It is sometimes referred to as the "adder" configuration because of its resemblance to the voltage adder of Figure 5-12b. Its popularity stems from the fact that all of the control inputs can be referred to ground as well as the output voltage of the current follower (amplifier 3), which is proportional to the current. (However, the internal impedance of the control inputs must be low.)

To understand the operation of this circuit ignore for the time being the capacitors C, C_1, and C_f (which are used to stabilize the frequency response) and the components enclosed by the dotted rectangle (which are used for positive feedback compensation). The working electrode is maintained by amplifier 3 at virtual ground, which is the *current follower* configuration whose operation has been explained in the previous section. The output voltage of amplifier 3 is given by

$$E_o = -iR_f \tag{5-42}$$

so that E_o is proportional to the current flowing through the working electrode.

Amplifier 2 is a *voltage follower* such that the output E_f is equal to the input e_f. The potential e_f is the potential of the reference electrode measured with respect to the working electrode; because this is the reverse of the conventional definition of E_{cell},

$$E_f = e_f = -E_{cell} = -(E_{working} - E_{reference}) \tag{5-43}$$

This assumes that there is no uncompensated iR drop in the cell, which includes the solution, the leads to the working electrode, and the working electrode itself. The voltage follower draws only minute current through the

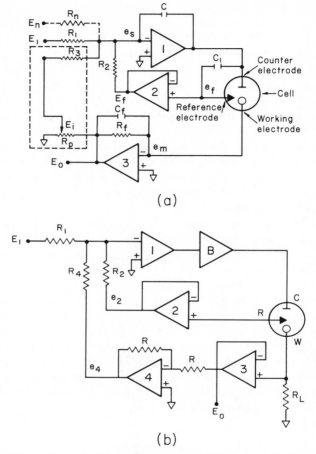

Figure 5-20. (*a*) "Adder" type operational amplifier potentiostat. Components in the dashed rectangle provide positive feedback *iR* compensaion. Amplifier 1, control amplifier; amplifier 2, voltage follower; amplifier 3, current follower; *C*, C_1, C_f, network stabilizing capacitors. (*b*) High current potentiostat. Amplifier 1, control amplifier; amplifier *B*, booster amplifier; amplifiers 2 and 3, voltage followers; amplifier 4, voltage inverter.

reference electrode so that *iR* drops in the reference electrode are negligible. If *iR* drops are present in the solution or in the electrode

$$E_f = -E_{\text{cell}} + iR_u \tag{5-44}$$

where R_u is the uncompensated resistance and *i* is the cell current (cathodic current taken as positive; anodic current taken as negative). The only electrically measurable quantity is the potential of the lead from the reference electrode measured with respect to the lead from the working electrode. At zero current, this is equal to $-E_{cell}$.

The control input(s), E_1 (up to E_n) and the voltage follower output E_f are added or summed through separate input resistors connected to the summing point of amplifier 1, the control amplifier. Amplifier 1 maintains its summing point at ground ($e_s = 0$) by virtue of the feedback loop that contains the counter electrode, reference electrode, and follower. The summing point is a current node where the algebraic sum of all the input currents is zero such that

$$\frac{E_n}{R_n} + \frac{E_1}{R_1} + \frac{E_f}{R_2} = 0 \tag{5-45}$$

If $R_1 = R_2 = R_n$, then

$$-E_f = E_1 + E_n \tag{5-46}$$

and in view of Equation 5-43

$$E_{cell} = E_1 + E_n \tag{5-47}$$

Simply stated, the control amplifier forces current to flow through the counter and working electrodes to enforce the condition described by Equation 5-47. For example, any tendency of the summing point to drift negative with respect to ground causes the control-amplifier output and the counter-electrode voltage to become more positive, E_{cell} to become more negative, and E_f to become more positive; all of which tend to cancel the negative drift at the summing point.

If the gain or frequency response of the amplifiers is inadequate, there will be potential-control errors, particularly if the control inputs change rapidly with time. Likewise if the voltage and current output of amplifiers 1 and 3 are inadequate to meet the demands, there will be an overload condition; neon lights or meters usually are incorporated in the circuit to detect overloads.

The basic circuit is versatile and permits as many control inputs as desired to be connected to the input of the control amplifier, each through its own resistor. To illustrate, a voltage ramp generator, based on the integrator configuration, can be connected to the input to provide a linearly changing voltage. The potential at which the voltage sweep starts can be varied by applying to a second input a variable voltage source to provide initial potential control. This is the basic configuration used in D-C polarography and linear potential-sweep voltammetry. A triangular waveform applied at the input enables one to do cyclic voltammetry.

If an integrator is connected to the output of the current follower (amplifier 3), the current may be integrated to yield the total number of coulombs passed through the cell. This is the basic configuration used for controlled potential coulometry.

For A-C polarography a sinusoidal input (generally of about 10-mV amplitude) is applied in addition to the voltage ramp and initial potential. Rectification of the output signal is necessary to record the A-C response (61).

1. Other Useful Potentiostat Configurations

Potentiostats of the type shown in Figure 5-20a have some drawbacks if a large current output is required because both amplifiers 1 and 3 must be able to supply current that flows through the cell. High current output amplifiers are more expensive, although only amplifier 1 needs to have a high voltage output because the voltage drop across R_f can be maintained below 10 V by appropriate selection of R_f. This drawback is avoided in the circuit of Figure 5-20b by connecting the working electrode through a load resistor to ground. The potential-control error that would be created by the iR drop through the load resistor, R_L, is compensated by negative feedback through the follower and inverter to the input of the control amplifier (amplifier 1). In this circuit all of the amplifiers except the booster (amplifier B) can be inexpensive low voltage, low current types.

The booster amplifier functions to augment the current and voltage output of the control amplifier. It is usually required because the 10 V output typical of many solid-state amplifiers is not adequate for solutions of moderate-to-high resistance. The booster amplifier must be of the noninverting type to preserve the proper phase relationship in the feedback loop. A number of manufacturers of operational amplifiers sell current and/or voltage boosters; an auxiliary power supply for the booster usually is required. An inexpensive current-voltage booster and power supply recently has been described which can provide ± 100 V at ± 100 mA for a parts cost of approximately $30 (62). Even simpler transistorized current-voltage boosters also have been described (63).

Note that in the circuit of Figure 5-20b the adder configuration is used at the input of the control amplifier (which is analogous to that in the circuit of Figure 5-20a), and that the voltage E_o at the output of the follower in the circuit of Figure 5-20b is proportional to the current flowing in the cell ($E_o = iR_L$). Hence these two circuits are alike in most respects as shown in Table 5-8.

2. Positive Feedback Compensation

If uncompensated resistance is present, the cell is represented approximately by the equivalent circuit shown in Figure 5-21a. Here R_u represents the sum

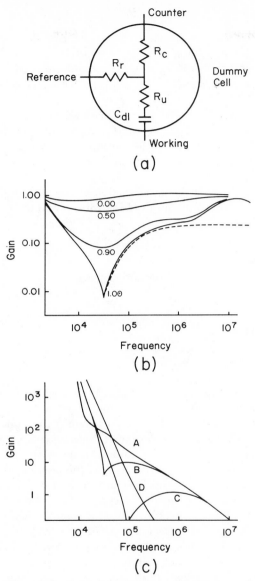

Figure 5-21. Potentiostat-cell frequency response. (*a*) Dummy cell. R_r, reference electrode resistance; R_c, counter electrode and cell resistance (compensated by positive feedback); R_u, total uncompensated resistance; C_{dl}, working electrode double layer capacitance. (*b*) Equivalent-cell transfer function for "dummy cell" as a function of frequency (several different values of β, the fraction of feedback compensation, are indicated). Total compensation corresponds to $\beta = 1.00$. $R_d = 25$ ohms; $R_u = 100$ Ω; $C_{dl} = 1.00$ μF. $R_rC_1 = 10^{-6}$ sec; $C = C_f = 0$. Solid line, rigorous calculation; dotted line, approximate calculation. (*c*) Bode plot of potentiostat-cell system open-loop gain for various values of R_rC_1 (Figure 5-20*a*). $R_1 = 25$ Ω; $R_u = 100$ Ω; $C_{dl} = 1.0$ μF; $\beta = 1.00$; $C = C_f = 0$. R_rC_1 equals (A) 10^{-5} sec; (B) 10^{-6} sec; (C) 10^{-7} sec; (D) 0.

of the uncompensated solution resistance and the electrode resistance (such as the internal resistance of the mercury thread in a dropping mercury electrode). These can be lumped together as a single resistance even though the internal electrode resistance and solution resistance are physically separated by the double-layer capacitance because a series R-C-R' circuit has an impedance indistinguishable from a series $C(R + R')$ circuit (64). C_{dl} represents the double-layer capacitance, R_C represents the solution and barrier (glass frit) resistance between the working electrode and the counter electrode that is compensated, and R_r represents the internal resistance of the reference electrode and probe.

In positive feedback compensation a fraction of the voltage output of the current follower is fed back to the input of the control amplifier. This is done by adjusting the tap of the potentiometer, R_p, in Figure 5-20a, so that

$$E_i = -iR_u \tag{5-48}$$

Summing the inputs at the control amplifier gives

$$E_f + E_i + E_1 + E_n = 0 \tag{5-49}$$

From Equations 5-44 and 5-48

$$E_{cell} = E_1 + E_n + iR_u - iR_u = E_1 + E_n \tag{5-50}$$

so that error due to uncompensated resistance is canceled.

There is the practical question of how to tell when adequate positive feedback compensation has been achieved. Although R_u, the uncompensated resistance, can be measured by a variety of techniques (65, 66) the average laboratory is not likely to have the equipment necessary to make these measurements in the manner described. For this reason, the methods employed by Whitson et al. (47) and by Garreau and Saveant (67) are of particular interest because they are in the context of cyclic voltammetry, a widely used technique. The first paper reports two simple methods that give reasonable agreement with each other. In the first method a cyclic voltammetric experiment is carried out in a potential region preceding the wave, where the current is due almost entirely to double-layer charging. The extent of compensation is increased until the onset of oscillation and then it is reduced slightly. In the second method a computation of R_u is made that is based on the known electrode geometry and the specific resistance of the solution and an equation similar in form to Equation 3-1. Measurements of the electrode geometry are made by photographing the hanging mercury-drop assembly and making measurements from enlarged prints. The reference electrode probe is located at a sufficient distance away so that its positioning is noncritical. In a comparison experiment for 0.08 F tetra-n-butylammonium perchlorate in acetonitrile, R_u was found to be 230 Ω by the first method and 234 Ω by the second method.

A detailed quantitative description of positive feedback operation is not at all simple, but the review by Smith (68) is recommended for its well-balanced and lucid discussion of positive feedback compensation. Smith emphasizes two points which should be kept in mind: (a) the ohmic resistance to be compensated must be determined accurately; and (b) if fast response is desired, an effort must be made to characterize the bandwidth limit of one's system.

3. Stabilization Techniques and Optimization of Potentiostat-Cell Response

The problem of how to obtain maximum bandwidth and optimum response from a cell-potentiostat system is too complex to discuss fully here. The reader is advised to study recent papers and reviews (68–71) to understand the different approaches to the analysis of stability and to obtain guidelines regarding the design of cells and stabilization techniques. In this discussion a few of the key ideas are introduced by use of the Bode diagram to represent graphically criteria for system stability.

Methods for evaluation of the stability of a feedback circuit rely on a knowledge of the open-loop gain of the system. The open-loop gain can be represented on a Bode diagram and in other ways (71). By use of the Bode-diagram approach the feedback circuit can be viewed as a series of individual elements (i.e., the potentiostat and the cell), each performing its open-loop operation on the signal as it passes around the feedback loop. The operation performed by each element is defined by its *transfer function*, the ratio of the output signal to the input signal (e_{out}/e_{in}), which is frequency dependent. The loop gain of the system is given by the product of the transfer functions of each element in the feedback loop. On a Bode diagram, the open-loop responses of each element may be plotted separately and added to give the total system open-loop response because the transfer functions are plotted on a logarithmic scale (70).

In assessing the stability of a potentiostat-cell system, a region of primary interest on its Bode plot (log open-loop gain versus log frequency) is the frequency where the gain becomes unity—the so-called unity gain crossover frequency of the system. The stability of the system is related to the slope of the gain curve at the unity gain crossover frequency; this slope in turn is related to the phase shift of the system (discussed in the section above on amplifier frequency response). When the phase shift becomes equal to or greater than 180° at unity gain, a net positive feedback results and the system is unstable. (Remember that the phase shift is referred to the normal phase shift of 180°.) This corresponds to a slope of -40 dB (or greater)/

decade of frequency at the unity gain crossover point. Any roll-off of gain less than -40 dB/decade indicates a nominally stable system. A -20 dB/decade roll-off means a 90° phase shift with a 90° phase margin of stability. In general, a phase margin of 40° to 60° is necessary for stable, critically damped transient response (72).

The transfer function of a cell can be measured (71,73) or calculated from the equivalent circuit of the cell (64). The potentiostat transfer function ordinarily can be calculated from the data that are supplied by the manufacturer. The measured cell transfer function is a useful tool for optimizing cell design and electrode geometry; Harrar and Pomernacki have presented a set of recommendations based on their measurements of cell transfer functions.

There are several reports in the literature that positive feedback compensation tends to produce instability and oscillation (68). Some understanding of why this might happen can be obtained by referring to Figure 5-21b which shows the equivalent-cell transfer function for the "dummy cell" of Figure 5-21a plotted as a function of frequency for various values to β, the fraction of feedback compensation. At close to 100% feedback compensation the cell develops a resonant response, which produces a steep roll-off of gain.

When the dummy cell is incorporated in the feedback loop of the potentiostat of Figure 5-20a, the open-loop response of the total system can be calculated from the gain characteristics of the amplifiers. This is plotted in Figure 5-21c for various values of the dummy-cell parameters and the stabilizing capacitor, C_1, which bridges the counter and reference electrode leads. In all the curves of this figure there is complete iR_u compensation ($\beta = 1.00$). The response of the system is strongly dependent on the $R_r - C_1$ time constant. With no stabilizing by-pass capacitance ($C_1 = 0$) the system would be unstable, as shown by curve D. Increasing C_1 slightly produces curve C, which still indicates system instability. However, a sufficiently large capacitor C_1 between the reference and auxiliary electrodes will stabilize the system as shown by the unit gain cross-over points of curves A and B of Figure 5-21c.

In the calculated open-loop response shown in Figure 5-21c the damping capacitance across the control amplifier (C) and current follower (C_f) have been set equal to zero without adverse effects for stable system operation. The addition of feedback capacitance around the control amplifier (C) initially has a strongly destabilizing effect and significantly decreases the bandwidth of the system (64). Stabilization of the system is achieved again when C is made quite large. Increasing C_f has a stabilizing effect but only in combination with a finite value of C_1. Hence the use of a by-pass capacitor, C_1, is a simple and efficient stabilizing element.

Experimental confirmation of this approach has been provided by a study of positive-feedback resistance compensation for cyclic voltammetry in acetonitrile (47). In this work, C_1 has been made just large enough to insure unconditional stability (a value of 2000 pF is typical).

C.   CURRENT CONTROL INSTRUMENTATION—THE GALVANOSTAT

One of the simplest ways to obtain a constant current of a few milliamperes or less through a cell is to connect a high voltage battery or power supply in series with the cell and a large resistor to limit the current. Constant current sources especially designed for coulometric titrations also are available (Leeds and Northrup, Sargent-Welch) as well as power supplies which operate in the constant-current mode. However, the operational amplifier provides an elegant and inexpensive way to obtain a controlled current source.

The basic operation of a galvanostat is like that of a single-amplifier potentiostat. First, a resistor is placed in series with the cell in the current carrying feedback loop. The iR drop across this resistor is then compared with the potential supplied by the arbitrarily adjustable signal source. Through use of a negative feedback loop that contains the cell, the amplifier forces current to flow through the cell in such a way as to reduce the difference between the iR drop and signal-generator voltage to zero. A readout device, M, measures the potential of the working electrode with respect to the reference electrode, but it plays no role in controlling the current in the circuit.

Many different circuit configurations are possible, as with the single-amplifier potentiostat (58); three of these are shown in Figure 5-22. In the circuit of Figure 5-22a the working electrode is maintained as a virtual ground by means of the feedback loop that contains the counter electrode and working electrode. The cell current that flows through the resistor generates a potential drop, $E = iR$. Because one end of the signal generator is grounded and one end of the resistor is a virtual ground, the voltages in this loop can be summed to obtain

$$iR + E_{sg} = 0 \qquad (5\text{-}51)$$

where E_{sg} is the signal-generator output voltage. Therefore the current will be given by

$$i = \frac{-E_{sg}}{R} \qquad (5\text{-}52)$$

A cathodic current through the working electrode requires a negative signal-generator output (measured with respect to ground).

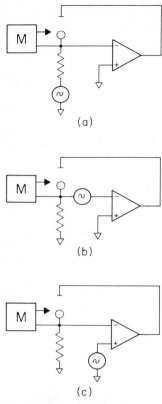

Figure 5-22. Circuit diagrams for single-amplifier galvanostats.

In this circuit the source resistance of the signal generator must be low; otherwise, a current-control error will be created by the iR drop in the signal source because the cell current flows through it. One input of the readout device can be grounded because the working electrode is a virtual ground. The amplifier also may be chopper stabilized.

In the circuit of Figure 5-22b, the signal generator is relocated so that negligible current flows through it. However, the signal generator and potential readout device must both be floating. The amplifier may be chopper stabilized.

The circuit of Figure 5-22c requires a differential amplifier, but the signal generator can have one terminal grounded. As in the previous circuit, the voltage readout device must have a differential (floating) input.

The qualitative characteristics of these circuits and the multiamplifier

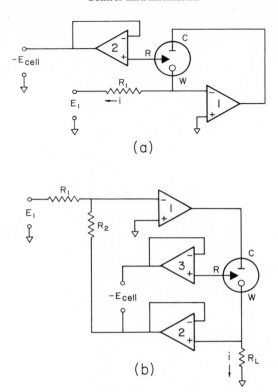

Figure 5-23. (*a*) Galvanostat with limited current capacity; (*b*) high current galvanostat.

circuits of Figure 5-23 are summarized in Table 5-9. The multiamplifier circuits have found practical use in chronopotentiometry and controlled current coulometry (see Chapters 8 and 9). The circuit of Figure 5-23*a* resembles that of Figure 5-22*a*; a follower has been added whose output voltage is equal to $-E_{cell}$ because its voltage (measured with respect to ground) is the potential of the reference electrode with respect to the working electrode (which is a virtual ground). This circuit also suffers from the same drawback that the cell current must be supplied by the reference voltage source. However, such a limitation usually does not create a problem for currents of a few milliamperes. (The source resistance should be less than 0.1% of the resistance connected to the summing point to prevent current-control error.) In the case of constant-current coulometric titrations, there may be need for much larger currents and the circuit of Figure 5-23*b* is better adapted for this purpose.

Table 5-9 Characteristics of Galvanostats

		Figure Number			
	5-22a	5-22b	5-22c	5-23a	5-23b
(a) Is the circuit free of current control error caused by iR drop in the signal generator?	No[a]	Yes	Yes	No[a]	No[a]
(b) Can stabilized (single-ended) amplifiers be used?	Yes	Yes	No	Yes[b]	Yes[b]
(c) Can one end of the signal generator be grounded?	Yes	No	Yes	Yes	Yes
(d) Can the potential measuring device have one input grounded?	Yes	No	No	Yes	No

[a] Yes, if the internal resistance of the signal source is negligible compared to the current determining resistor.
[b] Except for follower amplifiers, which require a differential input.

272

For the latter circuit the current flowing through the cell is supplied by the control amplifier (amplifier 1) which may be augmented with a booster. By means of the feedback loop, which contains the cell and follower amplifier 2, control amplifier 1 maintains its summing point at ground potential. The current through the cell flows through the load resistor, R_L, to generate a voltage drop, iR_L, which is fed by the follower (amplifier 2) through R_2 to the summing point. There is a current node at the summing point such that

$$\frac{E_1}{R_1} + \frac{iR_L}{R_2} = 0 \tag{5-53}$$

If $R_1 = R_2$

$$i = \frac{-E_1}{R_L} \tag{5-54}$$

The current will be determined by the reference voltage input and the load resistance, R_L. However, note that the current drawn from the voltage source is given by E_1/R_1, so that the iR drop in the voltage source is negligibly small if R_1 and R_2 are large enough.

For some experiments the ability to switch rapidly from potential to current control is useful (74,75). While some older designs allow the instrument function to be switched from potential control to current control, the switching sometimes is accompanied by severe transients. Roe has mentioned a method which is claimed to provide rapid switching (76). A current source is floated on the output of a potentiostat and connected to the working electrode. When the electrode is grounded, potential is controlled; but when the working electrode is disconnected from ground, the constant current prevails. Two new circuits have been described recently for rapid switching without transients (75,77).

D.   CONTROL OF CHARGE—THE COULOSTAT

In coulostatic experiments a fixed charge is injected into the working electrode (78). The method has been shown to be useful for measuring electrode kinetics (79), and to have some analytical utility (80), although as originally described it has not been widely used in analysis.

The technique, when applied to the dropping mercury electrode ("charge-step polarography"), is useful for rapid analysis of trace metals (81,82) and for the analysis of low concentrations of electroactive species in very dilute concentrations of supporting electrolyte (83).

The method has been recommended by Lauer as a way to charge the double layer rapidly. He has described an operational amplifier instrument for galvanostatic measurements which injects a precisely controlled charge followed by a constant current (84). Coulostatic charge-injection techniques also have been used to improve potentiostat rise times (85).

E.   DIGITAL CONTROL INSTRUMENTATION—MINICOMPUTERS

The power of the digital computer in data processing and simulation is well known. Clearly the use of small and relatively inexpensive on-line digital computers ("minicomputers") represents a major innovation in chemical instrumentation which is comparable to the introduction of operational amplifier methodology, but one that is likely to be of much vaster scope. At the same time, there is a much greater barrier to the use of minicomputers by the person who is not knowledgeable in digital techniques because of the complexities of interfacing the computer and the experiment. Indeed, the authors find themselves in this position. Fortunately, a number of books (86–88) and short reviews (89–91) are reducing this barrier but it is still formidable. The advice of most of those working in this area is that there is no substitute for "doing it." But how should one go about doing it?

To start, the reader should become familiar with basic number systems and digital codes, logical operations and digital circuits, and basic computer logic. This can be assisted by a variety of references (1,4,5,86,87,92). Practical experience with digital circuitry can be obtained by use of instrumentation packages like the Malmstadt/Enke Analog Digital Designer hardware manufactured by Heath/Schlumberger (Benton Harbor, Mich.). Finally, several university chemistry departments offer summer courses which provide first-hand experience in digital techniques and the interfacing of minicomputers (University of Illinois, Urbana; Michigan State University, East Lansing; Purdue University, Lafayette, Indiana; Virginia Polytechnic Institute, Blacksburg). Films on minicomputers and interfacing also are available on loan from Heath/Schlumberger.

A generalized interface for 16-bit computers has been described (93), and Heath/Schlumberger manufactures an interface package for use with the PDP-8 family of minicomputers manufactured by Digital Equipment Corp.

Naturally most manufacturers of minicomputers will provide considerable technical assistance to customers with interfacing problems. The trend is for greater standardization and in the future minicomputers and interfacing packages for a variety of tasks probably will be available "off the shelf." Still, the customer must be able to define his needs intelligently and he will be more successful at this task if he has some understanding of minicomputer architecture.

In this short chapter presentation of the fundamentals of digital-logic devices and computers is impossible, as is a meaningful description of the detailed circuitry associated with digital-controlled instrumentation. Therefore, only a few conceptual aspects are discussed; the reader should pursue any further interests in the cited references.

A block diagram of an on-line minicomputer-controlled experiment is shown in Figure 5-24. The computer (as purchased in "naked" form) consists of a central processor, a fast-core memory, and some means for the operator to communicate with the computer—Teletype keyboard, punched paper tape, or switches on the console. The central processor is the place where the arithmetic and logical operations take place; the programming instructions and output are stored in the core memory. There must be an interface between the Teletype and computer which translates the electrical

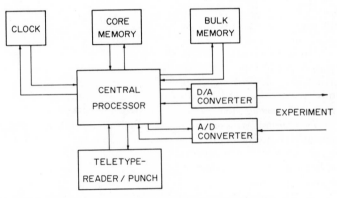

Figure 5-24.   Block diagram of on-line minicomputer-controlled experiment.

signals of the computer to the mechanical action of the Teletype and *vice-versa*. Likewise, there must be an interface between the computer and the experiment in order for the computer to be able to accept data from the experiment. Because the data from the cell are in analog form, the analog output must first be converted to digital form by an analog-to-digital converter. In order for the computer to act upon the cell, its digital output must be converted to analog form by a digital-to-analog converter. The sampling rate of the analog-to-digital converter is critical to proper performance of the system (94, 95), and a small handbook is available which gives practical advice on the selection of the optimum converter to meet the user's needs (96).

Table 5-10   Minicomputer Applications in Electrochemistry

Application	Reference
Double Potential Step Chronocoulometry. Under interactive Teletype keyboard control, the minicomputer initiates the experiment, acquires, digitizes, and stores the electrochemical data, tests for completion of a run, and performs a functional analysis on the stored data.	Lauer, Abel, and Anson, Ref. 99
Fast Sweep Derivative Polarography. Externally initiated sweep initiates a cycle of data acquistion by the computer. Externally filtered signal is differentiated by an analog differentiator to give the first or second derivative response, with the computer providing ensemble averaging of a large number of repetitive scans.	Perone, Harrar, Stephens, and Anderson, Ref. 100
Analysis of Fluctuating Current-time Curves with a Digital Polarograph. The instrument repetitively digitizes and applies an autocorrelation function to analyze current fluctuations in polarographic maxima.	Kugo, Umezawa, and Fujiwara, Ref. 101

Many experiments must be executed in real time; that is, the cell response as a function of time must be accurately known. In the minicomputer, there is often no compelling reason to operate in real time so that the sequence of events is controlled by the characteristic execution time—the time required to perform various arithmetic functions. Therefore, some form of "clock" generally is required to time the flow of data accurately and to provide for an orderly flow of control signals between the computer and the experiment, and between the computer and the operator. If the computer is connected to more than one experiment, provision must be made for priority interrupts— the computer stops whatever it is doing to receive some data input from the experiment—then resumes its tasks according to the assigned priorities. A recent review has described a number of different types of clocks; it also outlines the circumstances that dictate the choice of a particular kind of clock (97).

Electroanalytical Study of Flash Photolysis at the Hanging Mercury Drop Electrode. Operating in a continuous or time delay mode the computer initiates the flash, controls the electrode potential, digitizes and stores the data and provides a display output.	Kirschner and Perone, Ref. 102
General Purpose System for Computer Data Acquisition and Control. Sample and hold amplifiers will take X-Y data pairs with no time skew at rates of 10 kHz per X-Y pair, allowing the computer to acquire data, examine it, and then exert control on the experiment.	Ramaley and Wilson, Ref. 103
Computerized Electrochemical System for Pulse Polarography at a Hanging Mercury Drop Electrode. All experimental parameters are under control of the computer, with signal averaging employed to increase sensitivity.	Keller and Osteryoung, Ref. 104
Computerized Capillary Electrometer. Under control of the computer, the pressure required for birth of a drop is measured and individual data points are smoothed to give the electrocapillary curve.	Lawrence and Mohilner, Ref. 105

The system often requires bulk memory (slower than core memory) for storage of data and program instructions. This information is moved in and out of the core in large blocks as it is needed. The two most common forms of bulk storage are on magnetic discs and magnetic tape.

Finally, the computer must take over the task formerly carried out by human hands, throwing switches to stop and start the experiments and control auxilary devices. This usually requires relays or electrical switches, such as FET switches which can be operated by control signals from the computer. The reader unfamiliar with FET switching will find the applications described by O'Haver helpful in the use of these devices (98).

A number of applications of minicomputers for control of electrochemical instrumentation have been described in the literature (99–105); Table 5-10 provides a capsule summary of these examples. Careful reading of the references will give the reader a more concrete understanding of the mode of operation and capabilities of minicomputer systems.

Although most electrochemical applications employ analog control of the experiment, there are some examples of all digital control. An all digital bipolar instrument has been described which can function as a potentiostat through pulsed injection or extraction of charge to maintain a control potential (106). It sumultaneously functions as a current-to-digital converter which counts and sums these pulses as a function of time and allows the instrument to serve as an integrator. The name "bipolar digipotentiogrator" has been suggested for the instrument, which is claimed to have error limits of $\pm 0.01\%$ in the measurement of current. Although not providing very fast potential control response, it is of moderate cost and small size and makes extensive use of integrated circuits. It consumes about 1 W of power and delivers ± 10 V at ± 10 mA.

To build the devices which comprise a modern minicomputer is beyond the power of any chemical laboratory. Nevertheless the use of integrated circuits allows the construction of instruments which have some of the properties of minicomputer systems, including logic-control systems for switching, and small memories for storage of data (107). An example of this is the previously cited instrument for the measurement of solution resistance (38).

1. Hybrid Analog-Digital Instrumentation

There is a large gap between the simple operational amplifier control instrumentation that has been described and the minicomputer systems. Hybrid analog-digital systems, which cover this gap, use digital oscillators and logic circuitry for the control of FET switches that turn analog devices on and off. A simple and useful instrument in this category is shown in Figure 5-25 (108). It is basically a D-C polarograph which operates by stepping the potential rather than by continuously varying the potential. The step change is synchronized with the growth of the mercury drop, which is periodically dislodged by a mechanical "drop knocker." After each step change, there is a short delay to allow time for decay of the double-layer charging current. Then the Faradaic current is sampled by use of an integrator which has an R-C time constant that is equal to the period of the integration; this ensures that the output is the true average of the current over the integration period. The use of an integration system provides good signal-to-noise ratios. The output of the integrator (proportional to the average current) is transferred to a sequence of three sample-and-hold circuits (amplifiers $A5$, $A6$, and $A7$), with the samples in these amplifiers identified as sample 1, 2, and 3. After several cycles, the sampled currents of two adjacent potential steps are contained in $A4$ and $A5$, with the signal inverted when it is transferred from $A4$ to $A5$. The output of $A7$ is the

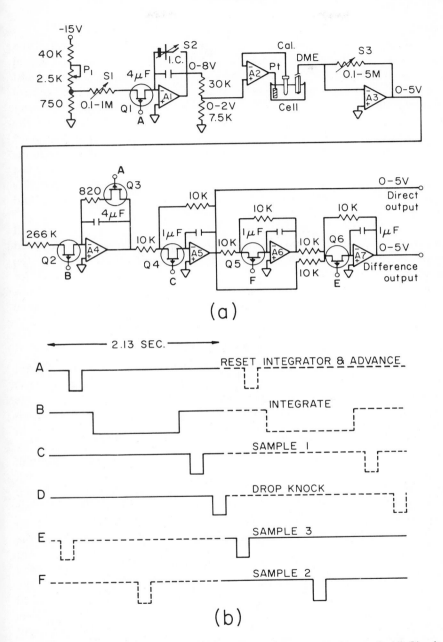

Figure 5-25. Hybrid analog-digital electronics D-C and differential polarograph. (*a*) Circuit diagram; (*b*) timing sequence of six FET switches.

279

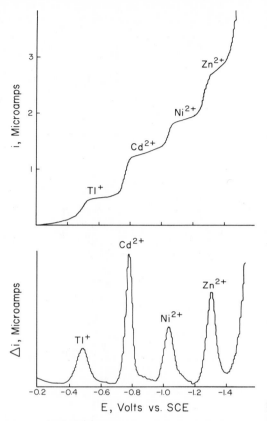

Figure 5-26. Direct current and differential polarograms produced by instrument of Figure 5-25.

sampled *sum* of the signals from $A4$ and $A5$; hence it is proportional to the current difference between two potential steps. Thus the instrument provides the average D-C current and the differential current, which amounts to the derivative of the average current. Amplifier $A1$ is an integrator which generates the potential steps by integration of input pulses, $A2$ is the control amplifier arranged in a conventional potentiostat configuration, and $A3$ is a current follower.

The two output modes for a test sample are shown in Figure 5-26. The averaging action tends to make the polarograms smoother and cleaner than conventional ones and permits higher sensitivities; these approach $10^{-6}\ M$ in the differential mode. For this instrument simple logic circuitry (detailed in Refs. 4 and 108) provides the timing sequence (shown in Figure 5-25b)

that controls the six FET switches. Although many other examples could be cited, this application has been chosen because it involves only moderate costs for the amplifiers and digital circuity, and can be built easily in most laboratories.

III. Noise Sources and Optimization of Signal-to-Noise Ratios

A. INTRODUCTION

Signals which represent an electrical or physical quantity are invariably accompanied by "noise"—fluctuations of the signal power. These fluctuations may be inherent in the physical process that is being measured or in the measuring system, or they may result from the environment. As an example, the potential of a glass electrode (which responds to the presence of hydrogen ions) will adopt an average equilibrium potential. If the electrode has a small active area and if the hydrogen ion activity is small, there will be only a limited number of hydrogen ions at the solution-electrode interface and their concentration will show microscopic fluctuations because of random thermal motions. This will give rise to corresponding fluctuations of the electrode potential which will limit the ultimate attainable precision of measurement. An excellent short review of noise sources and signal-to-noise optimization has been given by Coor (109).

B. SOURCES OF NOISE

The statistical and quantum properties of matter give rise to random fluctuations which sometimes are referred to as fundamental (unavoidable) noise. These generally have to do with the fact that there are fluctuations of random thermal origin and the fact that the current in a device is composed of discrete charges which move in an uncorrelated fashion.

1. Johnson, Nyquist, or Resistance Noise

The random motion of electrons produces fluctuations in the current in a resistor. The average of these fluctuations over a long period of time will be zero, but at any instant in time there is a net movement which gives rise to a thermal-noise voltage whose magnitude is given by Equation 5-55 of Table 5-11. The root-mean-square (rms) thermal-noise voltage in a resistor increases with the temperature, frequency bandwidth, and resistance. It appears at all frequencies at constant power per unit bandwidth; hence it is called "white" noise. The rms thermal noise in a 100-kΩ resistor at room

Table 5-11 Types of Noise and Their Frequency Dependence[a]

Type of Noise	Frequency Dependence	Magnitude	
Johnson or Resistance noise (thermal noise)	Constant power per unit bandwidth; "white noise"	Rms noise power $= 4kT\Delta f$	(5-55)
		Rms noise voltage $= (4kT\Delta fR)^{1/2}$	(5-56)
		Rms noise current $= \left(\dfrac{4kT\Delta f}{R}\right)^{1/2}$	(5-57)
Shot noise	Constant power per unit bandwidth; "white noise"	Rms noise power $= 2ei\Delta fR$	(5-58)
		Rms noise current $= (2ei\Delta f)^{1/2}$	(5-59)
Flicker noise	Increases with decreasing frequency; $1/f$ dependence	Dependent on type of device	
Environmental noise	Strongly peaked at certain frequencies, e.g., 60 Hz and higher harmonics	Dependent on environment and shielding	

$^a k = 1.38 \times 10^{-23}$ J/°K
T = temperature, °K
Δf = frequency bandwidth, Hz (sec^{-1})
f = frequency, Hz (sec^{-1})
e = electron charge, 1.60×10^{-19} C
i = D-C current, A
R = resistance, Ω

temperature is about 4 μV when measured with an instrument with a 10-kHz bandwidth. If the resistor forms the input to an amplifier with a gain of 1000, the noise at the output of the amplifier will be at least 4 mV. Hence thermal noise is not an insignificant problem.

Other sources of noise can be present in resistors besides Johnson noise (which appears across any resistor whether it is carrying current or not). However, wire-wound resistors are near the Johnson limit because microphonics (vibration) usually are the only other source of detectable noise. On the other hand, metal-film, deposited-carbon, and composition resistors exhibit voltage-dependent noise in increasing order. This apparently is due to the granular nature of the resistance elements and has a characteristic $1/f$ power-spectrum with a flicker-like appearance on an oscilloscope or meter. Good metal-film resistors are almost, but not quite, as good as wire-wound resistors and generally cost much less.

2. Shot Noise

In diodes (as well as photocells, transistors, and vacuum tubes) the passage of current is governed by a random single-event electron emission, while metallic conduction in a resistor is a large-scale correlated drift phenomenon with many discrete charges taking part at once. The shot noise which is generated in diode-like devices increases with the current that flows in the device, and is found in all active electronic amplifying systems. This noise determines in large measure the overall performance of an amplifying and measuring system. The shot noise in a vacuum tube will be much less than given by Equation 5-55 because the cloud of electrons around the cathode (space charge) introduces a smoothing action on the random conduction processes.

3. Flicker Noise

Flicker-effect noise is not completely understood, but seems to be associated with conduction processes in granular semiconductor material, or with cathode emission which is governed by the diffusion of clusters of barium atoms to the cathode surface. Almost all electronic elements exhibit flicker noise to some degree and their noise-power spectra differ widely. It generally dominates thermal or shot noise at frequencies below 100 Hz.

The characteristic $1/f$ dependence of flicker noise, which holds down to frequencies as low as can be measured, has disturbing implications for D-C measurements. The long-term drift in D-C transistor and tube amplifiers appears to be the manifestation of very low frequency flicker noise. Although in principle such noise might seem to be averaged to zero for a sufficiently long measurement period (as with "white" noise), the fact is that the flicker noise increases in proportion to the measurement period. This means that D-C *measurements must be avoided* if there is a signal-to-noise problem.

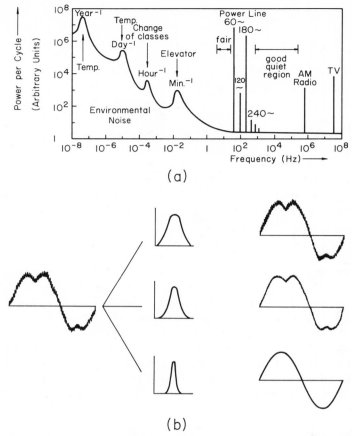

Figure 5-27. (*a*) Environmental noise power per unit frequency bandwidth as a function of frequency; (*b*) noise suppression as a function of the bandwidth of the processing system (center).

4. Environmental Noise

Noise from the environment finds its way into electrical measurements in a variety of ways. Sixty-hertz oscillations (and higher harmonics) from the power line, line-voltage variations (created by heavy loads such as an elevator motor), room-temperature fluctuations (or drafts created by foot-traffic in and out of the laboratory), and seasonal variations of temperature are all sources of noise. The noise power per unit frequency bandwidth is represented approximately by Figure 5-27*a*, which shows the sharp peaks of power-line noise sources and the broader peaks created by temperature fluctuations. In principle the system can be shielded (thermal insulation and electrical shielding) from these sources, but in practice they can be difficult to eliminate entirely.

5. Noise in Electrochemistry

Some recent work has attempted to describe the origin of noise in electrochemical systems. The first references to this topic are of Russian origin as are the first experimental measurements. In a short communication Barker (110) has cited this earlier work and has discussed the possible origin of flicker noise connected with the hydrogen evolution reaction on mercury.

Because noise consists of noncoherent random frequencies, the possibility exists that measurements of noise might give useful information. One group has shown that noise measurement should give information about the kinetics of homogeneous chemical reactions (111). Others have discussed the possibilities of measuring the A-C admittance of an electrode by use of a noise source or pseudo-random noise that consists of a number of discrete noncoherent frequencies (112). They also have shown that the approach allows the acquisition of the frequency response profile with great precision and rapidity (113,114). The term Fourier transform faradaic admittance has been used to describe the measurement, which is implemented with a minicomputer-controlled system.

C. NOISE FIGURE

Any electrical or physical measurement has a certain irreducible noise associated with it. Given a particular signal from the input source, S_i, with a noise, N_i, associated with it, one endeavors to transduce, amplify, and measure the characteristics of the signal with the least possible noise. To do this, the transducer and amplifier through which the signals are processed should degrade the initial S_i/N_i ratio as little as possible. A quantitative figure of merit for the system which processes the signal is the noise figure (NF) of the system, defined as

$$NF(\mathrm{dB}) = 10 \log \frac{S_i/N_i}{S_o/N_o} \tag{5-60}$$

where S_i/N_i is the ratio of the signal-to-noise *power* at the input and S_o/N_o the ratio at the output.

In specifying the NF of any element of the system, the signal source is assumed to have an internal resistance at 20°C and the thermal noise of the system, N_i, is taken as the irreducible baseline noise. Because there are other sources of noise (in addition to the Johnson noise generated by the source resistance) the *NF* of a signal processor or transducer is a complicated function of source impedance, frequency, and input-device characteristics. A perfect signal processing system will have a *NF* of 0 dB, while one that degrades the S/N of the signal source by a factor of two has a 3-dB *NF*. Amplifiers frequently are characterized by their measured *NF* values.

D.   EXTRACTION OF SIGNALS FROM NOISE

1. Noise Filters

One of the simplest ways to remove noise is to reduce the bandwidth of the processing system so that it passes only the frequency bandwidth of interest. This effect is shown qualitatively in Figure 5-27*b* for three different bandwidths. Reduction of the bandwidth can be accomplished by means of filters (passive or active) and frequency-selective and phase-selective amplifiers. A passive filter consists of simple resistance and capacitance or inductance

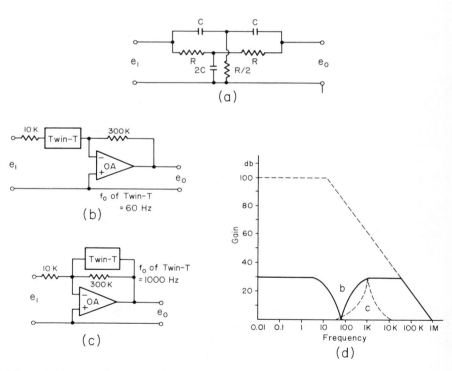

Figure 5-28. (*a*) Twin-T filter; (*b*) band-reject filter; (*c*) band-pass filter; (*d*) frequency responses of filters (*b*) and (*c*).

networks which can be used to reject or pass certain frequencies. The twin-T filter of Figure 5-28a is an example of a passive network which operates as a band-reject (notch) filter. In combination with an operational amplifier it can be used to form an active filter which acts either as a band-reject filter (Figure 5-28b) or as a bandpass filter (tuned amplifier) (Figure 5-28c). The frequency responses of these two filters are shown in Figure 5-28d as curves b

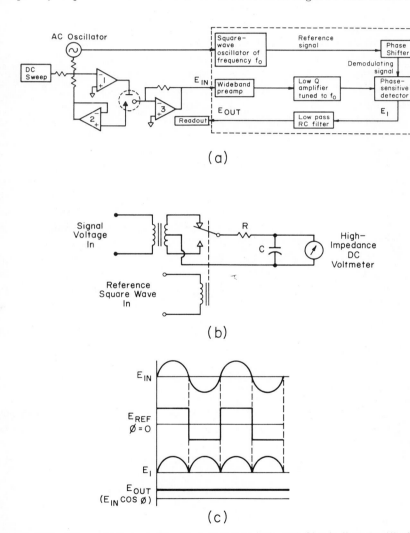

(a)

(b)

(c)

Figure 5-29.   Phase-selective amplifier for A-C polarography. (a) Circuit diagram; (b) phase-sensitive detector; (c) potential-time responses of various elements of circuit.

and c, respectively. Twin-T active filters have found many uses in electrochemical instrumentation, particularly in A-C polarography and derivative D-C polarography (68,115). Many other kinds of active filters are available which can provide low or high bandpass, or tuned-frequency response (116).

2. Lock-In or Phase-Selective Amplifiers

The principle of the phase-selective amplifier as applied to A-C polarography is shown in Figure 5-29. Here the A-C signal source is used to trigger a square wave which gates a phase-sensitive rectifier. Signals at other frequencies are averaged to zero, and even signals of the same frequency are attenuated if the phasing is incorrect; hence there is provision for phase shifting the detector. Such phase-sensitive detection gives an extremely narrow bandpass, and consequently a greatly improved S/N ratio. Phase-sensitive detectors have found a number of applications in A-C polarography and in other areas of chemical instrumentation (68,117–119).

In constructing an instrument every effort should be made to eliminate possible sources of noise. The techniques for doing this have been well-documented (120). In addition, attention should be paid to the grounding of the various components of the system. Figure 5-30a illustrates an example of poor practice in which the various ground points of the amplifier are connected to the ground return line (common) of the power supply at a number of places. This allows the generation of voltage drops between the ground points that can lead to potential-control errors. Figure 5-30b indicates how this can be avoided by bringing all ground returns to a single master ground point.

IV. Homemade and Commercial Instrumentation

A.   TO BUILD OR BUY

Unless one is engaged in advancing the state of the art in instrumentation, or exploiting a new technique for which no commerical instrumentation is available, the question of whether to build or buy is usually one of matching the needs and pocketbook of the individual with what is available on the market.

Fifteen years ago the instrumentation available for electrochemistry consisted mainly of D-C polarographs (of the two-electrode variety), potentiometers, pH meters, impedance bridges, and constant-current sources for coulometric titrations. However, since 1960 there has been a "construction boom" in electroanalytical laboratories with respect to instruments

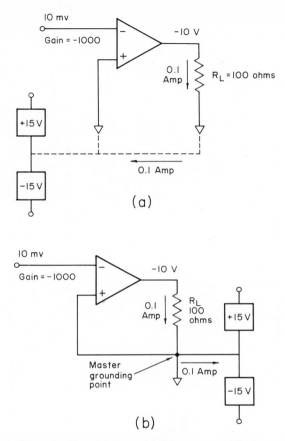

Figure 5-30. Amplifier grounding. (*a*) Example of poor system; (*b*) proper grounding.

that are based on operational amplifiers. With the seemingly limited market, few large instrument companies ventured into this new area at the time. The recent resurgence of interest in electrochemical applications to environmental problems has coincided with the appearance of several new instruments, and a number of large and small companies are now in the field (see for example the list published annually in *Science*, "Guide to Scientific Instruments," under "electrodes"and "polarographic analysis equipment"). In the rather specialized area of electrode kinetics, the need for fast-rise potentiostats stimulated their development and a number are now available commercially. A selected list of instruments is presented in Table 5-12 together with a few notes on their characteristics.

Table 5-12 A Selected List of Commercially Available Instrumentation for Voltammetric Analysis

Manufacturer	Functions Available[a]	Comments
A.R.F. Products, Inc.	1, 2e, 2g, 2h, 2i, 3a, 3c	Modular instrumentation designed primarily for teaching
Beckman Instruments, Inc.		
Electroscan 30	1, 2e, 2g, 2i, 2j, 2k, 2l, 3a, 3c	Integrator available separately
The Bendix Corp., Scientific Instruments and Equipment Division		
Davis Differential Cathode Ray Polarotrace 1660	2j	Readout on CRO tube; can be operated in subtractive mode with two cells and in derivative mode with preset voltage difference.
Chemtrix, Inc.		
Model SSP-2 three electrode polarographic analyzer	2j, 2k	Readout on storage oscilloscope Can be triggered by drop fall with adjustable sweep delay.
Environmental Science Associates, Inc.		
Anodic stripping analyzer	2l	Multiple cell instrument is available
Heath/Schlumberger		
EU-402 Polarography Lab	2i, 2l, 3a	± 10 mV stability after warmup. ± 50-V ± 1-mA output

290

Indiana Instrument and Chemical Corp.		
Controlled potential coulometer	2g	Based on the Oak Ridge designs of Kelley, Jones, and Fisher
Controlled potential and derivative voltammeter	2i	
McKee-Pedersen Instruments	1, 2e, 2f, 2g, 2h, 2i, 2j, 2k, 2l, 3a, 3c	Modular instrumentation for teaching and research
National Instrument Laboratories, Inc.		
N.I.L. Electrolab	1, 2e, 2g, 2i, 2j, 2k 2l, 3a, 3c	$\pm$40-V $\pm$1-A output
Princeton Applied Research		
Model 170	1, 2a-2g, 2i-2n 3a, 3c	$\pm$100-V $\pm$1-A output. Most versatile instrument on the market.
Sargent-Welch		
Model XVI polarograph	2h	Will record in derivative mode; iR Compensator available for three electrode circuit.

[a]1. Potentiometric measurements; 2. controlled potential: (a) small amplitude A-C polarography; (b) phase sensitive A-C polarography; (c) small amplitude pulse polarography; (d) derivative pulse polarography; (e) chronoamperometry; (f) chronocoulometry; (g) coulometry; (h) 2-electrode D-C polarography; (i) three-electrode D-C polarography; (j) three-electrode single sweep voltammetry; (k) three-electrode cyclic voltammetry; (l) anodic stripping voltammetry; (m) large amplitude pulse polarography; (n) current sampled D-C polarograhy; 3. controlled current: (a) chronopotentiometry; (b) current reversal chronopotentiometry; (c) constant current coulometry.

B. BUILDING

Although most of the circuits presented here have been simplified a great
deal for the sake of clarity (by omitting the power supply connections, bias
connections, control switches, etc.), the impression that these details are too
mysterious to be penetrated by the novice is not intended. On the contrary,
the construction of a versatile and reliable instrument for general
electrochemical use is not particularly difficult and requires no exceptional
skills in electronics. The experience is a particularly valuable way for
students to learn some instrumentation, can sometimes save considerable
money, and perhaps provide a capability not present in a commercial
instrument (at least at a price you can afford to pay). The beginner should
carefully study the detailed circuit diagrams of some of the instruments that
have been described recently in the literature (some of which are referred to
in Table 5-13). Such instruments can be duplicated down to the brand and
model of each amplifier, and the construction details provided by the
instrumentation group at Oak Ridge (Kelley, Jones, and Fisher) are a great
help to anyone who wants to build a polarograph. However, much simpler
instruments have been described which can easily be built by students with a
little assistance.

Those interested in the construction of instrumentation should obtain the
recent literature and catalogs from several manufacturers of operational
amplifiers; and then make the choice of the amplifiers, power supplies, and
other large components. Next plan a complete physical layout of the larger
components. Try to keep the signal path as straight as possible, avoid loops
and doubling back, and keep the input and output circuitry well separated.
Connections should be as short as possible without overcrowding. All large
components should be physically anchored to insulated tie points, or prin-
ted-circuit construction may be used.

Do not be afraid to experiment with the new amplifiers that continually
come on the market. The prices of amplifiers have decreased drastically in
the past few years, and integrated-circuit operational amplifiers are available which cost only a few dollars. Published circuits that are more than 5
years old may use amplifiers which are obsolete, or for which better and
cheaper substitutes are now available. Most amplifiers will require some
provision to bias out the voltage offset; the manufacturers specification sheet
will give the details of how this is to be done. By choosing low-drift or
stabilized amplifiers the bias adjustment will be infrequent and can be
located on the chassis next to the amplifier, rather than brought out to a
front panel. Low input current amplifiers must be used in long time-constant
integrators to minimize output drift.

For the present conditions switches, precision resistors, capacitors, and
chassis components probably will cost more than the amplifiers. These

Table 5-13 Selected Examples of Solid-State Voltammetric Instrumentation

Instrument	Comments	Ref.
A low cost, versatile electrochemical instrument	Will perform chronoamperometry, cyclic voltammetry, cyclic chronopotentiometry	S. C. Creason and R. F. Nelson, *J. Chem. Educ.*, **48**, 775 (1971).
Solid-state polarographic instrumentation	Uses digital switching function generator; provision for charging current compensation	R. Bezman and P. S. McKinney, *Anal. Chem.*, **41**, 1560 (1969).
Versatile solid-state potentiostat and amperostat	Cyclic voltammetry and chronopotentiometry with manual switching; coulometry	A. D. Goolsby and D. T. Sawyer, *Anal. Chem.*, **39**, 411 (1967).
A solid-state controlled potential D-C polarograph with first and second derivative modes	Uses active filters to obtain derivatives	H. C. Jones, W. L. Belew, R. W. Stelzner, T. R. Mueller, and D. J. Fisher, *Anal. Chem.*, **41**, 772 (1969).

components should be chosen with care to obtain optimum performance and service. In particular, switches should be of good quality because they are the first components to deteriorate in the atmosphere of a chemical laboratory. Where possible they should be of the enclosed type, or the whole chassis should be enclosed and sealed. This is especially important for integrating circuits where dirt and corrosion can lead to substantial leakage currents. The capacitors that are used for integrator circuits should be of high quality and contain a polystyrene dielectric. Polycarbonate- or Mylar-dielectric capacitors may be used in less critical integrator circuits with short time constants.

Wire-wound resistors of high precision (0.1% or better) are expensive, and careful thought should be given to decide wheter the uses to which the instrument will be put justify this expense. Good quality $\pm 1\%$ metal-film resistors may be used in many applications. Furthermore, an accurate digital ohmeter enables one to select or match resistors for critical applications (such as the input of an "adder" potentiostat, in voltage dividers, or current measuring resistors). Precision resistors in the high megohm range are quite expensive; their use can sometimes be avoided by means of a T-network (115, 121).

Construction on a totally enclosed metal chassis is preferred because it provides shielding and protection from the ravages of the laboratory atmosphere. The power supplies, with their 110 V A-C leads frequently are located external to the main chassis with the D-C power leads passed in through the chassis. The A-C leads should be tightly twisted together to help cancel the A-C pick-up from the alternating field. Careful attention should be paid to the grounding of the system by use of heavy gauge ground-return leads which are grounded at a single point. The use of shielded cables may be desirable for the external connections, but if the cable shield is connected to ground there will be a capacitance to ground, which may be a problem if high frequency response is desired. The shield on the follower input sometimes is connected to the output of the follower (driven shield) rather than to ground. This minimizes the capacitance at the follower input.

In large-instrument systems, modular construction may be preferred, even going so far as to make each unit (e.g., a follower, control amplifier, feedback resistors) a separate module on its own chassis. This provides maximum flexibility, but requires thorough familiarity with the circuit on the part of the user and some care in planning the interconnections of the modules to avoid a rat's nest of connecting wires.

C. BUYING

Assuming that the buyer can find what he wants at the right price, the choice of which instrument to buy will usually depend on the reputation of the manufacturer, gossip ("ask the man who owns one"), the advertised specifications of the instrument, and information about the instrument which may exist in the literature. Unfortunately there is no such thing as a "Consumer's Guide to Electrochemical Instrumentation," although a few experiences and candid opinions about new instruments have appeared in *Interface, a Newsletter of Electrochemical Science* (Department of Chemistry, Michigan State University, East Lansing, Mich. 48823).

1. Single-Purpose Versus Multiple-Purpose Instruments

While a large modular instrument system (with modules connected by internal switching or external patching) provides the maximum flexibility, some design compromises inevitably are necessary to obtain this flexibility (longer signal paths, more expensive switching, etc.). Furthermore, a multipurpose instrument often will be used for long periods of time in a single mode such that its versatility is unused. Also, a malfunction of a critical component (rare in a well-designed instrument) will make the whole instrument unusable. Finally, only one function at a time can be used on most multipurpose instruments and a single operator ties up the entire instrument; again, versatility is wasted.

One of the most successful multipurpose instruments is the Princeton Applied Research Model 170. Aside from the quality of its construction, one of the important ingredients of its success lies in the fact that it contains so many capabilities in one package. Yet it must be kept in mind when buying an instrument of this type that the functions can only be used one at a time and the same amount of money will buy (even from the same company) three or four single-purpose instruments, all of which can be used simultaneously.

Against these arguments may be weighed the economics of the multipurpose instrument (fewer power supplies, use of the same amplifiers for several different functions, etc.) and its usefulness in a teaching laboratory (where patching together modular elements may help the student to follow the signal path through the various modules).

Weighing all of these factors, a large industrial or graduate university laboratory probably will be better off if it builds or buys a number of less expensive single-purpose instruments than one large multipurpose instrument. The educational laboratory may find the modular multipurpose instrument with external patching useful for teaching purposes, while the smaller laboratory with few workers may find a multipurpose instrument cost-effective.

The minicomputer-controlled instrument may be seen as an example of maximum versatility. However powerful it may be, many of the uses to which it has been put could be accomplished more cheaply with simpler systems. The popularity of minicomputer systems in many electrochemical laboratories is partly due to the fact that the workers who employ them are engaged in advancing the state of the art of interfacing minicomputer systems. The immediate goal is not necessarily that of the maximum return of electrochemical information per dollar invested. Nevertheless, as manufacturing costs continue to decrease and packaged systems become increasingly available the minicomputer-controlled instrument ultimately will provide the maximum versatility, the greatest saving in human labor, and the lowest cost.

References

1. A. J. Diefenderfer, *Principles of Electronic Instrumentation*, W. B. Saunders Company, Philadelphia, 1972.

2. H. V. Malmstadt, C. G. Enke, and E. C. Toren, Jr., *Electronics for Scientists*, W. A. Benjamin, Inc., New York, 1962.

3. J. J. Brophy, *Basic Electronics for Scientists*, 2nd ed., McGraw-Hill Book Company, New York, 1972.

4. B. H. Vassos and G. W. Ewing, *Analog and Digital Electronics for Scientists*, Wiley-Interscience, New York, 1972.

5. R. D. Sacks and H. B. Mark, Jr., *Simplified Circuit Analysis—Digital-Analog Logic*, Marcel Dekker, Inc., New York, 1972.

6. D. M. Hunten, *Introduction to Electronics for Students of Physics and Engineering Science*, Holt, Rinehart and Winston, Inc., New York, 1964.

7. F. H. Mitchell, *Fundamentals of Electronics*, Addison-Wesley Publishing Company, Inc., Reading, Mass., 1959.

8. K. H. Goode, *J. Am. Chem. Soc.*, **44**, 26 (1922); **47**, 2483 (1925).

9. A. Hickling, *Trans. Faraday Soc.*, **38**, 27 (1942).

10. J. J. Lingane, *Electroanalytical Chemistry*, 2nd ed., Interscience Publishers, Inc., New York, 1958, Chap. 13.

11. G. L. Booman, *Anal. Chem.*, **29**, 213 (1957).

12. D. D. DeFord, Division of Analytical Chemistry, 133rd National Meeting, American Chemical Society, San Francisco, Calif., April 1958.

13. "Symposium on Operational Amplifiers," *Anal. Chem.*, **35**, 1770–1833 (1963).

14. "Symposium on Electroanalytical Instrumentation," *Anal. Chem.*, **38**, 1106–1148 (1966).

15. C. G. Enke, *Anal. Chem.*, **43**(1), 69A (1971).

16. P. W. Bridgman, *The Nature of Thermodynamics*, Harper Torchbooks, Harper and Bros., New York, 1961, pp. vii–xii.

17. D. Halliday and R. Resnick, *Physics for Students of Science and Engineering*, Part II, 2nd ed., John Wiley and Sons, Inc., New York, 1962, p. 854.

18. C. H. Page and P. Vigoureux, eds., *The International System of Units (SI)*, National Bureau of Standards Special Publication 330, April 1972.

19. J. Matisoo, *Anal. Chem.*, **41**(1), 83A (1969); **41**(2), 139A (1969); D. N Langenberg, D.J. Scalapino, and B. N. Taylor, *Sci. American*, **214**(5), 30(1966).

20. *Precision Electrochemical Measurements Utilizing the Cary Model 31 Vibrating Reed Electrometer*, Report 31-1, Applied Physics Corp., 1962 (Varian Instrument Division, Los Altos, Calif.).

21. S. M. Friedman, "H^+ and Cation Analysis of Biological Fluids in the Intact Animal," in *Glass Electrodes for Hydrogen and Other Cations*, G. Eisenman, ed., Marcel Dekker, Inc., New York, 1967, Chap. 16.

22. R. G. Bates, *Determination of pH—Theory and Practice*, 2nd ed., Wiley-Interscience, New York, 1973, p. 406.

23. M. J. D. Brand and G. A. Rechnitz, *Anal. Chem.*, **42**, 616 (1970).

24. M. J. D. Brand and G. A. Rechnitz, *Anal. Chem.*, **42**, 1659 (1970).

25. F. D. Tabbutt, *J. Chem. Educ.*, **39**, 611 (1962).

26. Ref. 1, p. 135.

27. Ref. 22, pp. 399–405.

28. Ref. 22, pp. 394–399.

29. G. W. Vinal, *Primary Batteries*, John Wiley and Sons, Inc., New York, 1950, Chap. 6.

30. Ref. 1, p. 325.

31. Ref. 3, p. 34.

32. Ref. 2, p. 353.

33. P. G. Cath and A. M. Peabody, *Anal. Chem.*, **43**(11), 91A (1971).

34. Ref. 2, p. 273.

35. J. Braunstein and G. D. Robbins, *J. Chem. Educ.*, **48**, 52 (1971).

36. G. Kortüm, *Treatise on Electrochemistry*, 2nd ed., Elsevier Publishing Company, Amsterdam, 1965, p. 202.

37. G. Jones and G. M. Bollinger, *J. Am. Chem. Soc.*, **53**, 411 (1931).

38. P. H. Daum and D. F. Nelson, *Anal. Chem.*, **45**, 463 (1973).

39. K. Schmidt, *Rev. Sci. Instrum.*, **47**, 671 (1966).

40. Ref. 1, p. 126.

41. T. Shedlovsky and L. Shedlovsky, "Conductometry," in *Techniques of Chemistry*, A. Weissberger, ed., Vol. 1, *Physical Methods of Chemistry*, Part IIA, *Electrochemical Methods*, Wiley-Interscience, 1971, Chapter III, p. 174.

42. D. W. Colvin and R. C. Propst, *Anal. Chem.*, **32**, 1858 (1960).

43. G. Knudson, L. Ramaley, and W. A. Holcombe, *Chem. Instrum.*, **1**, 325 (1969).

44. A. Ford and C. E. Meloan, *J. Chem. Educ.*, **50**, 85 (1973).

45. D. A. Aaker, *Anal. Chem.*, **37**, 1252 (1965).

46. I. G. McWilliam and H. C. Bolton, *Anal. Chem.*, **41**, 1755; 1762 (1969).

47. P. E. Whitson, H. W. VandenBorn, and D. H. Evans, *Anal. Chem.*, **45**, 1298 (1973).

48. J. G. Graeme, G. E. Tobey, and L. P. Huelsman, eds., *Operational Amplifiers—Design and Applications*, McGraw-Hill Book Company, New York, 1971.

49. *Applications Manual for Computing Amplifiers*, 2nd ed., George A. Philbrick Researches, Inc., Allied Drive at Route 128, Dedham, Mass. 02026, June 1966. (The Company is now known as *Teledyne Philbrick* at the same address.)

50. C. N. Reilley, *J. Chem. Educ.*, **39**, A853; A933 (1962).

51. L. P. Morgenthaler, *Basic Operational Amplifier Circuits for Analytical Chemical Instrumentation*,

2nd ed., 1968, McKee-Pedersen Instruments, P. O. Box 322, Danville, Calif. 94526.

52. *McKee-Pedersen Instruments Applications Notes*, Vol. 1–8, beginning 1966, McKee-Pedersen Instruments, P. O. Box 322, Danville, Calif. 94526.

53. R. R. Schroeder, "Operational Amplifier Instruments for Electrochemistry," in *Electrochemistry*, Vol. 2 of *Computers in Chemistry and Instrumentation*, J. S. Mattson, H. B. Mark, Jr., and H. C. MacDonald, Jr., eds., Marcel Dekker, Inc., New York, 1972, Chap. 10.

54. Ref. 5, Table 1, p. 4.

55. G. A. Korn and T. M. Korn, *Electronic Analog Computers*, 2nd ed., McGraw-Hill Book Company, New York, 1956, p. 415.

56. G. E. Roberts and H. Kaufman, *Table of Laplace Transforms*, W. B. Saunders Company, Philadelphia, 1966.

57. P. A. McCollum and B. F. Brown, *Laplace Transform Tables*, Publication No. 137, The Office of Engineering Research, Oklahoma State University, Stillwater, Okla. 74075, February 1964.

58. W. M. Schwarz and I. Shain, *Anal. Chem.*, **35**, 1770 (1963).

59. L. B. Anderson and C. N. Reilley, *J. Electroanal. Chem.*, **10**, 295 (1965).

60. C. N. Reilley, *Pure Appl. Chem.*, **18**, 137 (1968).

61. D. E. Smith, "AC Polarography and Related Techniques," in *Electroanalytical Chemistry*, Vol. 1, A. J. Bard, ed., Marcel Dekker, Inc., New York, 1966, pp. 102–121.

62. W. S. Woodward, T. H. Ridgway, and C. N. Reilley, *Anal. Chem.*, **45**, 435 (1973).

63. K. Lowe, *J. Chem. Educ.*, **47**, 846 (1970).

64. E. R. Brown, D. E. Smith, and G. L. Booman, *Anal. Chem.*, **40**, 1411 (1968).

65. E. R. Brown, H. L. Hung, T. G. McCord, D. E. Smith, and G. L. Booman, *Anal. Chem.*, **40**, 1424 (1968).

66. W. E. Thomas, Jr. and W. B. Schaap, *Anal. Chem.*, **41**, 136 (1969).

67. D. Garreau and J. M. Saveant, *J. Electroanal. Chem.*, **35**, 309 (1972).

68. D. E. Smith, *Crit. Rev. Anal. Chem.*, **2**, 247 (1971).

69. R. R. Schroeder and I. Shain, *Chem. Instrum.*, **1**, 233 (1969).

70. Ref. 53, pp. 325–347.

71. J. E. Harrar and C. L. Pomernacki, *Anal. Chem.*, **45**, 57 (1973).

72. W. C. Carter, *Instrum. Contr. Syst.*, **40**, 107 (January 1967).

73. D. T. Pence and G. L. Booman, *Anal. Chem.*, **38**, 1112 (1966).

74. J. T. Bowman and A. J. Bard, *Anal. Letters*, **1**, 533 (1968).

75. S. Bruckenstein and B. Miller, *J. Electrochem. Soc.*, **117**, 1040 (1970).

76. D. K. Roe, *Anal. Chem.*, **44**(5), 85R (1972).

77. Y. M. Sokolov, *Elektrokhimiya*, **6**, 1222 (1970).

78. P. Delahay, *Anal. Chim. Acta*, **27**, 90 (1962).

79. J. M. Kudirka, P. H. Daum, and C. G. Enke, *Anal. Chem.*, **44**, 309 (1972).

80. P. Delahay, *Anal. Chem.*, **34**, 1267 (1962).

81. M. Astruc and J. Bonastre, *Anal. Chem.*, **45**, 421 (1973).

82. M. Astruc, F. Del Rey, and J. Bonastre, *J. Electroanal. Chem.*, **43**, 113; 125 (1973).

83. J. M. Kudirka, R. Abel, and C. G. Enke, *Anal. Chem.*, **44**, 425 (1972).

84. G. Lauer, *Anal. Chem.*, **38**, 1277 (1966).

85. J. E. Davis and N. Winograd, *Anal. Chem.*, **44**, 2152 (1972).

86. B. Soucek, *Minicomputers in Data Processing and Simulation*, Wiley-Interscience, New York, 1972.

87. S. P. Perone and D. O. Jones, *Digital Computers in Scientific Instrumentation*, McGraw-Hill Book Company, New York, 1973.

88. J. S. Mattson, H. B. Mark, Jr., and H. C. MacDonald, Jr., *Computers in Chemistry and Instrumentation*, Vol. 2, Marcel Dekker, Inc., New York, 1972, Chap. 11–13.

89. R. E. Dessy and J. A. Titus, *Anal. Chem.*, **45**(2), 124A (1973).

90. J. W. Frazer, *Anal. Chem.*, **40**(8), 26A (1968).

91. G. Lauer and R. A. Osteryoung, *Anal. Chem.*, **40**(10), 30A (1968).

92. H. V. Malmstadt and C. G. Enke, *Digital Electronics for Scientists*, W. A. Benjamin, Inc., New York, 1969.

93. J. E. Davis and E. D. Schmidlin, *Chem. Instrum.*, **4**, 169 (1973).

94. Ref. 86, Chap. 8.

95. P. C. Kelly and G. Horlick, *Anal. Chem.*, **45**, 518 (1973).

96. D. H. Sheingold, ed., *Analog-Digital Conversion Handbook*, Analog Devices, Inc., Norwood, Mass. 02062, 1972.

97. B. K. Hahn and C. G. Enke, *Anal. Chem.*, **45**, 651A (1973).

98. T. C. O'Haver, *Chem. Instrum.*, **3**, 1 (1971).

99. G. Lauer, R. Abel, and F. C. Anson, *Anal. Chem.*, **39**, 765 (1967).

100. S. P. Perone, J. E. Harrar, F. B. Stevens, and R. E. Anderson, *Anal. Chem.*, **40**, 899 (1968).

101. T. Kugo, Y. Umezawa, and S. Fujiwara, *Chem. Instrum.*, **2**, 189 (1969).

102. G. L. Kirschner and S. P. Perone, *Anal. Chem.*, **44**, 443 (1972).

103. L. Ramaley and G. S. Wilson, *Anal. Chem.*, **42**, 606 (1970).

104. H. E. Keller and R. A. Osteryoung, *Anal. Chem.*, **43**, 342 (1971).

105. J. Lawrence and D. M. Mohilner, *J. Electrochem. Soc.*, **118**, 1596 (1971).

106. W. W. Goldsworthy and R. G. Clem, *Anal. Chem.*, **44**, 1360 (1972).

107. J. S. Springer, *Anal. Chem.*, **42**(8), 22A (1970).

108. B. H. Vassos, *Anal. Chem.*, **45**, 1292 (1973).

109. T. Coor, *J. Chem. Educ.*, **45**, A533; A583 (1968).

110. G. C. Barker, *J. Electroanal. Chem.*, **39**, 484 (1973).

111. M. Fleischmann and J. W. Oldfield, *J. Electroanal. Chem.*, **27**, 207 (1970).

112. B. J. Huebert and D. E. Smith, *Anal. Chem.*, **44**, 1179 (1972).

113. S. C. Creason and D. E. Smith, *J. Electroanal. Chem.*, **36**, Appl. 1 (1972).

114. S. C. Creason and D. E. Smith, *J. Electroanal. Chem.*, **40**, Appl. 1 (1972).

115. R. W. Stelzner, *Chem. Instrum.*, **2**, 213 (1969).

116. *Active Filters*, a collection of bound reprints from *Electronics*, 1971, McGraw-Hill Book Company, New York.

117. D. K. Means and H. B. Mark, Jr., *Chem. Instrum.*, **3**, 271 (1972).

118. T. C. O'Haver, *J. Chem. Educ.*, **49**, A131; A211 (1972).

119. G. M. Hieftje, *Anal. Chem.*, **44**(6), 81A; **44**(7), 69A (1972).

120. R. Morrison, *Grounding and Shielding Techniques* in *Instrumentation*, John Wiley and Sons, Inc., 1967.

121. Ref. 49, p. 24.

Six

POTENTIOMETRIC MEASUREMENTS

I. Introduction

Use of the potential of a galvanic cell to measure the concentration of an electroactive species developed later than a number of other electrochemical methods. In part this was because potentiometric measurement of ionic concentrations required the development of thermodynamics, and in particular its application to electrochemical phenomena. The work of J. Willard Gibbs (1) in the 1870s provided the foundation for potentiometry. However, the real keystone was the development of the well-known Nernst equation a decade later (2). With a quantitative relationship between potential and the concentration of an electroactive species, the method soon was applied to the detection of the equivalence point of a titration involving an oxidation-reduction reaction (3). Another milestone in the development of potentiometric measurements was the introduction of the hydrogen electrode for the measurement of hydrogen ion concentration (4); without question this was one of the major contributions of a great many made by Professor Hildebrand. The importance of potentiometric measurements for all phases of chemistry was established with the development of the glass electrode in 1919 (5). Potentiometric measurement of pH became the established method for monitoring solution acidity and continues to hold this position in modern chemistry.

At an early date Nernst also introduced the concept of potentiometry with polarized electrodes (6), which, together with the many other specialized forms of potentiometric measurements for a wide range of chemical systems, has been thoroughly discussed and reviewed in the definitive monograph by Kolthoff and Furman on the subject of potentiometric titrations (7). Unfortunately, this valuable work is out of print; it is still surprisingly current, although published 40 years ago.

II. Principles and Fundamental Relations

Potentiometric measurements are based on thermodynamic relationships and more particularly the Nernst equation which relates potential to the concentration of electroactive species. For our purposes it is most convenient to consider the redox process occurring at a single electrode, although two electrodes are always essential for an electrochemical cell. However, by considering each electrode individually, the two electrode processes are easily combined to obtain the entire cell process. Furthermore, confusion can be minimized if the half reactions for electrode processes are written in a consistent manner. Here, these are always reduction processes with the oxidized species being reduced by n electrons to give a reduced species.

$$Ox + ne^- \rightarrow Red \tag{6-1}$$

For such a half reaction the free energy is given by the relation

$$\Delta G = \Delta G^0 + RT \ln\left(\frac{[Red]}{[Ox]}\right) \tag{6-2}$$

where $-\Delta G$ indicates the tendency for the reaction to go to the right; R is the gas constant and in the units appropriate for electrochemistry has a value of 8.317 J; T is the temperature of the system in °K; and the logarithmic terms in the bracketed expression represent the activities (effective concentrations) of the electroactive pair at the electrode surface. The free energy of this half reaction is related to the electrode potential, E, by the expression

$$-\Delta G = nFE; \qquad -\Delta G^0 = nFE^0 \tag{6-3}$$

The quantity ΔG^0 is the free energy of the half reaction when the activities of the reactant and product have values of unity and is directly proportional to the standard half cell potential for the reaction as written. It also is a measure of the equilibrium constant for the half reaction assuming the activity of electrons is unity

$$-\Delta G^0 = RT \ln K \tag{6-4}$$

An extensive summary of E^0 values is presented in the classic monograph by Latimer (8); some of the most important couples are presented in Table 6-1.

Table 6-1 Electromotive Series for Aqueous Solutions at 25°C and 1 atm

Reaction	$E°$, V vs. NHE
$Li^+ + e^- \rightarrow Li$	-3.845
$Rb^+ + e^- \rightarrow Rb$	-2.925
$Cs^+ + e^- \rightarrow Cs$	-2.923
$K^+ + e^- \rightarrow K$	-2.925
$Ba^{2+} + 2e^- \rightarrow Ba$	-2.900
$Sr^{2+} + 2e^- \rightarrow Sr$	-2.890
$Ca^{2+} + 2e^- \rightarrow Ca$	-2.870
$Na^+ + e^- \rightarrow Na$	-2.714
$Mg^{2+} + 2e^- \rightarrow Mg$	-2.370
$Be^{2+} + 2e^- \rightarrow Be$	-1.850
$Al^{3+} + 3e^- \rightarrow Al$	-1.660
$S + 2e^- \rightarrow S^{2-}$	-0.920
$Co(CN)_6^{3-} + e^- \rightarrow Co(CN)_6^{4-}$	-0.830
$Se + 2e^- \rightarrow Se^{2-}$	-0.780
$Zn^{2+} + 2e^- \rightarrow Zn$	-0.763
$Fe^{2+} + 2e^- \rightarrow Fe$	-0.440
$Cr^{3+} + e^- \rightarrow Cr^{2+}$	-0.410
$Cd^{2+} + 2e^- \rightarrow Cd$	-0.402
$Ti^{3+} + e^- \rightarrow Ti^{2+}$	-0.370
$In^{3+} + 3e^- \rightarrow In$	-0.342
$Tl^+ + e^- \rightarrow Tl$	-0.335
$Co^{2+} + 2e^- \rightarrow Co$	-0.277
$Ni^{2+} + 2e^- \rightarrow Ni$	-0.250
$In^+ + e^- \rightarrow In$	-0.203
$V^{3+} + e^- \rightarrow V^{2+}$	-0.200
$Sn^{2+} + 2e^- \rightarrow Sn$	-0.136
$Pb^{2+} + 2e^- \rightarrow Pb$	-0.126
$Sn^{4+} + 2e^- \rightarrow Sn^{2+}$	$+0.154$
$Cu^{2+} + e^- \rightarrow Cu^+$	$+0.167$
$Cu^{2+} + 2e^- \rightarrow Cu$	$+0.337$
$Fe(CN)_6^{3-} + e^- \rightarrow Fe(CN)_6^{4-}$	$+0.356$
$O_2 + 2H_2O + 4e^- \rightarrow 4OH^-$	$+0.401$
$Cu^+ + e^- \rightarrow Cu$	$+0.521$
$I_3^- + 2e^- \rightarrow 3I^-$	$+0.535$
$I_2 + 2e^- \rightarrow 2I^-$	$+0.536$
$H_3AsO_4 + 2H_3O^+ + 2e^- \rightarrow H_3AsO_3 + 3H_2O$	$+0.559$
$Te^{4+} + 4e^- \rightarrow Te$	$+0.560$

Table 6-1   (*Continued*)

Reaction	$E°$, V vs. NHE
$MnO_4^- + 2H_2O + 3e^-$ $\rightarrow MnO_2 + 4OH^-$	$+0.587$
$Fe^{3+} + e^- \rightarrow Fe^{2+}$	$+0.771$
$Hg_2^{2+} + 2e^- \rightarrow 2Hg$	$+0.789$
$Ag^+ + e^- \rightarrow Ag$	$+0.799$
$2Hg^{2+} + 2e^- \rightarrow Hg_2^{2+}$	$+0.905$
$V(OH)_4^+ + 2H_3O^+ + e^-$ $\rightarrow VO^{2+} + 5H_2O$	$+1.000$
$Br_2 + 2e^- \rightarrow 2Br^-$	$+1.066$
$MnO_2 + 4H_3O^+ + 2e^- \rightarrow Mn^{2+} + 6H_2O$	$+1.236$
$Tl^{3+} + 2e^- \rightarrow Tl^+$	$+1.250$
$Au^{3+} + 2e^- \rightarrow Au^+$	$+1.290$
$Cl_2 + 2e^- \rightarrow 2Cl^-$	$+1.360$
$Cr_2O_7^{2-} + 14H_3O^+ + 6e^-$ $\rightarrow 2Cr^{3+} + 21H_2O$	$+1.360$
$ClO_3^- + 6H_3O^+ + 6e^- \rightarrow Cl^- + 9H_2O$	$+1.450$
$PbO_2 + 4H_3O^+ + 2e^- \rightarrow Pb^{2+} + 6H_2O$	$+1.456$
$Au^{3+} + 3e^- \rightarrow Au$	$+1.500$
$Mn^{3+} + e^- \rightarrow Mn^{2+}$	$+1.510$
$Ce^{4+} + e^- \rightarrow Ce^{3+}$	$+1.610$
$Pb^{4+} + 2e^- \rightarrow Pb^{2+}$	$+1.690$
$Co^{3+} + e^- \rightarrow Co^{2+}$	$+1.842$
$F_2 + 2e^- \rightarrow 2F^-$	$+2.850$

When Equations 6-2 and 6-3 are combined, the Nernst expression for the half reaction is obtained, which relates the half cell potential to the effective concentrations of the electroactive species

$$E = E^0 - \left(\frac{RT}{nF}\right)\ln\left(\frac{[\text{Red}]}{[\text{Ox}]}\right)$$

$$= E^0 + \left(\frac{RT}{nF}\right)\ln\left(\frac{[\text{Ox}]}{[\text{Red}]}\right) \tag{6-5}$$

The activity of a species is indicated as the symbol of the species enclosed in brackets. This quantity is equal to the concentration of the species times a mean activity coefficient

$$[M^{a+}] = a_{M^{a+}} = \gamma_\pm C_{M^{a+}} \tag{6-6}$$

Although there is no straighforward and convenient method for evaluating activity coefficients for individual ions, the Debye-Hückel relationship permits an evaluation of the mean activity coefficient, $\gamma_\pm$, for ions at low concentrations (usually below 0.01 M)

$$\log\gamma_\pm = -0.509z^2 \frac{\sqrt{\mu}}{\left(1+\sqrt{\mu}\right)} \tag{6-7}$$

where z is the charge on the ion and μ is the ionic strength which is given by the expression

$$\mu = \frac{1}{2}\sum C_i z_i^2 \tag{6-8}$$

The reaction of an electrochemical cell always involves a combination of two redox half reactions such that one species oxidizes a second species to give the respective redox products. Thus the overall cell reaction can be expressed by a balanced chemical equation

$$a\mathrm{Ox}_1 + b\mathrm{Red}_2 \rightarrow c\mathrm{Red}_1 + d\mathrm{Ox}_2, \; K_{\mathrm{equil}} \tag{6-9}$$

However, electrochemical cells are most conveniently considered as two individual half reactions, whereby each is written as a reduction in the form indicated by Equations 6-1 through 6-5. When this is done and values for the appropriate quantities are inserted, a potential can be calculated for each half-cell electrode system. Then that half-cell reaction with the more positive potential will be the positive terminal in a galvanic cell and the electromotive force of that cell will be represented by the algebraic difference between the potential of the more positive half cell and the potential of the less positive half-cell.

$$E_{\mathrm{cell}} = E_{(\text{more positive})} - E_{(\text{less positive})}$$

$$= E_1 - E_2 \tag{6-10}$$

By inserting the appropriate forms of Equation 6-5 into Equation 6-10 an overall expression for the cell potential is obtained

$$E_{\mathrm{cell}} = E_1^0 - E_2^0 + \left(\frac{RT}{nF}\right)\ln\left(\frac{[\mathrm{Ox}]_1{}^a[\mathrm{Red}]_2^b}{[\mathrm{Ox}]_2{}^d[\mathrm{Red}]_1{}^c}\right) \tag{6-11}$$

The equilibrium constant for the chemical reaction expressed by Equation 6-9 is related to the difference of the standard half-cell potentials by the relation

$$\ln K_{\mathrm{equil}} = \left(\frac{nF}{RT}\right)(E_1^0 - E_2^0) \tag{6-12}$$

To apply potentiometric measurements to the determination of the concentration of electroactive species a number of conditions have to be met. The basic measurement system must include an indicator electrode which is capable of monitoring the activity of the species of interest, and a reference electrode which gives a constant, known half-cell potential to which the indicator electrode potential can be referred. The voltage resulting from the combination of these two electrodes must be measured in a manner that minimizes the amount of current drawn by the measuring system. For low impedance electrode systems a conventional potentiometer is satisfactory. However, electrochemical measurements with high impedance electrode systems, and in particular the glass-membrane electrode, require the use of an exceedingly high input impedance measuring instrument (usually an electrometer amplifier with a current drain of less than 10^{-12} A). Because of the logarithmic nature of the Nernst equation, the measuring instrumentation must have considerable sensitivity. For example, a one-electron half reaction at 25°C gives a voltage change of 59.1 mV for a tenfold change in concentration of the electroactive species being measured. Another important point is that the potential response is directly dependent on the temperature of the measuring system. Thus if the correct temperature is not used in the Nernst expression, large absolute errors can be introduced in the measurement of the activity for electroactive species.

III. Electrode Systems

The indicating electrodes for potentiometric measurements traditionally have been categorized into three separate classes. "First-class" electrodes include those electrode systems which provide a direct response to the ion or species being measured; thus the primary electrode reaction includes the species being measured. Such electrodes give a direct response according to the Nernst equation for the logarithm of the activity of the species. The details of electrode fabrication and their characteristics are discussed in Chapter 2. An extensive authoritative treatment is provided by Ives and Janz (9).

Electrodes classified as "second-class" electrode systems are those in which the electrode is in direct contact with a slightly soluble salt of the electroactive species such that the potentiometric response is indicative of the concentration of the inactive anion species. Thus the silver-silver chloride electrode system, which is representative of this class of electrodes, gives a potential response that is directly related to the logarithm of the chloride ion activity, even though it is not the electroactive species. This is true because the chloride ion concentration, through the solubility product, controls the activity of the silver ion which is measured directly by the potentiometric silver electrode system.

"Third-class" electrodes are really a specialized case of second-class electrodes. They consist of the metal being in direct contact with a slightly soluble salt of the metal which is then used to monitor the activity of an electroinactive metal ion in equilibrium with a more soluble salt which includes the same anion as the electrode-salt system. For example, the concentration of calcium ions in equilibrium with solid calcium oxalate may be monitored using a silver-silver oxalate electrode system. The concentration of calcium ion affects the concentration of oxalate ion, which in turn controls the concentration of silver ion; the latter monitored by the potentiometric silver electrode system. A second and extremely useful example of a third-class electrode is the mercury-mercury(II) EDTA electrode system (10,11), which is used as a sensing system for the potentiometric titration of electroinactive metal ions with EDTA. Because the mercury(II) EDTA complex is one of the most stable of those encompassing the common divalent metal ions it is the least dissociated system. Hence when calcium ion is titrated with EDTA, the concentration of calcium ion controls the equilibrium concentration of the EDTA anion in solution, which in turn directly controls the free concentration of mercuric ion. The latter is monitored by a mercury electrode system and gives a direct measure of the calcium ion concentration. This type of system can be applied to most of the divalent ions that form moderately strong complexes with EDTA. It is extremely useful in developing potentiometric titration systems for mixtures of these ions.

Because any potentiometric electrode system ultimately must have a redox couple (or an ion exchange process in the case of membrane electrodes) for a meaningful response, the most common form of potentiometric electrode systems involves oxidation-reduction processes. Hence to monitor the activity of ferric ion, an excess of ferrous ion is added such that the concentration of this species remains constant to give a direct Nernstian response for the activity of ferric ion. For such redox couples the most common electrode system has been the platinum electrode. This tradition has come about primarily because of the historic belief that the platinum electrode is totally inert and involves only the pure metal as a surface. However, during the past decade it has become evident that platinum electrodes are not as inert as long believed and that their potentiometric response is frequently dependent on the history of the surface and the extent of its activation. The evidence is convincing that platinum electrodes, and in all probability all metal electrodes, are covered with an oxide film that changes its characteristics with time. Nonetheless, the platinum electrode continues to enjoy wide popularity as an "inert" indicator of redox reactions and of the activities of the ions involved in such reactions.

For many systems the gold electrode is as satisfactory as the platinum electrode. Both rhodium and palladium have been used for specialized

Table 6-2   Potentiometric Electrode Systems

Electrode	Couple	Application
$Pt/H_2, H^+$	$2H^+ + 2e^- = H_2$	pH; P_{H_2}
$Pt/M^{m+}, M^{(m-n)+}$	$M^{m+} + ne^- = M^{(m-n)+}$	$[M^{m+}]$; $[M^{(m-n)+}]$
M/M^{n+}	$M^{n+} + ne^- = M$	$[M^{n+}]$
$Ag/AgX/X^-$	$AgX + e^- = Ag + X^-$	$[X^-]$
$Hg/Hg_2Cl_2/Cl^-$	$Hg_2Cl_2 + 2e^- = 2Hg + 2Cl^-$	$[Cl^-]$
$Ag/Ag_2C_2O_4/CaC_2O_4/C_2O_4{}^{2-}$	$Ag_2C_2O_4 + 2e^- = 2Ag + C_2O_4{}^{2-}$	$[C_2O_4{}^{2-}]$; $[Ca^{2+}]$
	$Ca^{2+} + C_2O_4{}^{2-} = CaC_2O_4$	
$Hg/HgY^{2-}, M^{n+}$	$HgY^{2-} + 2e^- = Hg + Y^{4-}$	$[Y^{4-}]$; $[M^{n+}]$
	$M^{n+} + Y^{4-} = MY^{(n-4)}$	$Y = EDTA$

systems as "inert" potentiometric electrodes.

For those redox couples involving a metal ion plus the metal the logical electrode system is the metal itself. In other words, if the measured quantity is to be cupric ion a practical indicating electrode is a piece of copper metal. All second-class electrodes involve an active metal in combination with an insoluble compound or salt. Thus the silver-silver chloride electrode actually is a silver-silver ion electrode system which incorporates the means of controlling the silver ion concentration through chloride ion concentration. A related form of this is the antimony electrode which involves antimony and its oxide (an adherent film on the surface of the antimony metal electrode) such that the activity of antimony ion is controlled by the pH of the solution. This is an illustration of a second-class electrode being used to monitor hydrogen ion concentration (or more properly hydroxide ion concentration).

On occasion tungsten wire electrodes are used to monitor a redox reaction. These actually are oxide covered systems which respond to the activity of the species making up a redox couple; however, the response usually does not obey the Nernst equation.

A number of the most common potentiometric electrode systems and their application are summarized in Table 6-2. Additional information is available in the biennial reviews of *Analytical Chemistry* within the section on potentiometric measurements.

One of the most important and extensively used indicating electrode systems is the glass membrane electrode employed to monitor hydrogen ion activity. Although it was developed in 1919 it did not become popular until reliable electrometer amplifiers were developed in the 1930s. Figure 6-1 gives

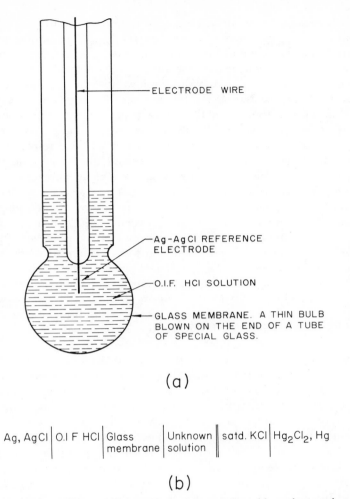

ELECTRODE WIRE

Ag-AgCl REFERENCE
ELECTRODE

O.I.F. HCl SOLUTION

GLASS MEMBRANE. A THIN BULB
BLOWN ON THE END OF A TUBE
OF SPECIAL GLASS.

(a)

Ag, AgCl | 0.l F HCl | Glass membrane | Unknown solution ‖ satd. KCl | Hg$_2$Cl$_2$, Hg

(b)

Figure 6-1. Glass electrode and its cell schematic in association with a reference electrode.

a schematic representation of this electrode and indicates that the primary electrode system is a silver-silver chloride (or mercury-mercurous chloride) electrode in contact with a known and fixed concentration of hydrochloric acid (usually about 0.1 F). When the outside surface of the glass membrane is exposed to an ionic solution a response for the hydrogen ion activity is obtained which follows the Nernst expression. Although there has been considerable debate concerning the mechanism of response for the glass electrode, the current thinking explains it in terms of an ion exchange process involving the hydroxyl groups on the surface of the glass. Thus the population of protons on the outside surface of the membrane affects the population on the inside of the membrane thereby generating a membrane potential which is indicated by the silver-silver chloride electrode. At one time there was a belief that hydrogen ions actually penetrated the glass membrane from the outside to the interior. However, experiments with labeled systems establish that this is not true. Further support for the ion exchange mechanism is provided by the realization that glass electrodes are not specific for hydrogen ions but only give a selective response. In other words, other metal ions, in particular sodium and lithium ion, cause a response from glass electrodes through an equivalent ion exchange process.

During the past decade the unwanted response of glass electrodes to metal ions, particularly alkali metal ions, has prompted the development of specialized glass membranes that have an enhanced selective response for metal ions. Electrodes are now available that give a selective response for sodium ion, potassium ion, silver ion, and the alkaline earth ions. Prototype electrodes currently are being developed which appear to give a fairly selective response for calcium ion and magnesium ion. With additional understanding of the response mechanism and the characteristics of glass membranes a whole class of selectively responsive glass membrane electrodes probably will be developed during the next decade. Details of such electrodes are given in Chapter 2.

A parallel approach during the past 5 to 10 years has been the development of ion exchange membranes that give a selective response to specific cations and anions. Because of the extensive literature on ion exchange resins and the improved control in the preparation of polymeric resin membranes this form of electrode shows great promise and has had extensive application in recent years. Again, Chapter 2 provides a detailed discussion.

A related form of the membrane electrode is the inorganic crystal electrode. Of the many being considered in current work the lanthanum fluoride (LaF_3) electrode for monitoring fluoride ion concentration has received the most attention and has become the standard sensor for fluoride concentrations in solution. It represents a combination of a membrane electrode with ion exchange characteristics and an incomplete form of a second-class electrode in which lanthanum fluoride is the insoluble material giving a response indicative of the free fluoride ion concentration in the test sample. This and a number of other prototype inorganic crystal sensing electrodes are discussed in Chapter 2.

In general a necessary part of a potentiometric measurement is the coupling of a reference electrode to the indicating electrode. The ideal reference electrode has a number of important characteristics including a reproducible potential, a low temperature coefficient, the capacity to remain unpolarized when small currents are drawn, and inertness to the sample solution. If the reference electrode must be prepared in the laboratory a convenient and reproducible system is desirable.

Although the standard half-cell reactions are all referenced to the hydrogen electrode, this is an exceedingly awkward reference electrode. It has been selected because it falls in the middle of the most common half reactions using water as a solvent and because with rigorous care highly reproducible potentials can be duplicated by equally careful workers in other laboratories. Furthermore, because it consists of a platinized platinum electrode over which hydrogen gas is passed in combination with a known activity of hydrogen ion, it can be combined in a number of cases with other half cells without a liquid junction. These three factors undoubtedly justify its selection as the ultimate reference electrode for fundamental measurements. Figure 2-8 illustrates a common form of the hydrogen electrode. The electrode reaction is a typical redox half reaction which includes the oxidized and reduced forms of hydrogen. By controlling the hydrogen partial pressure at a fixed level this becomes an indicating electrode for hydrogen ion. Conversely by controlling the activity of hydrogen ion in the sample solution one might suppose that this electrode would indicate partial pressures of dissolved hydrogen. Although in principle this is true the platinized platinum surface gives a very slow response to changes in hydrogen partial pressure; this is a major reason for using a platinized electrode when its main function is to monitor hydrogen ion activity.

For most potentiometric measurements either the saturated calomel reference electrode or the silver-silver chloride reference electrode are used. These electrodes can be made compact, are easily produced, and provide reference potentials that do not vary more than a few millivolts. The discussion in Chapter 2 outlines their characteristics, preparation, and temperature coefficients. The silver-silver chloride electrode also finds application in nonaqueous titrations, although some solvents cause the silver chloride film to become soluble. In addition electrodes involving zinc and silver couples have been proposed for use in nonaqueous solvents. From our own experience, aqueous reference electrodes are as convenient for nonaqueous systems as any of the prototypes that have been developed to date, except when there is a need to rigorously exclude water. This is true although they involve a liquid junction between the aqueous electrolyte system and the nonaqueous solvent system of the sample solution. The use of conventional reference electrodes does cause some difficulties if the electrolyte of the reference electrode is insoluble in the sample solution. Hence the use of a calomel electrode saturated with potassium chloride in conjunction with a sample solution containing perchlorate ion can cause erratic measurements due to the precipitation of potassium perchlorate at the junction.

Such difficulties normally can be eliminated by using a double junction which inserts another inert electrolyte solution between the reference electrode and the sample solutions (e.g., a sodium chloride solution).

For measurement of redox couples, a frequently overlooked but convenient reference electrode is a conventional glass pH electrode. Assuming the sample solution system contains a constant level of acidity this provides an extremely inert and stable reference electrode which is completely indifferent to most redox species. Naturally such a reference electrode requires the use of an electrometer amplifier such as is contained in pH meters.

Where potentiometric measurements are used to monitor a titration rather than being used as an absolute measure of a constituent's activity, a number of other monitoring systems based on potentiometric principles are possible. For example, if a platinum electrode is coupled with a tungsten wire sharp potential breaks are generally observed which coincide with the equivalence point for a redox titration. This undoubtedly comes about because the platinum electrode behaves in a reversible Nernstian fashion while the tungsten electrode tends to be irreversible and unresponsive to changes in the activity of redox ions. Many other bimetallic pairs have been used but the platinum-tungsten combination is the most popular and provides a highly inert electrode pair. Another approach for monitoring potentiometric titrations is to obtain a differential type measurement by using two identical electrodes. For example, a pair of platinum electrodes with one of them shielded from the sample solution when an increment of titrant is added results in a differential potential step per increment, which becomes a maximum at the equivalence point. This can be accomplished by inserting one of the platinum electrodes in a shielded compartment such that mixing with the sample solution is slow. An alternative form is to immerse the tip of the buret in the sample solution and have one of the electrodes inserted in the bore of the tip itself. An elegant version of the differential titration system is the Pinkhoff-Treadwell method which uses as a reference electrode an indicator electrode in a solution with the exact composition that will exist at the equivalence point of the titration. This is connected by a salt bridge to the sample solution where an identical indicator electrode is used to monitor the sample. The measuring circuit is simply a sensitive galvanometer or voltmeter which monitors the signal and indicates when it becomes zero at the equivalence point. If current is being measured, the direction of current flow will change at the equivalence point. Extremely precise results are possible under ideal conditions with this very simple measuring system. Figure 6-2 illustrates a differential titration system which utilizes a medicine dropper.

Another approach to differential measurements is the use of two identical

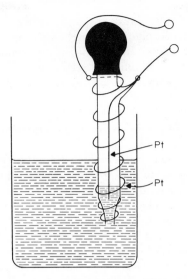

Figure 6-2. Differential potentiometric electrode system for titrations. After each addition of titrant and reading the solution in the dropper is exchanged with the bulk solution.

electrodes in the sample solution with a small polarizing current passed between them, usually 5 or 10 μA. Most modern pH meters have provision for a polarizing current built into them and only require the appropriate connections to be made at the back of the meter. With dual polarized electrodes (using a constant current) there generally is a sharp change in potential at the equivalence point. This results because either a reversible couple is destroyed during the course of the titration such that a much higher potential is generated at one of the electrodes or there is an absence of a reversible couple for one of the electrodes until the equivalence point is reached. A common application of the dual polarized potentiometric indicating system is in the Karl Fischer titration. Up to the equivalence point, there is a large potential on the indicating meter which decreases sharply at the equivalence point because an excess of iodine is introduced into the sample solution. The $I_2 - I^-$ couple is reversible and the anode clearly has plenty of iodide available to be oxidized to iodine prior to the equivalence point. However, the cathode reaction cannot be the reduction of iodine to iodide because none is available prior to the equivalence point. Hence to pass the small current some other species must be reduced; in general this is the solvent which requires a high potential. When an excess of iodine is introduced into the sample solution, the potential drops to almost zero to provide a dramatic indication of the equivalence point. Undoubtedly dual polarized potentiometric endpoint detection systems should be used more widely than they have been. Unfortunately predictability is limited and one must try each system to establish whether the method is applicable. Other examples of their application are discussed in Chapter 9.

IV. Application to Potentiometry

Although potentiometric measurements frequently are applied as a means of endpoint detection in potentiometric titrations the purpose of this volume is not to review the many titration procedures that have been developed utilizing potentiometric measurements. However, a brief summary of the species that can be monitored through the use of potentiometric measurements will illustrate the potential applications.

Although all potentiometric measurements (except those involving membrane electrodes) ultimately are based on a redox couple the method can be applied to oxidation-reduction processes, acid-base processes, precipitation processes, and metal ion-complexation processes. Measurements which involve a component of a redox require that either the oxidized or reduced conjugate of the species to be measured must be maintained at a constant and known activity at the electrode. If the goal is to measure the activity of silver ion in a solution, then a silver wire coupled to the appropriate reference electrode makes an ideal potentiometric indicating system. Likewise if the goal is to monitor ferric ion concentration with a platinum electrode a known concentration of ferrous ion must be present in the sample solution such that potential changes only are dependent on the ferric ion concentration.

Table 6-1 summarizes a number of redox couples that are well-behaved in aqueous solutions and provide a means for monitoring the indicated species by potentiometric measurements. This can be either in the form of monitoring a titration or as a direct absolute measurement of activity. Although the tabulations of Latimer (8) infer that the listing should be much more comprehensive, most of the couples tabulated in Latimer are not well-behaved in an electrochemical sense and do not provide a Nerstian response under normal laboratory conditions. The vast majority of the data tabulated by Latimer is based on other than electrochemical measurements.

To date our knowledge of redox systems in nonaqueous solvents is severely limited. Undoubtedly a number of well behaved systems are amenable to potentiometric measurements. However, data are just beginning to appear in the literature which can serve as an indication of where potentiometric techniques might appropriately be applied.

The major application of the potentiometric method is for acid-base measurements, both in aqueous and nonaqueous solvent systems. Although the glass electrode is universally the most common indicating electrode system for such measurements, many other electrodes have been developed. However, except in extremely specialized circumstances none of these provides the reliability and precision that is afforded by the glass electrode. In the absence of interfering substances the quinhydrone electrode (an equimolar combination of quinone and hydroquinone with a gold foil electrode) provides a simple monitoring system for the measurement of pH up to pH 8. However, the presence of oxidizing or strongly reducing ions in the sample system will interfere, as is true for the hydrogen gas electrode and most other systems that are an alternative to the glass electrode. Reference to Table 6-1 indicates that a number of redox couples involve hydrogen ion. Any of these can be used if the oxidized and reduced species are introduced into the sample solution in controlled amounts. For such conditions the response of the electrode will be dependent on the hydrogen ion activity. The iodate-iodide couple is an example of such a system.

For adverse conditions (either in terms of temperature or vibration) the antimony electrode has proven useful, particularly for industrial processes with extreme environmental problems. The electrode is not particularly reliable for precise measurements but its simple form (consisting of antimony metal imbedded in an insulating material) allows pH measurements under such adverse conditions. The principle of the electrode is based on a half reaction whereby the metal and its metal oxide are both insoluble and the electrode's response is dependent on hydrogen ion activity

$$Sb_2O_3 + 6H^+ + 6e^- \rightarrow 2Sb + 3H_2O \qquad (6\text{-}13)$$

pH measurements in nonaqueous solvents almost without exception use the glass electrode in combination with an appropriate reference electrode, frequently the silver-silver chloride electrode. In general the response of the glass electrode follows the Nernst expression in nonaqueous solvents and is an accurate representation of the changes in activity of hydrogen ion. Unfortunately few if any standard buffers are available for calibrating the pH meters that are used with glass electrodes for nonaqueous measurements. Thus nonaqueous pH measurements are only meaningful for monitoring the course of an acid-base titration or relative to some reference measurement made within the individual laboratory. Little if any confidence can be attached to absolute pH measurements in nonaqueous systems.

A particularly important application of the glass electrode is the carbon dioxide electrode. This is a self-contained system with a glass electrode and a concentric silver-silver chloride electrode enclosed by a CO_2 permeable membrane. The latter holds a thin film of bicarbonate solution in contact

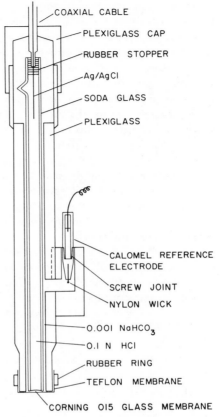

COAXIAL CABLE

PLEXIGLASS CAP

RUBBER STOPPER

Ag/AgCl

SODA GLASS

PLEXIGLASS

CALOMEL REFERENCE
ELECTRODE

SCREW JOINT

NYLON WICK

0.001 NaHCO₃

0.1 N HCl

RUBBER RING

TEFLON MEMBRANE

CORNING OI5 GLASS MEMBRANE

Figure 6-3. Potentiometric CO_2 electrode. Response proportional to pCO_2 ($-\log[CO_2]$).

with the glass membrane, which provides a junction to the silver-silver chloride reference electrode. The electrode, which is illustrated schematically by Figure 6-3, has found extensive application in monitoring pCO_2 levels in blood and probably will find increasing application in other systems requiring continuous measurement of partial pressures of carbon dioxide. The electrode response is based on the reaction

$$CO_2 + H_2O \rightarrow H^+ + HCO_3^- \qquad (6\text{-}14)$$

such that changes in the partial pressure of carbon dioxide cause an attendant change in the concentration of hydrogen ion which provides a direct potentiometric response for the partial pressure of carbon dioxide. Other monitoring electrode systems should be possible which are based on similar processes. For example, an ammonium electrode might well be developed with the converse of the reactions indicated for the CO_2 electrode. Thus an ammonium ion electrolyte would be used such that that changes in pH would be proportional to changes in the partial pressure of ammonia.

$$NH_3 + H_2O \rightarrow NH_4^+ + OH^- \tag{6-15}$$

Second-class electrodes, that is those whose response is dependent on the change in concentration of an anion which gives an insoluble salt with the metal ion of the indicator electrode, provide a general means for monitoring concentrations of anions. Table 6-3 summarizes those half reactions which

Table 6-3   Electromotive Series for Electrodes of the Second Class at 25°C and 1 atm

Reaction	$E°$, V vs. NHE
$PbO + H_2O + 2e^- \rightarrow Pb + 2OH^-$	-0.578
$PbSO_4 + 2e^- \rightarrow Pb + SO_4^{2-}$	-0.276
$AgI + e^- \rightarrow Ag + I^-$	-0.152
$Hg_2I_2 + 2e^- \rightarrow 2Hg + 2I^-$	-0.040
$AgBr + e^- \rightarrow Ag + Br^-$	$+0.071$
$AgSCN + e^- \rightarrow Ag + SCN^-$	$+0.095$
$HgO + H_2O + 2e^- \rightarrow Hg + 2OH^-$	$+0.097$
$Hg_2Br_2 + 2e^- \rightarrow 2Hg + 2Br^-$	$+0.140$
$AgCl + e^- \rightarrow Ag + Cl^-$	$+0.222$
$Hg_2Cl_2 + 2e^- \rightarrow 2Hg + 2Cl^-$	$+0.268$
$AgN_3 + e^- \rightarrow Ag + N_3^-$	$+0.292$
$Hg_2SO_4 + 2e^- \rightarrow 2Hg + SO_4^{2-}$	$+0.614$
$PbO_2 + 4H_3O^+ + SO_4^{2-} + 2e^- \rightarrow PbSO_4 + 6H_2O$	$+1.685$

are well-behaved electrochemically and provide a means for the potentiometric monitoring of anion species. This table also includes a tabulation of redox reactions which are useful for monitoring the concentration of ligands that are capable of complexing metal ions. Consideration of these indicates one of the difficulties with absolute potentiometric measurements as a measure of metal ion activity. If one is concerned purely with the actual activity of free metal ion these measurements are meaningful. However, if the measurement is taken to represent the total metal ion content of the solution, both as a free ion and as its various complexes, then highly erroneous conclusions can be made. Thus the ability to monitor the concentration of ligands should be recognized as a pitfall if one does not take account of this in the use of potential measurements for monitoring concentrations of metal ions.

A recent and rapidly developing extension of potentiometry is in the area of membrane-type indicator electrodes. These include specialized glass electrodes that respond to ions other than the hydrogen ion, ion exchange membranes, and single inorganic crystal membranes. Each year the selectivity and reliability are improved for this important class of electrode. In particular the development of ion exchange membranes that provide selective response for a number of anions has made new areas of analysis amenable to potentiometric measurements. This has been particularly important for the biomedical field where nondestructive, highly specific potentiometric measurements are desirable. Furthermore, the potentiometric method, because of its continuous nature, is particularly attractive to those concerned with *in vivo* monitoring of biological substances.

Table 2-11 summarizes the presently available electrodes categorized as glass, ion exchange membrane, and crystal membrane. These electrodes can be used either for direct potentiometric measurements of ionic activity after calibration of the Nernst expression for the particular electrode or they may be used to monitor a potentiometric titration when a selected reaction involving the monitored ion is available. Table 2-11 also indicates the common interfering ions. Many instrument companies currently are endeavoring to develop potentiometric membrane electrodes for direct monitoring of biologically important ions in the body fluids.

Because potentiometry (through the Nernst equation) gives a response that is proportional to the logarithm of the activity of the electroactive ion, the accuracy and precision are more limited than for many methods which give a directly proportional response. Thus for a one-electron redox process an order of magnitude change in activity gives a potential change of 59.1 mV (at room temperature), a 10% change in activity gives a change of 2.5 mV, and a 1% change in activity gives only a 0.25-mV change in potential. This has prompted many efforts to improve the accuracy of potentiometry and has lead to the development of differential methods of various types. One of the most effective of these is the method of Professor Howard Malmstadt which utilizes two identical indicator electrodes with one of them immersed in the sample solution and the second immersed in a reference solution containing pure solvent and an inert electrolyte. To make the measurement a sensitive voltage measuring device is attached to the two indicator electrodes and a standarized solution of the constituent species is added to the reference half-cell with a precision micrometer syringe until the indicated voltage reaches a null point. By knowing the concentration of the added constituent and the volume it is possible to compute the concentration

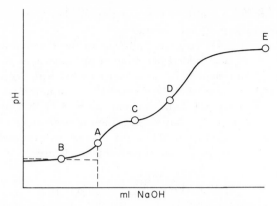

Figure 6-4. Titration curve for a polyprotic acid titrated with NaOH

of the constituent ion in the unknown solution. This approach allows very dilute sample systems to be analyzed and the level of precision frequently is higher than by direct potentiometric measurements.

Potentiometry has found extensive application over the past half century as a means of evaluating various thermodynamic parameters. Although this is not the major application of the technique today it still provides one of the most convenient and reliable approaches to the evaluation of thermodynamic quantities. In particular the activity coefficients of electroactive species can be evaluated directly through the use of the Nernst equation, assuming the species being measured represents a reversible electrochemical reaction. Thus if an electrochemical system is used without a junction potential and with a reference electrode that has a well-established potential then potentiometric measurement of the constituent species at a known concentration provides a direct measure of its activity. This provides a direct means for evaluating the activity coefficient (assuming the standard potential is known accurately for the constituent half reaction). If the standard half-reaction potential is not available, it must be evaluated under conditions where the activity coefficient can be determined by the Debye-Huckel equation.

Another important application is the use of potentiometric measurements for the evaluation of thermodynamic equilibrium constants. In particular the dissociation constants for weak acids and weak bases in a variety of solvents are evaluated conveniently with a pH electrode measuring system. The most precise approach for evaluating this type of equilibrium constant is to perform an acid-base titration such that the titration curve can be recorded. Figure 6-4 illustrates the titration curve for a polyprotic acid which has two breaks; actually it is for orthophosphoric acid and illustrates the principles of evaluating such constants.

Point A on the titration curve is half-way to the first equivalence point with one-half mole of base added per mole of phosphoric acid. If the pH at this point were greater than pH 5 and less than pH 9 the concentration of H_3PO_4 and of $H_2PO_4^-$ ion could be assumed to be equal. However, because the pH is approximately pH 3 the concentration of H_3PO_4 at this point is less than that of the $H_2PO_4^-$ ion by an amount equivalent to about twice the hydrogen ion concentration. Thus the correct expressions at the half-equivalence point for the equilibrium concentrations are

$$[H_2PO_4^-] = \frac{N_{OH^-}V_{OH^-}}{v + V_{OH^-}} + [H^+];$$

$$[H_3PO_4] = \frac{N_{OH^-}V_{OH^-}}{v + V_{OH^-}} - [H^+] \qquad (6\text{-}16)$$

and

$$K_1 = \frac{[H^+][H_2PO_4^-]}{[H_3PO_4]}$$

$$= [H^+]\left[\frac{N_{OH^-}V_{OH^-}}{v + V_{OH^-}} + [H^+]\right] \Bigg/ \left[\frac{N_{OH^-}V_{OH^-}}{v + V_{OH^-}} - [H^+]\right] \qquad (6\text{-}17)$$

where N_{OH^-} is the normality of the base, V_{OH^-} its volume, and v the initial sample volume.

At point B on the titration curve one and one-half moles of base have been added per mole of phosphoric acid and the pH indicates that the concentration of $H_2PO_4^-$ ion equals the concentration of HPO_4^{2-} ion. Thus the correct expression for the equilibrium constant at this point is

$$K_2 = \frac{[H^+][HPO_4^{2-}]}{[H_2PO_4^-]} = [H^+] \qquad (6\text{-}18)$$

At point D on the titration curve two and one-half moles of base per mole of phosphoric acid have been added. Again, it is erroneous to assume that the PO_4^{3-} ion concentration equals the HPO_4^{2-} ion concentration at this half-equivalence point because a significant fraction of the base added beyond the second equivalence point (C on Figure 6-4) has not reacted with the HPO_4^{2-} ion. Thus the proper expression for the third dissociation constant for phosphoric acid is given by the relations

$$[PO_4{}^{3-}] = \frac{N_{OH^-}[V_{OH^-} - (V_{OH^-})_c]}{V_{OH^-} + v} - [OH^-] \qquad (6\text{-}19)$$

$$[HPO_4{}^{2-}] = \frac{N_{OH^-}[V_{OH^-} - (V_{OH^-})_c]}{V_{OH^-} + v} + [OH^-] \qquad (6\text{-}20)$$

$$K_3 = \frac{[H^+][PO_4{}^{3-}]}{[HPO_4{}^{2-}]}$$

$$= [H^+]\left[\frac{N_{OH^-}[V_{OH^-} - (V_{OH^-})_c]}{V_{OH^-} + v} - [OH^-]\right] \Bigg/$$

$$\left[\frac{N_{OH^-}[V_{OH^-} - (V_{OH^-})_c]}{V_{OH^-} + v} + [OH^-]\right] \qquad (6\text{-}21)$$

Similar expressions can be developed for other weak acids and weak bases to permit the evaluation of their dissociation constants. The constants normally are evaluated at points on the titration curve where the pH is changing slowly relative to the added titrant; equivalence points are to be avoided because of the significant experimental errors that are possible at this point on the titration curve.

Obviously one could measure the pH of a known concentration of a weak acid and obtain a value of its hydrogen ion activity which would permit a direct evaluation of its dissociation constant. However, this would be a one point evaluation and subject to greater errors than by titrating the acid halfway to the equivalence point. The latter approach uses a well-buffered region where the pH measurement represents the average of a large number of data points. Similar arguments can be made for the evaluation of solubility products and stability constants of complex ions. The appropriate expression for the evaluation of solubility products again is based on the half-equivalence point of the titration curve for the particular precipitation reaction (Ag^+ represents the titrant)

$$Ag^+ + Cl^- \rightarrow \underline{AgCl} \qquad (6\text{-}22)$$

$$[Cl^-] = \frac{N_{Cl^-}V_{Cl^-}}{V_{Cl^-} + v} + [Ag^+] \qquad (6\text{-}23)$$

$$K_{sp} = [\text{Ag}^+][\text{Cl}^-] = [\text{Ag}^+]\left[\frac{N_{\text{Cl}^-}V_{\text{Cl}^-}}{V_{\text{Cl}^-} + v} + [\text{Ag}^+]\right] \tag{6-24}$$

In the use of potentiometry for the evaluation of stability constants for complex ions the expressions can become extremely complicated if multi-equilibria are present. For a simple one-to-one complex a direct potentiometric titration curve again provides the most satisfactory route to an accurate evaluation of the constant. The curve looks similar to that for an acid-base titration and the appropriate point to pick is the half-equivalence point. If the complex is extremely stable, then the amount of free metal ion at this point on the titration curve (ligand titrated with metal ion) is sufficiently low that it can be disregarded. If not it must be handled in a way similar to the first point on the titration curve for phosphoric acid. Assuming it is a stable complex, at the first half-equivalence point the concentration of complexed metal ion will be equivalent to that of the free ligand. The potential will give a direct measure of the free metal ion and allow the stability constant for the complex to be evaluated at the half-equivalence point

$$[\text{Ag(en)}^+] = \frac{N_{\text{Ag}^+}V_{\text{Ag}^+}}{V_{\text{Ag}^+} + v} - [\text{Ag}^+] \tag{6-25}$$

$$[\text{en}] = \frac{N_{\text{Ag}^+}V_{\text{Ag}^+}}{V_{\text{Ag}^+} + v} + [\text{Ag}^+] \tag{6-26}$$

and

$$K_f = \frac{[\text{Ag(en)}^+]}{[\text{Ag}^+][\text{en}]} \tag{6-27}$$

For multistep complexation reactions and for ligands that are themselves weak acids extremely involved calculations are necessary for the evaluation of the equilibrium expression from the individual species involved in the competing equilibria. These normally have to be solved by a graphical method or by computer techniques. Their discussion at this point is beyond the scope of this book. However, those who are interested will find adequate discussions in the many books on coordination chemistry, chelate chemistry and the study and evaluation of the stability constants of complex ions (10–12). The general approach is the same as outlined here; namely, that a titration curve is performed in which the concentration or activity of the substituent species is monitored by potentiometric measurement.

Table 6-4 Formal Potentials

Half-Reaction	E', V vs. NHE	Solution Composition
$Ag^+ + e^- = Ag$	$+0.792$	$1\ F\ HClO_4$
	$+0.770$	$1\ F\ H_2SO_4$
$Ag^{2+} + e^- = Ag^+$	$+2.000$	$4\ F\ HClO_4$
	$+1.930$	$4\ F\ HNO_3$
$AgI + e^- = Ag + I^-$	-0.137	$1\ F\ KI$
$H_3AsO_4 + 2H^+ + 2e^- = HAsO_2 +$		
$2H_2O$	$+0.577$	$1\ F\ HCl\ or\ HClO_4$
$AuCl_2^- + e^- = Au + 2Cl^-$	$+1.110$	$1\ F\ Cl^-$
$AuCl_4^- + 2e^- = AuCl_2^- + 2Cl^-$	$+0.930$	$1\ F\ HCl$
$Ce(IV) + e^- = Ce(III)$	$+0.060$	$2.5\ F\ K_2CO_3$
	$+1.280$	$1\ F\ HCl$
	$+1.700$	$1\ F\ HClO_4$
	$+1.600$	$1\ F\ HNO_3$
	$+1.440$	$1\ F\ H_2SO_4$
$Co^{3+} + e^- = Co^{2+}$	$+1.850$	$4\ F\ HNO_3$
	$+1.820$	$8\ F\ H_2SO_4$
$Co(en)_3^{3+} + e^- = Co(en)_3^{2+}$	-0.200	$0.1\ F\ en + 0.1\ F\ KNO_3$
$Cr(III) + e^- = Cr(II)$	-0.260	Satd. $CaCl_2$
	-0.400	$5\ F\ HCl$
	-0.370	$0.1{-}0.5\ F\ H_2SO_4$
$Cr(CN)_6^{3-} + e^- = Cr(CN)_6^{4-}$	-1.130	$1\ F\ KCN$
$CrO_4^{2-} + 2H_2O + 3e^- = CrO_2^- + 4OH^-$	-0.120	$1\ F\ NaOH$
$Cr_2O_7^{2-} + 14H^+ + 6e^- = 2Cr^{3+} +$		
$7H_2O$	$+0.930$	$0.1\ F\ HCl$
	$+1.000$	$1\ F\ HCl$
	$+1.080$	$3\ F\ HCl$
	$+0.840$	$0.1\ F\ HClO_4$
	$+1.025$	$1\ F\ HClO_4$
	$+0.920$	$0.1\ F\ H_2SO_4$
	$+1.150$	$4\ F\ H_2SO_4$
$Cu(CN)_3^{2-} + e^- = Cu + 3CN^-$	-1.000	$7\ F\ KCN$
$CuCl_3^- + e^- = Cu + 3Cl^-$	$+0.178$	$1\ F\ HCl$
$Cu(II) + e^- = Cu(I)$	$+0.010$	$1\ F\ NH_3 + 1\ F\ NH_4^+$
	$+0.520$	$1\ F\ KBr$
	$+0.300$	$0.1\ F\ py + 0.1\ F\ pyH^+$
$Cu(C_2O_4)_2^{2-} + 2e^- = Cu + 2C_2O_4^{2-}$	$+0.060$	$1\ F\ K_2C_2O_4$
$Cu(EDTA)^{2-} + 2e^- = Cu + EDTA^{4-}$	$+0.130$	$0.1\ F\ EDTA$, pH 4–5
$Eu(III) + e^- = Eu(II)$	-0.430	$0.1\ F\ HCOOH$
	-0.920	$0.1\ F\ EDTA$, pH 6–8
$Fe(III) + e^- = Fe(II)$	$+0.710$	$0.5\ F\ HCl$

Table 6-4 (*Continued*)

Half-Reaction	E', V vs. NHE	Solution Composition
	+0.640	5 F HCl
	+0.530	10 F HCl
	−0.680	10 F NaOH
	+0.735	1 F HClO$_4$
	+0.010	1 F K$_2$C$_2$O$_4$, pH 5
	+0.460	2 F H$_3$PO$_4$
	+0.680	1 F H$_2$SO$_4$
	+0.070	0.5 F Na$_2$Tart, pH 5–6
$Fe(CN)_6^{3-} + e^- = Fe(CN)_6^{4-}$	+0.560	0.1 F HCl
	+0.710	1 F HCl
	+0.720	1 F HClO$_4$
$FeO_4^{2-} + 2H_2O + 3e^- = FeO_2^- + 4OH^-$	+0.550	10 F NaOH
$Fe(EDTA)^- + e^- = Fe(EDTA)^{2-}$	+0.120	0.1 F EDTA, pH 4–6
$2H^+ + 2e^- = H_2$	+0.005	1 F HCl or HClO$_4$
$Hg_2^{2+} + 2e^- = 2Hg$	+0.776	1 F HClO$_4$
$I_3^- + 2e^- = 3I^-$	+0.545	0.5 F H$_2$SO$_4$
$2ICl_2^- + 2e^- = I_2 + 4Cl^-$	+1.060	1 F HCl
$IrCl_6^{2-} + e^- = IrCl_6^{3-}$	+1.020	1 F HCl
$Mn(CN)_6^{4-} + e^- = Mn(CN)_6^{5-}$	−1.080	1 F NaCN
$Mn(III) + e^- = Mn(II)$	+1.500	7.5 F H$_2$SO$_4$
	−0.244	1.5 F NaCN
$Mo(IV) + e^- = Mo(III)$	+0.100	4.5 F H$_2$SO$_4$
$Mo(CN)_8^{3-} + e^- = Mo(CN)_8^{4-}$	+0.800	0.25 F KBr, KCl, or KNO$_3$
$Mo(V) + 2e^- = Mo(III)$(green)	−0.250	2 F HCl
(red)	+0.110	2 F HCl
$Mo(VI) + e^- = Mo(V)$	+0.530	2 F HCl
	+0.700	8 F HCl
	+0.500	2 F KSCN + 1 F HCl
$NO_3^- + 3H^+ + 2e^- = HNO_2 + H_2O$	+0.920	1 F HNO$_3$
$NbO^{3+} + 2H^+ + 2e^- = Nb^{3+} + H_2O$	−0.340	2–6 F HCl or 1.5–3 F H$_2$SO$_4$
$Nb(V) + e^- = Nb(IV)$	−0.210	12 F HCl
$Ni(CN)_4^{2-} + e^- = Ni(CN)_4^{3-}$	−0.820	1 F KCN
$Np(IV) + e^- = Np(III)$	+0.140	1 F HCl
	+0.155	1 F HClO$_4$
$Np(V) + e^- = Np(IV)$	+0.739	1 F HClO$_4$
$NpO_2^{2+} + e^- = NpO_2^+$	+1.140	1 F HCl
	+1.137	1 F HClO$_4$
$Os(IV) + e^- = Os(III)$	+0.350	2 F HBr
$Os(VI) + 2e^- = Os(IV)$	+0.660	0.5 F HCl
	+0.840	5 F HCl

Table 6-4 *(Continued)*

Half-Reaction	E', V vs. NHE	Solution Composition
	$+0.970$	9 F HCl
$Os(VIII) + 4e^- = Os(IV)$	$+0.790$	5 F HCl
$Pb(II) + 2e^- = Pb$	-0.320	1 F NaOAc
$PbO_3^{2-} + H_2O + 2e^- = PbO_2^{2-} + 2OH^-$	$+0.210$	8 F KOH
$PbSO_4 + 2e^- = Pb + SO_4^{2-}$	-0.290	1 F H$_2$SO$_4$
$Pd^{2+} + 2e^- = Pd$	$+0.987$	4 F HClO$_4$
$PdBr_6^{2-} + 2e^- = PdBr_4^{2-} + 2Br^-$	$+0.990$	1 F KBr
$PdI_6^{2-} + 2e^- = PdI_4^{2-} + 2I^-$	$+0.480$	1 F KI
$Po(IV) + 4e^- = Po$	$+0.600$	1 F HCl
	$+0.800$	1 F HNO$_3$
$Po(IV) + 2e^- = Po(II)$	$+0.700$	1 F HCl
$PtBr_6^{2-} + 2e^- = PtBr_4^{2-} + 2Br^-$	$+0.640$	1 F NaBr
$PtCl_6^{2-} + 2e^- = PtCl_4^{2-} + 2Cl^-$	$+0.720$	1 F NaCl
$PtI_6^{2-} + 2e^- = PtI_4^{2-} + 2I^-$	$+0.390$	1 F NaI
$Pt(SCN)_6^{2-} + 2e^- = Pt(SCN)_4^{2-} + 2SCN^-$	$+0.470$	1 F NaSCN
$Pu(IV) + e^- = Pu(III)$	$+0.400$	1 F HOAc + 1 F NaOAc
	$+0.970$	1 F HCl
	$+0.500$	1 F HF
	$+0.920$	1 F HNO$_3$
	$+0.590$	0.6 F H$_3$PO$_4$ + 1 F HCl
	$+0.750$	1 F H$_2$SO$_4$
$PuO_2^{2+} + 4H^+ + 2e^- = Pu^{4+} + 2H_2O$	$+1.050$	1 F HCl
	$+1.040$	1 F HClO$_4$
$PuO_2^{2+} + e^- = PuO_2^+$	$+0.916$	1 F HClO$_4$
p-Benzoquinone + 2H$^+$ + $2e^-$ = hydro- quinone	$+0.696$	1 F HCl or HClO$_4$
$Re^+ + 2e^- = Re^-$	-0.230	0.4–2 F H$_2$SO$_4$
$ReCl_6^{2-} + e^- = ReCl_4^- + 2Cl^-$	-0.250	1 F HCl
$Re(V) + 2e^- = Re(III)$	$+0.140$	2 F NaCN
$Rh(IV) + e^- = Rh(III)$	$+1.430$	0.5 F H$_2$SO$_4$
$Rh(VI) + 2e^- = Rh(IV)$	$+1.500$	0.1 F H$_2$SO$_4$
$Ru(CN)_6^{3-} + e^- = Ru(CN)_6^{4-}$	$+0.800$	0.05 F H$_2$SO$_4$
$Ru(IV) + e^- = Ru(III)$	$+0.910$	0.5 F HCl
	$+0.860$	2 F HCl
$SO_4^{2-} + 4H^+ + 2e^- = SO_2 + 2H_2O$	$+0.070$	1 F H$_2$SO$_4$
$SbO_2^- + 2H_2O + 3e^- = Sb + 4OH^-$	$+0.675$	10 F KOH
$Sb(V) + 2e^- = Sb(III)$	$+0.750$	3.5 F HCl
	$+0.820$	6 F HCl
$SbO_3^- + H_2O + 2e^- = SbO_2^- + 2OH^-$	-0.589	10 F NaOH
$SnCl_4^{2-} + 2e^- = Sn + 4Cl^-$	-0.190	1 F HCl

Table 6-4   (*Continued*)

Half-Reaction	E', V vs. NHE	Solution Composition
$Sn(IV) + 2e^- = Sn(II)$	+0.140	1 F HCl
	+0.130	2 F HCl
$Te(IV) + 4e^- = Te$	+0.560	2 F HCl
$TiOCl^+ + 2H^+ + 3Cl^- + e^- = TiCl_4^- + H_2O$	−0.090	1 F HCl
	+0.240	6 F HCl
$Ti(IV) + e^- = Ti(III)$	−0.050	1 F H_3PO_4
	−0.150	5 F H_3PO_4
	−0.240	0.1 F KSCN
	−0.010	0.2 F H_2SO_4
	+0.120	2 F H_2SO_4
	+0.200	4 F H_2SO_4
$Tl(III) + 2e^- = Tl^+$	+0.890	0.1 F HCl + 0.9 F HClO$_4$
	+0.780	1 F HCl
$U(IV) + e^- = U(III)$	−0.640	1 F HCl
	−0.630	1 F HClO$_4$
$UO_2^{2+} + 4H^+ + 2e^- = U(IV) + 2H_2O$	+0.410	0.5 F H_2SO_4
$UO_2^{2+} + e^- = UO_2^+$	+0.062	0.1 F Cl$^-$
$V(EDTA)^- + e^- = V(EDTA)^{2-}$	−1.020	0.001–0.02 F EDTA
$V(III) + e^- = V(II)$	−0.217	0.1–1 F NH$_4$SCN
$VO^{2+} + 2H^+ + e^- = V^{3+} + H_2O$	+0.360	1 F H_2SO_4
$W(V) + 2e^- = W(III)$ (green)	+0.100	12 F HCl
(red)	−0.200	12 F HCl
$W(V) + e^- = W(IV)$	−0.300	12 F HCl
$W(VI) + e^- = W(V)$	+0.260	Concd. HCl
$Yb^{3+} + e^- = Yb^{2+}$	−1.150	0.1 F NH$_4$Cl

Potentiometry also is a direct means of evaluating the standard potential for half reactions (E^0) and has been applied for appropriate reversible systems in the evaluation of a number of the standard potentials that are routinely tabulated. Such measurements require corrections for activity coefficients or extrapolation of the data to infinite dilution. Again direct measurements in which equal molar concentrations of the oxidized or reduced species are introduced into the system provide a simple approach to such evaluations and are as precise as those obtained by less direct methods. However, E^0 values also can be extracted from potentiometric titration data. For example, in the titration of Fe^{2+} ion with Ce^{4+} ion the Fe^{3+} ion concentration equals the Fe^{2+} concentration at the half-equivalence point; the half-reaction potential, assuming activities are equal to concentrations, is given directly by the potential of the indicator electrode relative to the reference electrode. If the latter is a hydrogen electrode the measured potential is equal to the E^0 for the Fe^{3+}/Fe^{2+} couple. As indicated at the beginning of this chapter the evaluation of the E^0 for a half reaction provides a direct measure of the free energy for the half reaction relative to the free energy for the reduction of hydrogen ion to hydrogen gas. Likewise, a combination of any pair of E^0 values or of the free energy values permits the evaluation of the equilibrium constant and the standard free energy for a redox reaction.

From a practical standpoint it is often useful to have the observed potential in the medium of measurement for the condition of equal concentrations of the oxidized and reduced species of a half reaction. Such potentials are known as formal potentials, E', rather than standard potentials, and are not purely thermodynamic quantities. The term formal potential comes from the tradition of having the supporting electrolyte at a one formal concentration. However, other stated solution conditions also are included in many listings. Thus the indicated potential is what one would expect at the half equivalence point under actual titration conditions. In other words, activity corrections have not been made and it is only a practically useful quantity. Table 6-4 summarizes a number of formal potentials for commonly encountered half reactions.

References

1. *Collected Works of J. Willard Gibbs*, Yale University Press, New Haven, Conn., 1948.
2. W. Nernst, *Z. Phys. Chem.*, **4**, 129 (1889).
3. R. Behrend, *Z. Phy. Chem.*, **11**, 466 (1893).
4. J. H. Hildebrand, *J. Am. Chem. Soc.*, **35**, 847 (1913).
5. F. Haber and Z. Klemeciewicz, *Z. Phys. Chem.*, **67**, 385 (1919).
6. W. Nernst and E. S. Merriam, *Z. Phys. Chem.*, **52**, 235 (1905).
7. I. M. Kolthoff and N. H. Furman, *Potentiometric Titrations*, 2nd ed., John Wiley and Sons, New York, 1931.
8. W. M. Latimer, *The Oxidation States of the Elements and Their Potentials in Aqueous Solutions*, 2nd ed., Prentice-Hall, Inc., New York, 1952.
9. D. J. G. Ives and G. J. Janz, *Reference Electrodes*, Academic Press, New York, 1961.
10. C. N. Reilley and R. W. Schmid, *Anal. Chem.*, **30**, 947 (1958).
11. C. N. Reilley, R. W. Schmid, and D. W. Lamsen, *Anal. Chem.*, **30**, 953 (1958).
12. F. J. C. Rossotti and H. Rossotti, *The Determination of Stability Constants*, McGraw-Hill Book Company, New York, 1961, pp. 127–170.

Seven

CONTROLLED POTENTIAL
METHODS

I. Introduction—Control of Potential and Measurement
of Current

With the formulation of the laws of electrolysis by Michael Faraday in 1834 the basis for relating electrolysis currents to chemical quantities was established. Although the concept of electrolysis was known prior to then, its utility in terms of chemical analysis depended on a quantitative relationship between current and equivalents of substance. Because an electrolysis current always necessitates mass transfer to or away from the electrode the formulation of equations for diffusion by Fick was an important event in developing quantitative relationships (1). With the laws of electrolysis and diffusion established Professor Heyrovsky combined these in a preferred form to provide a practical analytical method, namely, polarography (2). His real contribution beyond combining the important concepts of Faraday and Fick was to realize that a reproducible and continuously renewed electrode surface was essential for electrochemistry to be a reliable analytical tool. Another important factor was the realization that a diffusion controlled current could best be established by using a combination of a large nonpolarized electrode with a small working electrode. Thus the dropping mercury electrode became an important part of the effectiveness of polarography. Ilkovic developed the equation relating diffusion currents with the parameters involved in the dropping mercury electrode (3) and thus provided an understanding of the variables affecting the current response for polarographic analysis.

329

Because polarography rapidly provides information about the best conditions for performing an electrolysis with well-defined products, electrolysis conditions frequently are adjusted to take advantage of polarographic data. Thus the technique of controlled potential electrolysis often makes use of a mercury pool as a working electrode. By stirring this pool vigorously and reproducing the conditions of the polarographic data, the control potential and supporting electrolyte conditions can be idealized to obtain the maximum yield.

An as yet undeveloped application of electrochemistry is practical electrochemical synthesis based on the vast amount of electrochemical data from polarography as well as solid-electrode voltammetry. Although the latter has serious limitations because of the changing character of solid electrodes, they provide the ability to go to much more positive potentials than are possible with the mercury electrode. This has caused continuing interest in solid microelectrodes as well as in macroscopic electrolysis using such electrodes.

Although solid-electrode voltammetry has been practiced almost from the date of the discovery of polarography, the complex nature of the diffusion current with time has precluded effective applications of this other than to an extremely specialized group of applications. However, with modern electronic instrumentation the principles of voltage-sweep voltammetry and cyclic voltammetry recently have been developed. The success has been such that this has become one of the most important research tools for electrochemists concerned with the kinetics and mechanisms of electrochemical processes. These important contributions by Professors Nicholson and Shain (4,5) rely, as have all electrochemical kinetic developments, on the pioneering work by Eyring, Glasstone, and Laidler (6).

A number of specialized electrode systems have been developed in the last decade that provide practical extensions of the principles of voltammetry. These include the rotating disc electrode as well as the ring-disc configuration. The latter is particularly effective for studying pre- and postchemical processes that are coupled with the electrochemical reaction. Because with such systems the limiting current is proportional to the rate of rotation of the electrode system, experimental conditions can be extended well beyond those possible with a stationary electrode. Another useful development has been an electrode system which involves extremely closely spaced electrodes (of the order of a few microns apart in the most refined design) such that the electrolysis products of one electrode rapidly diffuse to the second electrode where they can be reelectrolyzed. This configuration also provides the possibility of almost instantaneously electrolyzing the contents in the narrow space between the electrodes.

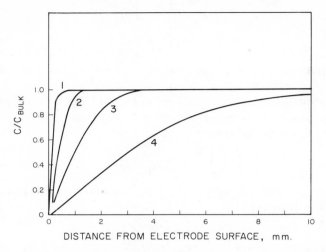

Figure 7-1. Concentration-distance curves for different periods of linear diffusion to a planar electrode surface. Diffusion times: 1, 10 sec; 2, 100 sec; 3, 1000 sec; and 4, 10,000 sec. Data for a diffusion coefficient, D, of 1×10^{-5} cm^2/sec.

II. Principles and Fundamental Relations

The basic approach in controlled potential methods of electrochemistry is to control in some manner the potential of the working electrode while measuring the resultant current, usually as a function of time. Figure 7-1 illustrates a planar electrode surface exposed to a solution with an electroactive species at a bulk concentration of C. When a potential sufficient to electrolyze the electroactive species completely is applied to the electrode at ($t=0$), the concentration at the electrode surface is reduced to zero. Passage of current requires material to be transported to the electrode surface as well as away from it. Thus relationships must be developed which involve the flux and diffusion of materials; this is appropriately accomplished by starting with Fick's second law of diffusion

$$\frac{dC_{(x,t)}}{dt} = \frac{D\,d^2C_{(x,t)}}{dx^2} \tag{7-1}$$

where D represents the diffusion coefficient, C the concentration of the electroactive species at a distance x from the electrode surface, and t the amount of time that the concentration gradient has existed. Through the use of La Place transforms Equation 7-1 can be solved to give a relationship for concentration in terms of the parameters x and t

$$C_{(x,t)} = C \operatorname{erf}\left(\frac{x}{D^{1/2}t^{1/2}}\right) \tag{7-2}$$

Figure 7-1 indicates concentration gradients as a function of time that result from this expression.

By taking the derivative of Equation 7-2 for the proper boundary condition, namely at the electrode surface, the diffusion gradient at the electrode surface is expressed by the relation

$$\left(\frac{dC}{dx}\right)_{x=0} = \frac{C}{\pi^{1/2}D^{1/2}t^{1/2}} \tag{7-3}$$

This flux of material crossing the electrode boundary can be converted to current by the expression

$$i = nFAD\left(\frac{dC_{(0,t)}}{dx}\right) \tag{7-4}$$

where n is the number of electrons involved in the electrode reaction, F the faraday, and A the area of the electrode. When Equation 7-3 is substituted into this relation a complete expression for the current resulting from semiinfinite linear diffusion is obtained (the Cottrell equation)

$$i = \frac{nFACD^{1/2}}{\pi^{1/2}t^{1/2}} \tag{7-5}$$

This relationship holds for any electrochemical process involving semiinfinite linear diffusion and is the basis for a variety of electrochemical methods, including polarography, voltammetry, and controlled potential electrolysis. The diffusion coefficient is directly related to the mobility, δ, of the species by the relation

$$D = \frac{RT\delta}{6.02 \times 10^{23}} \tag{7-6}$$

This expression can be extended for the diffusion coefficients of ionic species and thereby be related to the equivalent conductance, λ. For the condition of infinite dilution this gives the relationship

$$D_i^0 = \frac{RT\delta_i^0}{6.02 \times 10^{23}} = \frac{RT\lambda_i}{z_i F^2}$$

$$= 2.67 \times 10^{-7}\lambda_i / z_i \ \mathrm{cm^2/sec} \ \mathrm{at} \ 25°C \tag{7-7}$$

where z_i is the charge of the ionic species.

Equation 7-5 is the basic relationship used for solid electrode voltammetry with a preset initial potential on the plateau region of the current-voltage curve. Its application requires that the electrode configuration be such that semiinfinite linear diffusion is the controlling condition for the mass transfer process.

The most important and extensively studied form of voltammetry has been and continues to be polarography. Its unique characteristic is that it uses a dropping mercury electrode, such that the electrode surface is continuously renewed in a well-defined and regulated manner to give reproducible effective electrode areas as a function of time. The diffusion current equation (Equation 7-5) can be extended to include a dropping mercury electrode by appropriate substitution for the area of the electrode. Thus the volume of the drop for a dropping mercury electrode is given by the relationship

$$V = \frac{4}{3}\pi r^3 = \frac{mt}{d} \tag{7-8}$$

where r is the radius of the drop of mercury, m the mass flow rate of mercury from the orifice of the capillary, t the life of the drop, and d the density of mercury under the experimental conditions. When this equation is solved for r and the latter is substituted into the equation for the area of a sphere, an expression for the area of the dropping mercury electrode drop as a function of the experimental parameters is obtained.

$$A = (4\pi)^{1/3} 3^{2/3} m^{2/3} t^{2/3} d^{-2/3} \tag{7-9}$$

This then can be substituted into Equation 7-5 to give a calculated diffusion current for the dropping mercury electrode.

$$(i_t)_{\text{calc}} = 464 n C D^{1/2} m^{2/3} t^{1/6} \tag{7-10}$$

Actually, experimental tests indicate that the constant in Equation 7-10 is too small by a factor of $\sqrt{7/3}$. We now realize that this $\sqrt{7/3}$ quantity is not empirical but is the appropriate contribution due to the growth of the mercury drop into the solution away from the capillary orifice. Thus the correct diffusion current expression for a dropping mercury electrode is

$$i = 706 n C D^{1/2} m^{2/3} t^{1/6} \tag{7-11}$$

which gives the current at any time up to the life time of the drop. If the drop time, t_d, is substituted the well-known Ilkovic equation results

$$i_d = 706 n C D^{1/2} m^{2/3} t_d^{1/6} \tag{7-12}$$

In the original work on polarography well-damped ballistic galvanometers were used to record the current because potentiometric recorders were not available. The characteristics of a ballistic galvanometer are such that it gives the average current for a dropping mercury electrode as used in

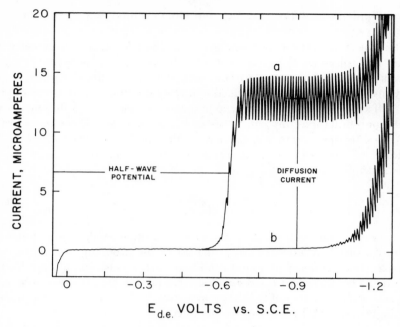

Figure 7-2. Polarograms for (a) $0.5\,\text{mF}\,Cd^{2+}$ ion, in 1 F HCl and (b) 1 F HCl alone.

polarography (see Figure 7-2). Integrating Equation 7-11 over the life of a drop permits the average current to be evaluated and is given by Equation 7-13.

$$\bar{i}_d = \tfrac{6}{7}i_d = 607nCD^{1/2}m^{2/3}t_d^{1/6} \qquad (7\text{-}13)$$

It is noteworthy that much of the polarographic literature is tabulated on the basis of Equation 7-13. In spite of this almost all modern recording polarographs use potentiometric strip chart recorders which have time constants such that the maximum of the polarographic oscillations is the instantaneous current. Furthermore, the mode of response of a potentiometric recorder is such that the mean of the oscillation is not equal to the average current. This is true because ballistic galvanometers obey Hooke's law (in terms of rate of movement) while strip chart recorders use constant-speed drive motors for the pen. Thus unless a ballistic galvanometer is used for recording current, Equation 7-12 should be used for polarographic data with the envelope of the maxima of the polarographic current oscillations being used for the measurement of current.

Although the experimental conditions for diffusion controlled current may be in effect for a polarographic measurement, it does not necessarily follow that the resultant current is controlled purely by diffusional processes. One of the most convenient ways to test whether this is true is to vary the height of the mercury column. The fluid flow characteristics of a capillary with a hydrostatic head are such that the diffusion current is directly proportional to the square root of the height of the column (small corrections for the surface tension of mercury on glass and for the hydrostatic back pressure of the water-immersed portion of the capillary are necessary for the most precise measurements)

$$i_d = \text{constant} \times h^{1/2} \qquad (7\text{-}14)$$

Thus if three or four different heights of a mercury column are used for the dropping mercury electrode the resulting current can be tested by Equation 7-14. If kinetic or catalytic complications are present the current will not adhere to this relationship.

Because of the extensive amount of data available from the polarographic literature, advantage of this should be taken when performing macroscopic electrolyses. In particular, controlled potential electrolysis at a mercury pool can be approached with predictable success on the basis of available polarographic information for the system of interest. An electrolysis can be accelerated by maximizing the electrode surface area and minimizing the thickness of the diffusion layer. However, the same electrode material must be used as in polarography. Thus a conventional approach in controlled potential electrolysis is the use of a mercury pool stirred as vigorously as possible with a magnetic stirring bar to minimize the concentration gradient. Under such conditions the decay of the current as well as the decay of the concentration of the electroactive species is given by the relation

$$\frac{i_t}{i_{t=0}} = \frac{C_t}{C_{t=0}} = \exp\left[\left(\frac{-DA}{V\Delta x}\right)t\right] \qquad (7\text{-}15)$$

where V is the volume of the solution to be electrolyzed and Δx is the thickness of the concentration gradient. Thus the current and concentration decay exponentially. Under idealized conditions 90% of the electroactive species will be electrolyzed in approximately 20 min. Increases in the temperature as well as in the electrode area relative to the solution volume will accelerate the rate of electrolysis.

For systems where kinetic, adsorptive, or catalytic effects complicate a simple diffusion controlled process, polarography provides a means of verifying the existence of these complications. Hence if there is a prechemical process which is kinetically limiting then the observed current will be less than that anticipated from Equation 7-12. Furthermore, tests based on Equation 7-14 will indicate that the current is more or less independent of the height of the mercury column. With adsorptive effects the diffusion current tends to be larger than would be anticipated from Equation 7-12, unless the rate of adsorption becomes a limiting part of the process. Again, such currents will not obey Equation 7-14. Currents resulting from catalytic processes will take many different forms and cannot be predicted *a priori*. However, these systems will not in general obey Equations 7-12 and 7-14.

The polarographic current-potential wave is illustrated by Figure 7-2 and can be expressed by the Nernst equation for reversible electrochemical processes. However, it is more convenient to express the concentrations at the electrode surface in terms of the current, i, and diffusion current, i_d. Because i_d is directly proportional to the concentration of the electroactive species in the bulk and i at any point on the curve is proportional to the amount of material produced by the electrolysis reaction, these quantities can be directly related to the concentration of the species at the electrode surface. Hence by such substitution the Nernst equation takes the form

$$E = E_{1/2} + \left(\frac{RT}{nF}\right)\ln\left[\frac{(i_d - i)}{i}\right]; \qquad E_{3/4} - E_{1/4} = \frac{0.056}{n} \text{ at } 25°C \quad (7\text{-}16)$$

For the reduction of a simple solvated metal ion to its amalgam $E_{1/2}$ is given by

$$E_{1/2}(s) = E_s^{\,0} + \left(\frac{RT}{nF}\right)\ln\left(\frac{\gamma_{ion}D_a^{\,1/2}}{\gamma_a D_{ion}^{\,1/2}}\right) \qquad (7\text{-}17)$$

where γ_{ion} is the activity coefficient for the ion and γ_a is the activity coefficient for the amalgamated species. The diffusion coefficients for the amalgamated and ionic species also are a part of this expression. The standard reduction potential is for reduction of the ion to the amalgamated species. These expressions also hold for the reduction of an ion to a lower oxidation state, but require that the appropriate value be used. For well-behaved polarograms, the number of electrons in the reduction process can be determined by taking the difference between the three-fourths-wave potential and the one-fourth-wave potential (Equation 7-16). Equation 7-16 also can be extended to conditions where the electroactive species is a complexed metal ion. For such conditions the half-wave potential for the complexed species, $E_{1/2(c)}$, is related to that for the uncomplexed species, $E_{1/2(s)}$, by the relationship

$$E_{1/2(c)} = E_{1/2(s)} + \left(\frac{RT}{nF}\right)\ln K_{diss} - p\left(\frac{RT}{nF}\right)\ln(\gamma_x C_x) \qquad (7\text{-}18)$$

where K_{diss} is the dissociation constant for the complex, p the number of ligands per metal ion, γ_x the activity coefficient of the ligand, and C_x the concentration of the ligand. A similar relationship holds for the reduction of a complexed species to another ionic complexed species

$$E_{1/2(c)} = E_{1/2(s)} + \left(\frac{RT}{nF}\right)\ln\left(\frac{K_{diss}^{ox}}{K_{diss}^{red}}\right) - (p-q)\left(\frac{RT}{nF}\right)\ln(\gamma_x C_x) \qquad (7\text{-}19)$$

where q represents the number of ligands per reduced ion. This expression indicates that only the ratio of dissociation constants can be determined; some independent means must be used to evaluate one of them.

Unfortunately, a large number of substances reduced at the dropping mercury electrode do not behave reversibly. In other words, they do not behave according to equilibrium thermodynamics. For a totally irreversible process (one in which the kinetics for the back reaction are essentially equal to zero) the current-potential relationship takes the form

$$E = E_{1/2} + \frac{0.916RT}{\alpha n_a F}\ln\left[\frac{(i_d - i)}{i}\right] \qquad (7\text{-}20)$$

where α (with values between 0 and 1) is the transfer coefficient which indicates the symmetry of the potential energy function for the transition state and n_a is the number of electrons involved in the rate determining step of the reduction process. For such a system the half-wave potential does not represent the standard potential but is related to the kinetic parameters for the reduction process by the expression

$$E_{1/2} = -0.241 + \frac{RT}{\alpha n_a F}\ln\left(\frac{1.35 k_{f,h}^{0} t_d^{1/2}}{D^{1/2}}\right) \qquad (7\text{-}21)$$

where $k_{f,h}^{0}$ is the heterogeneous forward rate constant for the reduction process at a potential of 0.00 V versus the standard hydrogen electrode and t_d is the drop time for the dropping mercury electrode. A simplified form of Equation 7-20 can be used to evaluate the quantity αn_a

$$E_{3/4} - E_{1/4} = \frac{-0.052}{\alpha n_a} \qquad \text{at } 25°C \qquad (7\text{-}22)$$

whereby the potential at the three-fourths wave-height minus that at the one-fourth wave-height is measured.

During the past decade interest has developed in accelerating the measurement process for voltammetric studies. Whereas in classical polarography voltage is scanned at the rate of approximately $\frac{1}{10}$ V/min,

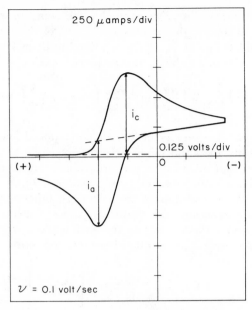

Figure 7-3. Linear voltage-sweep voltammogram with reversal of sweep direction to give a cyclic voltammogram. Initial sweep direction to more negative voltage.

voltages can be scanned up to rates as fast as 100 V/sec. For such accelerated scan rates, whether a dropping mercury electrode or a solid electrode is used, the electrode has essentially a finite and fixed area which does not change during the course of the scan.

Figure 7-3 illustrates the shape of a voltage-sweep voltammogram with an electrode of fixed area. When the reduction process is reversible the peak current is given by the relation

$$i_p = 0.4463 n F A (Da)^{1/2} C \qquad (7\text{-}23)$$

with

$$a = \frac{nF\nu}{RT} = \frac{n\nu}{0.026} \qquad \text{at } 25°C \qquad (7\text{-}24)$$

where ν is the scan rate in volts per second. Thus in terms of the adjustable parameters the peak current is given by

$$i_p = 2.67 \times 10^5 n^{3/2} A D^{1/2} C \nu^{1/2} \qquad \text{at } 25°C \qquad (7\text{-}25)$$

For a reversible process the peak potential can be related to the polarographic half-wave potential, $E_{1/2}$, by the expression

$$E_p = E_{1/2} - 1.11\left(\frac{RT}{nF}\right) = E_{1/2} - \left(\frac{0.0285}{n}\right) \quad \text{at } 25°C \quad (7\text{-}26)$$

Because of the dynamic nature of voltage sweep voltammetry irreversible processes give a distinctly different expression for the peak current from those for reversible systems,

$$i_p = 3.01 \times 10^5 n[\alpha n_a]^{1/2} A D^{1/2} C \nu^{1/2} \quad \text{at } 25°C \quad (7\text{-}27)$$

where n_a represents the number of electrons in the rate controlling step and α is the transfer coefficient (normally with a value between 0.3 and 0.7). The latter two quantities can be evaluated by taking the difference between the peak potential and the half peak potential. This difference is related by

$$E_p - E_{p/2} = -1.857\left(\frac{RT}{\alpha n_a F}\right) = -1.857\left(\frac{0.026}{\alpha n_a}\right) \quad \text{at } 25°C \quad (7\text{-}28)$$

An alternative approach is to scan the voltammogram at two different rates. Under these conditions α and n_a may be evaluated by the expression

$$(E_p)_2 - (E_p)_1 = \left(\frac{RT}{\alpha n_a F}\right)\ln\sqrt{\frac{\nu_1}{\nu_2}} \quad (7\text{-}29)$$

The peak current for a reversible reduction may also be expressed in terms of a heterogeneous rate constant, k_s, and the peak potential by the relation

$$i_p = 0.227 nFACk_s \exp\left[\left(\frac{-\alpha n_a F}{RT}\right)(E_p - E')\right] \quad (7\text{-}30)$$

where k_s represents the heterogeneous rate constant of the electrode process for the condition that the electrode has a potential equal to the formal potential for the electrode process. A somewhat simpler form of this expression is possible if the instantaneous current of the wave is measured at a point which is less than one-tenth that of the peak current. For this condition the measured current is related to the kinetic parameters by the expression

$$i = nFACk_s \exp\left[\left(\frac{-\alpha n_a F}{RT}\right)(E - E_i)\right] \quad (7\text{-}31)$$

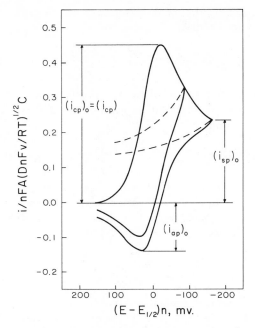

Figure 7-4. Method for measurement of peak currents and peak-current ratios of cyclic voltammograms.

$$\frac{i_{ap}}{i_{cp}} = \frac{(i_{ap})_0}{(i_{cp})_0} + \frac{0.485(2_{sp})_0}{(i_{cp})_0} + 0.086$$

where E is the potential at the measured current and E_i is the potential at which the scan was initiated.

An extension of voltage-sweep voltammetry is called cyclic voltammetry and involves reversing the triangular scan after the peak of the reduction process has been passed. Thus the voltage is scanned negatively beyond the peak and then reversed in a linear positive sweep. Such a technique provides even more information concerning the properties and characteristics of the electrochemical process and also gives insight into any complicating side processes such as pre- and postchemical reactions as well as kinetic considerations.

For a reversible process the ratio of the peak current for the cathodic process relative to the peak current for the anodic process is equal to unity. To measure the peak current for the anodic process the extrapolated baseline going from the foot of the cathodic wave to the extension of this cathodic current beyond the peak must be used as a reference, as illustrated by Figure 7-3. If a postchemical process destroys the product before the reverse scan occurs then the ratio of the cathodic peak current to the anodic peak current will be greater than unity. Another approach to measuring peak-current ratios is illustrated by Figure 7-4.

For the condition

$$|E_\lambda - E_{p/2}| \geqslant \frac{0.141}{n} \tag{7-32}$$

where E_λ is the extent of the voltage sweep, the difference in the peak potentials between the anodic and cathodic processes of a reversible reaction is given by the relationship

$$(E_p)_a - (E_p)_c = \frac{0.0595}{n} \qquad \text{at } 25°C \tag{7-33}$$

Again, this provides a very rapid and convenient means for establishing the number of electrons involved in the electrochemical reaction.

Cyclic voltammetry is of particular value for the study of electrochemical processes that are limited by finite rates of electron transfer. In the last ten years Professors Nicholson and Shain (4,5) have derived quantitative relationships which allow the evaluation of the kinetic parameters for such rate-limited processes using cyclic voltammetry. A particularly useful function for such measurements is given by the relation

$$\psi = \frac{\gamma^\alpha k_s}{\sqrt{\pi a D_{ox}}} = f(n\Delta E_p) \tag{7-34}$$

where

$$\gamma = \left(\frac{D_{ox}}{D_{red}}\right) \tag{7-35}$$

and the quantity a is given by Equation 7-24. Values of the function in Equation 7-34 are given in Table 7-1. Thus by adjusting the cyclic scan rate such that the peak potential difference times the number of electrons involved in the electrochemical process approaches 100 mV, a sensitive measure of the kinetic parameters is possible. This clearly is one of the most convenient methods of evaluating rapid heterogeneous electron-transfer rate constants. Furthermore, cyclic voltammetry is one of the most reliable electrochemical approaches for elucidating the nature of electrochemical processes and for providing insight into the nature of any complicating processes beyond the electron-transfer reaction.

Table 7-1   Analytical
Function Values for Cyclic
Voltammetry of Irreversible Processes
$(\alpha = 0.5)$

$(n\Delta E_p)(\text{mV})$	Ψ
61	20.00
63	7.00
64	6.00
65	5.00
66	4.00
68	3.00
72	2.00
84	1.00
92	0.75
105	0.50
121	0.35
141	0.25
212	0.10

Recent work has extended this method to the study of the chemical kinetics preceding and following the electron transfer process, as well as for the study of various adsorption effects which occur at the electrode surface. However, these are sufficiently complicated that those interested should consult the original treatments by Professors Shain and Nicholson and the more recent ones of Saveant (7).

In addition to the forms of voltammetry discussed so far, recent interest has developed in the use of a solid electrode of fixed area under conditions where the diffusion layer is constant. This has led to the development of the rotating disc and ring-disc electrodes (8). By rotation of a disc, the electrode diffusion layer becomes fixed such that the current is constant as a function of time and does not decay according to the indicated relationship in Equation 7-5. Voltammetry with such an electrode system gives a current-potential wave which is analogous to a polarogram and follows the relationship

$$E = E_{1/2} + \left(\frac{RT}{nF}\right)\ln\left[\frac{(i_{\text{lim}} - i)}{i}\right] \tag{7-36}$$

for reversible processes, where i_{lim} is the limiting current (in mA) on the plateau. The magnitude of the latter is given by the relationship

$$i_{lim} = 0.62 nFAD^{2/3}C\omega^{1/2}\nu^{-1/6} \tag{7-37}$$

where ω is the angular velocity of the disc ($\omega = 2\pi$ rps), ν the kinematic viscosity, and the other quantities have their usual meaning. The method provides the advantage of having the current increase with the rate of rotation as long as the process is diffusion controlled. Thus increasing the rate of rotation until the limiting current ceases to obey Equation 7-37 provides a measure of the electron-transfer kinetics. Furthermore, by having a concentric pair of electrodes the product from the disc electrode, which is produced at a given potential, is conveyed centrifically to the ring electrode. The latter usually is controlled at a different potential such that the product can be monitored. Furthermore, because the relationship between the ring current and disc current has been quantitatively established, the ring-disc electrode provides a means of measuring the kinetics of postchemical reactions of electrode products.

III. Methodology

With the exception of the indicating electrode systems the instrumentation for polarography and voltammetry is basically the same. The unique feature for polarography is the dropping mercury electrode which has a number of advantages but also introduces certain difficulties that must be overcome if reliable data are to be obtained. The advantages of this electrode are that it provides a continually renewed surface of liquid mercury which eliminates electrode contamination. Furthermore, the high hydrogen overvoltage on mercury permits a number of ions to be reduced before solvent or proton ions are reduced. Thus it extends the range of cathodic processes that may be studied. Because of the repetitive formation of the drop of mercury at the end of a capillary tube, sufficient stirring is provided such that the maxima of the current oscillations remain constant at a given potential and do not decay with time. This is in contrast to a static microelectrode whose current is inversely proportional to the square root of time, due to semiinfinite linear diffusion.

If a solid microelectrode is used in a static position the voltammetric current-voltage curve will give a peak just past the half-wave potential after which the current will decay as a function of the square root of time. In the preceding section relations are given which relate the peak current to the concentration as well as to the rate of voltage scan. By agitation or rotation of the solid electrode conditions can be obtained to yield a current plateau, similar to the polarographic wave obtained with a dropping mercury

344   Controlled Potential Methods

electrode, which will be directly proportional to the concentration of the active species (with the current constant and independent of time). In the early days of voltammetry this form of electrode had the wire mounted such that its axis was perpendicular to the axis of rotation; commercial equipment used a 600-rpm motor to rotate this electrode with contact being made through a mercury pool at the center of the axis of rotation. In recent years rotating disc voltammetry has utilized a disc electrode made out of platinum or other metal which is sealed in the end of glass or plastic and of a larger area; frequently as large as a half-inch in diameter. With careful control of the geometry, conditions can be developed such that the limiting current on the voltammetric plateau is directly proportional to the concentration as well as proportional to the square root of the rate of rotation. Under these conditions the limiting current follows the relationship of Equation 7-37.

To summarize, voltammetric electrodes have taken many special forms; the necessary conditions are that they be of a geometry and be used in such a way that the current that flows is diffusion controlled. Thus the electrodes can be in the form of discs, cylinders, spheres, or any other shape that fulfills the necessary experimental conditions. The disc electrode provides the means for giving a linear diffusion controlled response over a wide dynamic range; in contrast, cylindrical and spherical electrodes have only a limited set of conditions that yield such a response. The use of closely spaced parallel pairs of electrodes is a specialized case and is more related to controlled potential electrolysis than to voltammetry. However, conditions can be adjusted such that the principles of voltammetry apply; namely, that the current is diffusion controlled.

In voltammetry the second necessary part of the electrode system is an adequate reference electrode. Such an electrode must have a well-established and invariant potential which is not affected by the passage of significant quantities of current. During the first decade of polarography the major interest was in its use for measuring diffusion currents rather than for evaluating kinetic and thermodynamic quantities. Thus a mercury pool at the bottom of the electrolysis cell frequently served as a reference electrode. This system was subject to a number of problems including a variable potential, as well as the introduction of electrolysis products that might ultimately interfere with the sample system being monitored at the indicator electrode. Today most voltammetric measurements utilize a saturated calomel electrode (SCE) connected to the sample solution by a large salt bridge. Convenient electrode and cell systems for polarography and voltammetry are discussed in Chapters 2 and 3.

When voltammetry is carried out using nonaqueous solvents, the problems of an adequate reference electrode are compounded. To date the most common reference electrode has been the mercury pool, because of its convenience rather than because of its reliability. With the advent of sophisticated electronic voltammetric instrumentation more reliable reference electrodes have been possible, especially if a three electrode system is used. Thus variation of the potential of the counter electrode is not a problem if a second noncurrent carrying reference electrode is used to monitor the potential of the sensing electrode. If three-electrode instrumentation is used any of the conventional reference electrodes common to potentiometry may be used satisfactorily. Our own preference is a silver chloride electrode connected to the sample solution by an appropriate noninterfering salt bridge. The one problem with this system is that it introduces a junction potential between the two solvent systems which may be quite large. However, such a reference system is reproducible and should ensure that two groups of workers can obtain the same results.

Because polarographic and voltammetric measurements assume that mass transfer occurs only by diffusion, any electrostatic field effects that will attract or repulse the electroactive species must be avoided. The significant electrostatic fields that exist between the indicator electrode and the reference electrode must be dissipated by introducing an inert supporting electrolyte. Normally a one-hundred fold excess of supporting electrolyte relative to the sample species is a minimum requirement if migration currents are to be avoided. Almost any strong electrolyte can serve as the supporting electrolyte in voltammetry but over the past decade approximately a dozen such systems have evolved as meeting most of the needs of practicing polarographers. These are summarized in Table 7-2 and indicate the range of media that are available for voltammetric studies. With the exception of perchlorate media almost all supporting electrolyte systems have some tendency to complex metal ions. Thus the half-wave potential of such ions will be affected significantly by the composition of the supporting electrolyte. Furthermore, if the electroactive species has any acidic or basic properties then the acidity of the supporting electrolyte will also have a direct effect on the potential at which many species will be reduced. Clearly the effect on the half-wave potential of the supporting electrolyte can be used to advantage for the identification and selective detection of species in mixtures. A number of tabulations have been published which summarize the voltammetric behavior of most of the common substances in the presence of a variety of supporting electrolytes. By far the most convenient of these is the one compiled by Professor Louis Meites (9).

Because oxygen is electroactive at almost all electrodes, it is a serious interference in voltammetric and electrochemical measurements. For this reason means must be provided for deaerating the sample solution before meaningful voltammetric measurements are made. Although for many years further purification of tank nitrogen was necessary to remove traces of

Table 7-2   Examples of Polarographic Supporting Electrolytes

Electrolyte	Effective Voltage Range, V versus SCE
1 F KCl	$+0.1$ to -2.1
1 F HCl	$+0.1$ to -1.4
12 F HCl	-0.4 to -1.0
1 F NH$_3$ and 1 F NH$_4$Cl	-0.1 to -1.8
1F NaOH	-0.1 to -2.0
1 F NaF	$+0.2$ to -2.0
1 F KCN	-0.6 to -1.9
7.3 F H$_3$PO$_4$	$+0.1$ to -1.3
1 F KSCN	-0.3 to -1.6
1 F Na$_3$ (citrate) and 0.1 F NaOH	-0.2 to -1.8
0.5 F Na$_2$ (tartrate), pH 9	0.0 to -1.8
2 F HOAc and 2 F NH$_4$OAc, 0.01% gelatin	$+0.3$ to -1.3
0.3 F N(C$_2$H$_4$OH)$_3$ and 0.1 F KOH	-0.3 to -1.3
0.1 F Pyridine and 0.1 F pyridinium chloride	$+0.2$ to -1.2

oxygen, today prepurified nitrogen is sufficiently inexpensive and available to eliminate this problem. For alkaline solutions sodium sulfite provides a convenient means for directly eliminating any interfering oxygen.

A final precaution is that the deaeration of organic solvents is far more difficult than aqueous solutions; this is primarily because of the much greater solubility of oxygen in such solvents relative to water.

During the past decade increasing interest in the use of electrochemical methods for organic and biochemical systems has promoted the application of nonaqueous solvents for voltammetric and polarographic measurements. A major problem is to find a system such that an adequate amount of supporting electrolyte may be introduced and dissociated to prevent migration currents. For some time the most popular solvent systems involved mixtures of alcohols and water, as well as mixtures of dioxane and water. However, with improved instrumentation and the use of three electrode systems pure nonaqueous solvents have become popular. Thus polar materials such as acetonitrile, dimethylsulfoxide, and dimethylformamide have sufficiently high dielectric constants to make them useful nonaqueous solvents for electrochemical measurements, including voltammetric studies. A number of electrolytes are soluble in these solvents, and are sufficiently dissociated to make effective supporting electrolytes. Each of them must be purified to avoid interferences; this has been outlined in Chapter 4. With such solvents the most popular supporting electrolytes are lithium chloride, lithium perchlorate, and various tetraalkylammonium salts, including tetraethylammonium perchlorate, tetramethylammonium chloride, and tetrabutylammonium iodide.

If the half-wave potentials or the peak potentials in polarography and voltammetry are to be used for identification purposes as well as for the evaluation of thermodynamic quantities, they must represent the actual electrode potential relative to the reference electrode. Unfortunately, with the passage of current and the existence of a finite resistance in all electrochemical cells, the recorded current-potential curve includes the iR drop across the cell if a two-electrode form of instrumentation is used. For most electrochemical measurements the upper limit for polarographic currents is approximately 20 μA. Assuming a cell resistance of 200 Ω, the recorded half-wave potential would be in error by -4 mV as a result of iR losses. Clearly this approaches the limits of accuracy of most instrumentation and is negligible. On the other hand, use of a nonaqueous solvent may result in a cell resistance of 20,000 Ω, which for a 20-μA current would give an iR drop of -0.4 V, and thereby preclude meaningful identification as well as thermodynamic data without making precise corrections. Because corrections of this order of magnitude are almost impossible with any degree of precision, the practical alternative for high resistant systems is employment of a three-electrode system, whereby the indicator electrode potential is referred to a noncurrent carrying reference electrode. Through the use of operational amplifiers for the construction of electronic polarographs and potentiostats such three electrode instrumentation is commonplace and straightforward in its construction. However, complete compensation demands judicious placement of the nonworking reference electrode because the iR losses represent a field gradient between the indicator electrode and the current-carrying counter electrode. Thus placement of the reference electrode must be as close as physically possible to the surface of the indicator electrode. To do this directly is awkward and unnecessary. A more convenient arrangement is the use of a Luggin capillary, which is normally constructed out of glass or plastic tubing. Figure 7-5 illustrates such a capillary as well as its appropriate placement in an electrochemical cell. The principle of this device is that the electric field gradient represented by the iR loss is not transmitted through the glass or plastic tubing because it is an insulator. Hence if the tip of the Luggin capillary is within a millimeter or so of the surface of the indicator electrode, then the electrolyte within the Luggin capillary does not contribute any iR loss because current is not flowing and the potential at the electrode surface is effectively referenced to that of the reference electrode within the Luggin capillary.

The instrumentation for voltammetry normally provides a means for scanning the applied potential or the potential of the indicator electrode

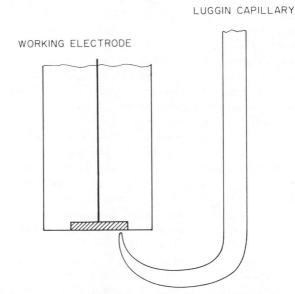

LUGGIN CAPILLARY

WORKING ELECTRODE

Figure 7-5. Luggin capillary and its placement relative to the working electrode.

with a motor driven slide wire or by electronic means. Prior to the advent of recording instrumentation manual instruments used a hand-adjusted potentiometer to obtain a point by point scan, with the current monitored by a ballistic galvanometer. Today there is an increasing tendency to use operational amplifiers such that the voltage is electronically swept with an R-C circuit, thereby eliminating the need for a motor-driven slide wire and its inherent problems of noise and wear. Chapter 5 discusses the form of a number of such instruments. Traditionally, the scan rates in polarography and voltammetry have been of the order of 1 V/10 min. This evolved in part because of the limitations in recording devices during the early years of polarography. However, there is a tendency to increase scan rates and many modern instruments record 1 V polarograms in 3 to 5 min. With the advent of voltage-sweep and cyclic voltammetry this acceleration of scan rate has progressed to the point where it is not unreasonable to scan with rates as high as 100 V sec.

Because of the availability of recording potentiometric strip-chart recorders the voltammetric current is almost invariably recorded on such devices. This requires that the current be passed through a standard resistor such that it is converted to a potential. The impedance of the recording device must be sufficiently high to avoid the introduction of additional errors. The important point is that in a two-electrode system the potential generated by the current measuring resistance is a part of the applied potential and represents another error in addition to the iR potential drop of the cell. Thus with the use of a 100-mV recorder a current causing a full-scale deflection would introduce a 100-mV error if the applied voltage is assumed to be equal to the potential of the indicator electrode relative to the reference electrode. Again, this problem can be avoided through the use of three-electrode instrumentation such that the voltage drop of the current measuring resistor is compensated in the feedback loop. This is discussed in Chapter 5.

Much of the recent instrumentation for classical voltammetry can be extended to the more sopisticated voltage sweep and cyclic techniques. The same basic elements are necessary; namely, a voltage source, a means of sweeping the voltage, a current measuring system, and some type of recording device. However, mechanical voltage scanning systems are incapable of giving the variable and rapid scan rates that are necessary. Thus electronic voltage scanning systems are necessary, usually by use of R-C circuits in association with operational amplifiers such that appropriate adjustment of the resistance into the amplifier will provide a wide range of scan rates. Ideally scan rates from approximately 0.1 V/min up to 100 V/sec are needed. Also, adjustment of the starting potential and the terminal potential for a voltage sweep is extremely convenient. In the case of cyclic voltammetry there is a need to be able to reverse the sweep at variable points along the potential scan. For moderate scan rates this can be accomplished manually, at more rapid scan rates microswitches are useful, and at the most rapid rates transistor switching circuits are necessary. In general voltage-sweep and cyclic voltammograms are most conveniently recorded with an X-Y recorder that has a slewing speed of at least 10 in./sec. For the fastest scan rates an oscilloscope is necessary to obtain an accurate record of the current-voltage curve. A particularly convenient form of such instrumentation has provision for single sweeps as well as for retention of the trace (so-called memory oscilliscopes). The general features of these devices are discussed in Chapter 5.

Another specialized form of current voltammetry involves the use of either a rotating disc or a ring-disc indicating electrode. With this type of electrode the current is directly proportional to the square root of the rate of rotation if it is a diffusion controlled process. To obtain complete adherence to the square-root relationship a hydrodynamically sound design for the electrode is essential (8). Figure 7-6 illustrates the geometric features that have been found to give reliable performance for rotation rates as high as 10,000 rpm.

The ring-disc electrode represents a specialized variant of the rotating disc electrode and involves a concentric ring electrically insulated from the disc

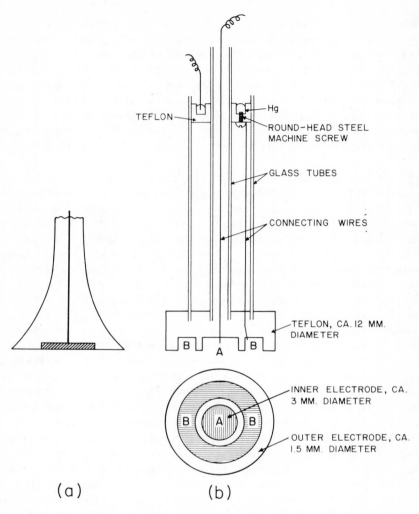

TEFLON

Hg

ROUND-HEAD STEEL
MACHINE SCREW

GLASS TUBES

CONNECTING WIRES

TEFLON, CA. 12 MM.
DIAMETER

B   A   B

INNER ELECTRODE, CA.
3 MM. DIAMETER

B   A   B

OUTER ELECTRODE, CA.
1.5 MM. DIAMETER

(a)   (b)

Figure 7-6. Construction of rotating disc and ring-disc electrodes. (*a*) Hydrodymic disc electrode for high rotation speeds. (*b*) Side and bottom views of a ring-disc electrode with wells for carbon paste or press-fit noble metals.

electrode. The advantage of this form of the electrode is that products produced by the disc electrode may be monitored at the ring electrode. Thus the potential of the ring can be controlled at a level which is different from that of the disc electrode to give a specific response for a transient or unstable product. Although this requires a separate potentiostat from that controlling the potential of the disc electrode, the ring-disc electrode has found extensive application for the study of the kinetics and mechanisms of postchemical reactions. Figure 7-6b indicates the details of construction.

Another recent specialized form of electrode configuration is known as the thin-layer electrolysis system (10). With this arrangement two electrodes are placed sufficiently close to each other that the product at one diffuses rapidly to the second electrode where it may be monitored. Figure 7-7 illustrates one form of this system in which a precision micrometer is used as a support of one electrode and provides the means for adjusting the gap between the electrodes. In its simplest application this system provides the means for rapidly evaluating the electron stoichiometry of an electrode reaction. It also, like the ring-disc electrode system, provides a convenient means for monitoring the products of the opposite electrode reaction and for evaluation of the kinetics of postchemical reactions. With adequate electronic instrumentation the potential of both electrodes can be controlled independently such that one or the other will be controlling the magnitude of the electroly-

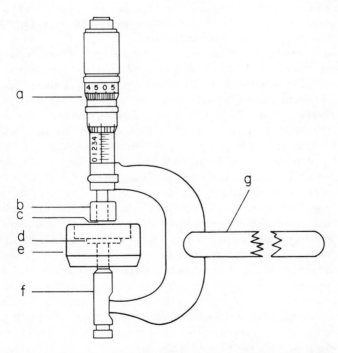

Figure 7-7. Micrometer thin-layer electrode system. (a) Outer thimble for readings to 1×10^{-4} in.; (b) Press-fit Teflon Collar; (c) Platinum face of micrometer spindle; (d) Flat, glass disc in Teflon cup against the face of the detachable anvil; (e) Machined Teflon cup; (f) Starrett No. 212 detachable anvil; (g) Rod for mounting cell assembly.

sis current. General control of the potential is accomplished with conventional potentiostats. A crucial feature in parallel-electrode electrolysis is the maintenance of a precisely parallel configuration. Obviously this becomes an ever more demanding specification as the spacing is made smaller and smaller. To date systems have been developed with gaps as small as 20 μ.

One of the oldest forms of electrochemistry involves the analysis of materials by electrodeposition. The principles are extremely simple but still provide some of the highest levels of precision possible in analysis. Usually the electrolysis potential is provided by either a D-C-motor generator or a battery with a crude potentiometer providing control of the applied potential. Provisions are not made for compensating the iR drop across the cell and selectivity usually is obtained by control of the acidity and of the supporting electrolyte of the electrolysis solution. A further degree of selectivity is provided by the working electrode and its overpotential for certain reactions. Thus mercury, because of its high overpotential for the reduction of hydrogen ion, provides a wide range of potentials for the reduction of most metal ions before the reduction of hydrogen ion interferes. In contrast, if platinum is used only electropositive ions such as copper(II) will be reduced without hydrogen ion interfering.

As an example, a sample containing a mixture of copper(II) and nickel (II) salts can be analyzed by first electrolyzing the sample solution under acidic conditions with platinum electrodes such that the copper is plated onto a platinum gauze electrode. Because the solution is acidic, hydrogen ion is reduced before nickel ion and there is no interference. After the electrolysis for copper is completed the electrolysis solution can be neutralized and made basic with ammonia. Having determined the copper and removed it from the platinum electrode the remaining basic electrolysis solution can be electrolyzed to plate nickel on the platinum electrode.

Classical electrodeposition has limited application today because of the development of more sophisticated methods. It is a slow and tedious process and there is a need for large samples. However, it still represents one of the most precise quantitative techniques for the determination of copper, cadmium, silver, and nickel. The best conditions for the determination of these and other ions by electrodeposition are summarized in Table 7-3.

The mercury pool electrode is especially attractive because of its renewable surface and because it allows the use of a stirring bar to vigorously stir the electrode-solution interface. Its major limitation relative to a platinum electrode is its high mass and the awkwardness of rinsing and of weighing a liquid electrode relative to a solid electrode.

The rate of electrodeposition is dependent on a number of factors and these are predictable to only a limited degree. However, the thickness of the diffusion layer must be minimized to obtain a rapid electrolysis. This is accomplished by vigorously stirring and by the use of electrodes with large surface areas. An increase in temperature enhances the rate of electrolysis because it increases the mobility of the electroactive species. The use of high ionic concentrations minimizes the iR drop between electrodes and also improves the electrolysis rate. The orientation and geometry of the electrodes is important to insure a uniform and adherent plate. Depolarizers frequently are introduced to prevent formation of interfering products from the counter electrode (see Table 7-3).

With the development of more sophisticated electronic instrumentation a higher degree of control has been made possible through the development of the potentiostat. The basic operating principle of a potentiostat involves a three-electrode assembly such that the working electrode's potential is monitored relative to a closely spaced reference electrode. In the simplest form this potential is controlled by manually adjusting the applied voltage across the working electrodes while visually monitoring the potential of the working electrode relative to the reference electrode. However, this form of controlled potential electrolysis is not only extremely tedious, but it is inadequate in terms of precise potential control. Even with the most "religious" attention the dexterity of an individual is such as to give an effective response time of at least 0.1 sec.

The next stage in the development of the potentiostat was to replace the human operator by a servomechanical system, whereby microswitches are placed on a recorder monitoring the working electrode potential such that when preset limits are approached these activate a motor-driven rheostat controlling the applied voltage. Again the mechanical nature and slow response time limit the applicability of this type of potentiostat. During the past decade electronic potentiostats, based to a large extent on operational amplifiers, have provided sophisticated instrumentation with response times of less than 1 msec, applied potential capabilities up to 100 V, and currents up to several amperes. Several commercial instruments based on these modern approaches are available and a design for such instrumentation is discussed in Chapter 5. In an ideal potentiostat one would like to have the ability to set the electrode potential to a precision of ± 1 mV over a potential range from $+3$ to -3 V, to have electronics capable of applying up to 100 V for high-resistance solutions while delivering currents up to at least 1 A, and to have the response time close to 1 msec.

The majority of controlled potential electrochemistry has been carried out at mercury pool electrodes. This is because the vast amount of reference data available from polarography permits conditions to be based on polarographic experience. Furthermore, the uniform and reproducible surface, and the high voltage for solvent reduction, make the mercury pool particularly attractive relative to solid electrodes. As with electrodeposition, con-

Table 7-3 Electrolysis Conditions for the Determination of Metal Ions

Metal Ion	Solution Conditions	Electrolysis Conditions	Interferences
Ag^+	200 ml of 1 F HNO_3, 0.5 F $NaNO_2$	2 A, 1 hr	As, Hg, Se, Te
Ag^+	10%KCN	2 A, 20 min	Au, Bi, Cd, Co, Cu, Hg, Ni, Zn
Au^{3+}	160 ml of 0.6 F HCl, 0.15 F $NH_2OH \cdot HCl$	2–3 A, 20 min	Bi, Cd, Cu, Hg, Pb, Pt, Sb, Sn, NO_3^-
Au^{3+}	100 ml of 0.2 F KCN	0.3–0.5 A, 30 min 30–50°C	Cu, Pd, Pt, Ag, Bi, Cd, Co, Hg, Ni, Zn
Bi^{3+}	200 ml of 0.5 F $HClO_4$ plus 5 ml satd. $N_2H_4 \cdot H_2SO_4$	1 A, 1 hr	Ag, As, Cd, Cu, Hg, Pb, Sb, Sn
Cd^{2+}	200 ml of 0.1 F H_2SO_4 plus 10 ml 0.1% gelatin (Cu/Pt cathode)	3 A, 40 min	Cu, Ag, Au, Bi, Hg, Pb, Pt, Sb, Sn
Co^{2+}	100 ml of 2 F NH_3, 0.6 F NH_4Cl, 0.015 F $NH_2OH \cdot HCl$	2–5 A, 45 min	Cu, Ni, Zn, Pd, Tl
Cu^{2+}	200 ml of 0.5 F H_2SO_4, 0.25 F HNO_3	2 A, 1 hr	Ag, As, Au, Bi, Hg, Mo, Pt, Pd, Sb, Se, Te, Sn, N-oxides, Cl^-
Fe^{2+}	100 ml of 0.2 F H_2SO_4, 0.4 F $(NH_4)_2 C_2O_4$	6 A, 30 min	Co, Cu, Mn, Ni
Fe^{3+}	100 ml of 1 F HCl, 1 F H_3PO_4 plus NH_3 to neutralize plus 15 ml NH_3 excess, diluted to 200 ml	2 A, 45 min	As, Co, Cu, Mo, Ni, Sb, W, Zn

Ion	Solution	Conditions	Separated from
Hg^{2+}	100 ml of 1.5 F HNO_3 (Au cathode)	1 A, 45 min	Ag, Au, Bi, Cd, Cu, Pb, Pt, Sb, Sn
Ni^{2+}	100 ml 6 F NH_3, 0.1 F $N_2H_4 \cdot 2HCl$ (Cu/Pt cathode)	0.5 A, 3 hr	Co, Cu, Fe, In, Pd, Tl, Zn, NO_3^-
Pb^{2+}	100 ml 1 F H_3PO_4 plus 10 ml 0.1% gelatin (Cu/Pt cathode)	0.4 A	Bi, Cu, Sb
Pb^{2+}	100 ml 1 F HNO_3	2.4 A, 1.5 hr, 70–80°C (on anode as PbO_2)	Fe, Mn, Tl
Pd^{2+}	150 ml 2.5 F NH_3 (Ag/Pt cathode)	0.05 A, 12 hr, (unstirred)	Co, Cu, Fe, Hg, In, Ni, Tl, Zn
Pt^{4+}	100 ml 0.05 F H_2SO_4 (Cu/Pt cathode)	0.01–0.03 A, 5 hr, 70°C (unstirred)	In, Ni, Tl, Zn
Rh^{3+}	100 ml 0.05 F H_2SO_4 (Ag/Pt cathode)	8 A, 15 min	Ag, Au, Bi, Cd, Cu, Hg, Pb, Pt, Sb, Sn, Tl
Sn^{2+}	200 ml 1 F H_2SO_4, 0.25 F $HClO_4$, 0.3 F HCl, 0.35 F $NH_2OH \cdot HCl$ (Cu/Pt cathode)	2 A	Ag, As, Au, Bi, Cd, Cu, Hg, Pb, Pt, Sb, Tl
Tl^+	180 ml of 0.5 F HNO_3, 0.05 F benzoic acid (Hg/Pt cathode)	5 A, 15 min, 45°C	Ag, As, Au, Bi, Cd, Hg, Pb, Pt, Sb, Sn, Mo
U^{6+}	100 ml of 0.5 F NH_4OAc, pH 6–7	0.2 A, 90°C	
Zn^{2+}	125 ml of 0.12 F KCN, 2.5 F NH_3 (Cu/Pt cathode)	3 A, 30 min	Ag, Au, Bi, Cd, Co, Cu, Hg, Ni
Zn^{2+}	100 ml of 0.04 F KNa (tartrate) $\cdot 4H_2O$ plus sufficient KOH to dissolve ppt. (Cu/Pt cathode)	0.3 A, 45 min	Bi, Cu, Fe, Pb, Sn

trolled potential electrolysis rates are dependent on an electrode area, stirring rates, solution volume, solution temperature, and supporting electrolyte. Assuming the diffusion layer is uniform and that the applied potential is such that one is on the diffusion plateau then the electrolysis obeys the relation

$$i_t = i_0 \exp\left[\left(\frac{-DA}{V\Delta x}\right)t\right] \tag{7-38}$$

where Δx is the thickness of the diffusion layer. Thus the electrolysis current for these conditions decays exponentially. Under the best obtainable conditions 99.9% of the material in a sample can be electrolyzed in approximately 20 min. Because of the nonlinear nature of the electrolysis current its

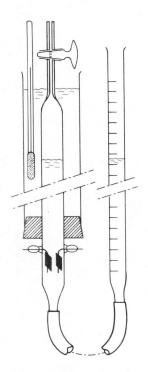

HYDROGEN–OXYGEN COULOMETER

Figure 7-8. The cell contains $1\,F\,H_2SO_4$ and platinized platinum electrodes. A thermometer monitors the temperature of the water jacket of the cell, and a conventional buret measures the volume of displaced electrolyte solution.

integration is not simple. To decrease the size of the samples and to avoid the necessity of weighing the electrode to determine the amount of substituent, a number of devices for integrating the electrolysis current have been developed. An example of a simple, but sensitive and accurate integrator is illustrated in Figure 7-8. This is based on the oxygen-hydrogen coulometer and relies on the stoichiometric electrolysis of acidified water to give a reproducible volume of gas under standard conditions per coulomb of electricity (11). Although cumbersome, it is capable of good precision when corrected for barometric pressure and the vapor pressure of water. Obviously, if controlled potential electrolysis is used for the conversion of a species from one ionic state to another it is essential to have some current integrating system because there is no means of weighing the amount of material electrolyzed. Through the use of standard resistors to convert the electrolysis current to a potential any integrating device capable of integrating potential-time curves can be used for current integration. Because of the extensive use of gas chromatography a number of devices are available today, including the ball-and-disc integrator and the sensitive voltage-to-frequency type of integrator. As discussed in Chapter 5, the most popular integrating device for electrochemistry consists of operational amplifiers in an R-C circuit. Assuming the use of operational amplifier instrumentation, this is a simple and inexpensive adjunct to provide for effective integration of potential-time curves.

IV. Application of Controlled Potential Methods

To date the most extensive application of electrochemical methods with controlled potential has been in the area of qualitative and quantitative analysis. Because a number of monographs have more than adequately reviewed the literature and outlined the conditions for specific applications, this material is not covered here. In particular, inorganic applications of polarography and voltammetry have been discussed in great detail in the classic monograph by Kolthoff and Lingane (12). Excellent compilations of polarographic half-wave potentials and diffusion current constants are presented in the monograph by Meites (9). The two-volume work by Kolthoff and Lingane also provides a good, but somewhat dated, review of organic and biological applications of polarographic methods. A more recent treatment is offered in the two-volume work by Kolthoff and Zuman (13). A brief summary of the best conditions for the polarographic determination of a number of inorganic and organic substances is presented in Table 7-4. This represents an extremely limited summary and it should be recognized that there are many other conditions which are satisfactory and which may provide the degree of selectivity needed for a specific analysis

Table 7-4 Optimum Conditions for Polarographic Determination of Inorganic and Organic Substances[a]

A. Inorganic Ions

Ion	Supporting Electrolyte	$E_{1/2}$ Versus SCE	I
Ag(I)	Dilute HNO_3, $NaNO_3$, $HClO_4$, or $NaClO_4$	>0	—
	1 F NaF, 0.01% gelatin	>0	2.35
	12 F HCl	>0	—
		-0.65	—
	0.1 F KNO_3	>0	2.50
As(III)	0.1 F NH_3, 0.1 F NH_4Cl	-1.71	—
	1 F Na_3 Cit, 0.1 F NaOH	(-0.31)	—
	1 F HCl, 0.0001% methylene blue	-0.43	6.04
		-0.67	12.00
	12 F HCl	>0	3.94
		-0.55	
	0.5 F KOH, 0.025% gelatin	(-0.26)	-3.82
	1 F HNO_3, 0.01% gelatin	-0.70	—
		-1.00	8.80
	7.3 F H_3PO_4	-0.46	—
		-0.71	—
	0.5 F H_2SO_4, 0.01% gelatin	-0.70	—
		-1.00	8.40
	1 F H_2 Tart, 1 F HCl	-0.40	4.32
		-0.67	
	1 F Na_2 Tart, 0.8 F NaOH	(-0.31)	-2.87
	0.1 F EDTA, pH 6–8	-1.60	—
Au(I)	0.1 F KCN	-1.46	—
	0.1 F KOH	-1.16	—
Au(III)	2 F HOAc, 2 F NH_4OAc, 0.01% gelatin	>0	—
	0.1 F KCN	>0	—
		-1.40	—
	1 F NaF, 0.01% gelatin	>0	—
Ba(II)	0.1 F LiCl or $N(C_2H_5)_4I$	-1.92	3.58
Bi(III)	2 F HOAc, 2 F NH_4OAc, 0.01% gelatin	-0.25	3.50
	0.25 F $(NH_4)_3Cit$, pH 5	-0.52	—
	1 F NaF, pH 0.7–2.1, 0.01% gelatin	-0.07	4.88
	1 F HCl, 0.01% gelatin	-0.09	5.23

[a]Parentheses indicate an anodic wave; $I = 607 n D^{1/2} = i_d / cm^{2/3} t^{1/6}$ and is based on average rather than maximum i_d values (multiplication by 7/6 will give appropriate I values for maximum diffusion currents). Negative I-values indicate anodic currents.

Table 7-4 (*Continued*)

A. Inorganic Ions

Ion	Supporting Electrolyte	$E_{1/2}$ Versus SCE	I
	12 F HCl	−0.45	—
	1 F HNO$_3$	−0.01	4.59
	0.7 F HClO$_4$	+0.02	—
	0.1 F KH phthalate	−0.23	—
	0.5 F H$_2$SO$_4$	−0.03	4.31
	0.5 F Na$_2$Tart, pH 4.5, 0.01% gelatin	−0.29	3.12
	0.3 F triethanolamine, 0.1 F KOH	−0.74	—
	0.1 F Na Gluconate, 1 F NaOH	−0.80	3.65
	0.1 F EDTA, 1 F K$_2$CO$_3$	−0.78	—
	0.1 F EDTA, 2 F NaOAc	−0.70	—
Br$^-$	0.1 F KNO$_3$	(+0.12)	—
BrO$_3$$^-$	0.1 F KCl	−1.78	—
	0.1 F BaCl$_2$ or CaCl$_2$	−1.53	—
	0.1 F KCl, 0.001–0.1 F LaCl$_3$	−0.82	—
	0.1 F H$_2$SO$_4$, 0.2 F KNO$_3$	−0.41	—
Cd(II)	2 F HOAc, 2 F NH$_4$OAc, 0.01% gelatin	−0.65	2.30
	1 F NH$_3$, 0.1 F KNO$_3$	−0.78	3.85
	1 F NH$_3$, 1 F NH$_4$Cl	−0.81	3.68
	5 F CaCl$_2$	−0.80	—
	0.1 F KCl or HCl	−0.60	3.51
	1 F KCl or HCl	−0.64	3.58
	0.5 F K$_3$Cit, pH 7	−0.71	—
	1 F Na$_3$Cit, 0.1 F NaOH	−1.46	—
	1 F KCN	−1.18	—
	1 F ethylenediamine, 0.1 F KNO$_3$, 0.01% gelatin	−0.93	3.13
	1 F NaF, 0.01% gelatin	−0.63	2.99
	1 F KI	−0.74	—
	0.1 F KNO$_3$	−0.58	3.53
	1 F KNO$_3$ or HNO$_3$	−0.59	—
	7.3 F H$_3$PO$_4$	−0.71	—
	0.1 F pyridine, 0.1 F KNO$_3$, 0.01% gelatin	−0.59	—
	0.5 F Na$_2$Tart, pH 4.5, 0.01% gelatin	−0.64	2.34
	0.5 F Na$_2$Tart, pH 8.8, 0.01% gelatin	−0.64	2.34
	2 F KSCN	−0.66	—
	0.3 F triethanolamine, 0.1 F KOH	−0.82	—

Table 7-4 (*Continued*)

A. Inorganic Ions

Ion	Supporting Electrolyte	$E_{1/2}$ Versus SCE	I
Ce(III)	0.1 F LiCl or N(CH$_3$)$_4$Br	-2.00	—
Ce(IV)	12 F HCl	>0	—
		-0.68	—
	7.3 F H$_3$PO$_4$	>0	—
	0.1 F solns. of strong acids	>0	—
Cl$^-$	0.1 F KNO$_3$	$(+0.25)$	—
ClO$^-$	Neutral 0.5 F K$_2$SO$_4$	$+0.08$	—
ClO$_2^-$	1 F KCl, 0.05 F LaCl$_3$	-1.02	—
	1 F NaOH	-1.00	—
CN$^-$	0.1 F KOH	(-0.36)	-3.00
C$_2$N$_2$	0.1 F NaOAc	-1.20	—
Co(II)	0.1 F NaOAc	-1.19	—
	1 F NH$_3$, 1 F NH$_4$Cl, 0.004% gelatin	-1.29	—
	5 F CaCl$_2$	-0.82	—
	0.1–1 F KCl or NaCl, 0.01% gelatin	-1.20	—
	1 F Na$_3$Cit, 0.1 F NaOH	-1.45	—
	1 F KCN	-1.45	—
	0.1 F ethylenediamine, 0.1 F KNO$_3$	(-0.46)	—
	1 F NaF, pH 3–6, 0.01% gelatin	-1.38	2.75
	1 F KOH	-1.43	—
	10 F NaOH	-1.54	—
	0.1 F KH phthalate	-1.24	—
	0.1 F pyridine, 0.1 F pyridinium chloride	-1.06	—
	0.1 F K$_2$SO$_4$	-1.21	—
	1 F K$_2$SO$_4$	-1.43	—
	1 F neutral or weakly acidic tartrate	-1.60	—
	1 F KSCN	-1.04	—
Co(III)	1 F K$_2$C$_2$O$_4$, 0.2 F NH$_4$OAc, 0.5 F HOAc, 0.02% gelatin	>0	1.38
	0.3 F EDTA, 0.3 F pyridine, 0.3 F pyridinium chloride	>0	—

Table 7-4   (*Continued*)

A.   Inorganic Ions

Ion	Supporting Electrolyte	$E_{1/2}$ Versus SCE	I
$Co(NH_3)_6{}^{3+}$	2 F NH$_3$, 2 F NH$_4$Cl	-0.28	—
		-1.30	—
	0.1 F KCl	-0.26	1.78
		-1.20	5.38
	5 F CaCl$_2$	-0.26	—
		-0.88	—
	0.5 F K$_3$Cit	-0.36	1.23
		-1.52	3.66
	10 F NaOH	-0.35	—
		-1.54	—
	0.1 F KNO$_3$	-0.24	1.74
		-1.21	5.36
	0.1 F K$_2$SO$_4$	-0.46	1.60
		-1.23	4.77
	0.1 F K$_2$Tart	-0.31	1.50
		-1.32	4.35
Cr(II)	Saturated CaCl$_2$	(-0.51)	-0.47
	1 F KCl, 0.005% gelatin	(-0.40)	-1.54
	1 F KCN, 0.001% gelatin	(-1.38)	—
	0.5 F HCOOH, pH 1.8, 0.005% gelatin	(-0.30)	-1.54
	0.1 F Na salicylate, 0.1 F NaOH, 0.005% gelatin	(-1.23)	-1.17
	0.01 F N(CH$_3$)$_4$Br or N(C$_2$H$_5$)$_4$Br, 0.005% gelatin	(-0.43)	-1.54
		-1.58	—
	1 F KSCN, 0.005% gelatin	(-0.80)	-1.64
Cr(III)	1 F NH$_3$, 1 F NH$_4$Cl,	-1.43	—
	0.004% gelatin	-1.71	—
	Saturated CaCl$_2$, 0.005% gelatin	-0.51	—
	1 F Na$_3$Cit, 0.1 F NaOH	-1.45	—
	1 F KCN	-1.38	1.55
	7.3 F H$_3$PO$_4$	-1.02	—
	0.1 F pyridine, 0.1 F pyridinium perchlorate, pH 5.4, 0.02% gelatin	-0.95	—
	0.05 F Na$_2$SO$_4$	-1.01	—
	0.001 F N(CH$_3$)$_4$Br or	-1.14	—
	N(C$_2$H$_5$)$_4$Br, 0.005% gelatin	-1.58	—
	1 F KSCN, 0.005% gelatin	-0.99	—

Table 7-4 (*Continued*)

A. Inorganic Ions

Ion	Supporting Electrolyte	$E_{1/2}$ Versus SCE	I
Cr(VI)	1 F NH$_3$, 1 F NH$_4$Cl, 0.1 F KCl	−0.20	—
		−1.60	—
	0.1 F KCl, 0.0003% sodium methyl red	−0.30	—
		−1.00	5.95
		−1.80	12.00
	1 F Na$_3$Cit, 0.1 F NaOH	−0.83	—
		−1.49	—
	1 F NaF, pH 9.3, 0.01% gelatin	−0.26	2.98
		−1.10	—
	12 F HCl	>0	—
		−0.61	—
	0.2 F NaOH	−0.97	—
	1 F NaOH	−0.85	5.72
	1.5 F NaOH, 3% mannitol	−0.83	—
Cs(I)	0.1 F N(CH$_3$)$_4$Cl or N(CH$_3$)$_4$OH	−2.09	—
Cu(I)	1 F NH$_3$, 1 F NH$_4$Cl	(−0.22)	—
		−0.50	—
	0.1 F KCl, 50% pyridine	−0.52	—
Cu(II)	2 F HOAc, 2 F NH$_4$OAc, 0.01% gelatin	−0.07	3.10
	1 F NH$_3$, 1 F NH$_4$Cl, 0.004% gelatin	−0.24	—
		−0.51	3.75
	1 F K$_2$CO$_3$, pH 9.5–11	−0.20	—
	5 F CaCl$_2$	−0.33	—
	0.1 F KCl or HCl, 0.01% gelatin	+0.04	3.23
	1 F KCl or HCl, 0.01% gelatin	+0.04	—
		−0.22	3.39
	0.5 F K$_3$Cit, pH 9–11	−0.37	—
	1 F Na$_3$Cit, 0.1 F NaOH	−0.50	—
	0.1 F ethylenediamine, 0.1 F KNO$_3$, 0.01% gelatin	−0.51	3.56
	0.1 F diethylenetriamine, 0.1 F KNO$_3$, 0.01% gelatin	−0.54	3.15
	1 F NaF, pH 5, 0.01% gelatin	0.00	3.48
	12 F HCl	>0	—
		−0.71	—
	1 F KOH or NaOH	−0.41	2.91
	1.5 F NaOH, 3% mannitol	−0.60	

Table 7-4   (*Continued*)

A.   Inorganic Ions

Ion	Supporting Electrolyte	$E_{1/2}$ Versus SCE	I
	0.1 F KNO$_3$	+0.02	3.41
	1 F KNO$_3$ or HNO$_3$, 0.01% gelatin	−0.01	3.24
	0.25 F (NH$_4$)$_2$C$_2$O$_4$, pH 10	−0.19	—
		−0.42	—
	1 F K$_2$C$_2$O$_4$, pH 5–10	−0.27	—
	0.1 F NaClO$_4$ or HClO$_4$	+0.01	—
	7.3 F H$_3$PO$_4$	−0.09	—
	0.1 F KH phthalate	−0.10	—
	0.1 F pyridine, 0.1 F pyridinium perchlorate, pH 5.4, 0.02% gelatin	+0.05	—
		−0.25	—
	0.1 F Na$_4$P$_2$O$_7$, pH 4.8, 0.0006% tropeoline 00	−0.11	2.51
	0.05 F Na$_4$P$_2$O$_7$, 0.1 F NaOH	−0.30	—
	0.5 F H$_2$SO$_4$, 0.01% gelatin	0.00	2.12
	0.25 F (NH$_4$)$_2$ tart, pH 10	−0.22	—
		−0.43	—
	0.5 F Na$_2$Tart, pH 4	−0.06	2.37
	0.5 F Na$_2$Tart, pH 10	−0.30	—
	0.5 F KNaTart, 1 F KOH	−0.52	—
	0.1 F KSCN	−0.02	—
		−0.39	—
	1 F KSCN	−0.62	—
	0.3 F triethanolamine, 0.1 F KOH	−0.53	—
	0.1 F EDTA, 1 F K$_2$CO$_3$, pH 9.5	−0.50	—
	0.1 F EDTA, 2 F NaOAc	−0.39	—
	0.25 F EDTA, pH 7	−0.41	2.83
	0.25 F EDTA, pH 9–11.3	−0.47	—
Eu(III)	0.1 F NH$_4$Cl	−0.67	1.47
	0.1 F EDTA, pH 6–8	−1.17	1.30
	0.1 F EDTA, 0.2 F NaOH	−1.22	1.50
Fe(II)	1 F NH$_3$, 1 F NH$_4$Cl	(−0.34)	—
		−1.49	—
	0.05 F BaCl$_2$ or 0.1 F KCl	−1.30	—
	1 F Na$_3$Cit, pH 7	(−0.34)	−0.78
	0.1 F KHF$_2$	+0.11	—
	1 F KOH, 8% mannitol	(−1.09)	—
		−1.55	—

Table 7-4   (*Continued*)

A.   Inorganic Ions

Ion	Supporting Electrolyte	$E_{1/2}$ Versus SCE	I
	0.2–1 F K$_2$C$_2$O$_4$, pH 5	(−0.24)	−1.37
	1 F NH$_4$ClO$_4$	−1.46	—
	0.5 F (NH$_4$)$_2$Tart, 1 F NH$_3$, 0.005% gelatin	(−0.62)	—
		−1.53	—
	0.5 F Na$_2$Tart, pH 5.6, 0.005% gelatin	(−0.17)	—
		−1.50	—
	1 F KSCN	−1.40	—
	0.04–0.4 F EDTA, pH 4	(−0.12)	−1.45
Fe(III)	5 F CaCl$_2$, "pH" 3.5	>0	—
		−1.20	—
	0.15 F Na$_3$Cit, pH 6	−0.18	—
	0.5 F Na$_3$Cit, pH 10	−0.85	—
	0.1 F KHF$_2$	−0.54	—
	1 F KF, pH 7	−0.74	—
		−1.44	—
	12 F HCl	>0	—
		−0.65	—
	1 F K$_2$ malonate, pH 8.2	−0.30	1.25
	3 F KOH, 3% mannitol	−1.12	—
		−1.74	—
	0.2 F K$_2$C$_2$O$_4$ or Na$_2$C$_2$O$_4$, pH 4, 0.005% gelatin	−0.24	1.50
	1 F H$_3$PO$_4$, 1 F NaH$_2$PO$_4$, 0.01% gelatin	−0.03	—
	7.3 F H$_3$PO$_4$	+0.06	—
	0.1 F Na$_4$P$_2$O$_7$, pH 10	−0.82	1.02
	0.1 F Na sulfosalicylate, 0.5 F NaBO$_2$, 0.4 F NaClO$_4$, pH 9	−1.50	—
	0.5 F (NH$_4$)$_2$Tart, 1 F NH$_3$, pH 9.7, 0.005% gelatin	−0.98	—
		−1.53	—
	0.1 F H$_2$Tart, pH 2.0	+0.12	1.55
	0.5 F Na$_2$Tart, pH 6, 0.005% gelatin	−0.19	1.11
		−1.52	—
	0.5 F Na$_2$Tart, pH 9, 0.005% gelatin	−1.20	—
		−1.73	—
	0.7 F KSCN, 0.02 F H$_2$SO$_4$, 0.004% gelatin	>0	1.48
	0.3 F triethanolamine, 1 F NH$_3$, 0.9 F NH$_4$Cl	−0.50	—
		−1.60	—
	0.3 F triethanolamine, 0.1 F KOH	−1.01	—

Table 7-4 (*Continued*)

A. Inorganic Ions

Ion	Supporting Electrolyte	$E_{1/2}$ Versus SCE	I
	0.1 F EDTA, 2 F NaOAc	-0.12	—
	0.04 F EDTA, pH 4.5–6.5	-0.14	1.46
	Dilute solns. of KCl, HCl, HClO$_4$, etc.	>0	—
Fe(CN)$_6^{3-}$	0.1 F KCl	>0	1.79
Ga(III)	1 F NH$_3$, 1 F NH$_4$Cl	-1.60	—
	0.1 F KCl	-1.10	—
	1 F KCN	-1.29	—
	0.001 F HCl	-1.20	—
Gd(III)	0.1 F KCl or LiCl, 0.01% gelatin	-1.75	3.70
Ge(II)	6 F HCl	-0.45	—
	0.1 F NH$_3$, 0.1 F NH$_4$Cl	-1.72	—
		-1.90	—
	0.5 F NH$_3$, 1 F NH$_4$Cl	-1.45	—
		-1.70	—
	0.1 F EDTA, pH 6–8, 10^{-4} F fuchsin	-1.30	—
H(I)	0.1–0.5 F KCl, LiCl, or NaClO$_4$	-1.58	5.60
Hg(I)	0.1 F HNO$_3$	>0	3.68
	1 F HCl, KCl, HClO$_4$, NaClO$_4$, or NaF	>0	—
Hg(II)	0.1 F HNO$_3$	>0	3.48
	2 F HOAc, 2 F NH$_4$OAc, 0.01% gelatin; 1 F NaF, 0.01% gelatin; dil. HCl, KCl, HClO$_4$, or NaClO$_4$	>0	—
In(III)	2 F HOAc, 2 F NH$_4$OAc, 0.01% gelatin	-0.71	3.70
	0.1 F KCl or HCl	-0.56	—
	1 F KCl	-0.60	—
	0.1 F KI	-0.53	—
	0.6 F H$_2$ Tart	-0.59	—
	1 F KOH or NaOH	-1.09	—
I$^-$	0.1 F KNO$_3$	(-0.03)	—
IO$_3^-$	1 F KCl, CsCl, LiCl, or NaCl	-1.16	—

Table 7-4 (*Continued*)

A. Inorganic Ions

Ion	Supporting Electrolyte	$E_{1/2}$ Versus SCE	I
	1 F HNO$_3$	0.00	—
	0.1 F biphthalate buffer, 0.1 F KCl, pH 3.2	-0.31	—
	0.1 F acetate buffer, 0.1 F KCl, pH 4.9	-0.50	—
	0.2 F phosphate buffer, pH 6.4	-0.79	—
	0.5 F Na$_2$B$_4$O$_7$, 0.2 F KNO$_3$, pH 9.2	-1.20	—
	0.1 F NaOH, 0.1 F KCl	-1.21	—
IO$_4^-$	0.16 F K$_2$SO$_4$, 1 F H$_2$SO$_4$	>0	—
		-0.12	—
	0.2 F HBO$_2$ + KOH, pH 10, 0.08% thymol	$+0.02$	—
K(I)	0.1 F N(C$_2$H$_5$)$_4$OH	-2.14	—
	0.1 F N(C$_2$H$_5$)$_4$OH in 50% ethanol	-2.10	1.70
Li(I)	0.1 F N(C$_4$H$_9$)$_4$OH	-2.33	—
	0.1 F N(C$_2$H$_5$)$_4$OH in 50% ethanol	-2.31	1.19
Mn(II)	1 F NH$_3$, 1 F NH$_4$Cl, 0.004% gelatin	-1.66	—
	5 F CaCl$_2$	-1.45	—
	1 F KCl	-1.51	—
	1.5 F KCN	-1.33	—
	1 F NaF, pH 2.5–7, 0.01% gelatin	-1.55	3.93
	1 F KOH or NaOH	-1.70	—
	1.5 F NaOH, 3% mannitol	(-0.38)	-0.64
	0.25 F Na$_2$Tart, 2 F NaOH	(-0.39)	-1.03
	1 F KSCN	-1.54	—
	0.3 F triethanolamine, 0.1 F KOH	(-0.5)	—
		-1.61	—
Mn(III)	0.4 F K$_4$P$_2$O$_7$, pH 2.3, 0.02% agar	>0	1.17
Mo(V)	0.2 F EDTA, pH 12	(-0.53)	—
Mo(VI)	0.03 F Na$_2$HPO$_4$, 0.9 F H$_3$Cit, 0.1 F KCl, pH 2.8	-0.23	—
		-0.58	—
	0.1 F Na$_3$Cit, pH 7	-1.11	—
	0.3 F HCl	-0.26	—
		-0.63	—

Table 7-4 (*Continued*)

A. Inorganic Ions

Ion	Supporting Electrolyte	$E_{1/2}$ Versus SCE	I
	2 F NH$_4$NO$_3$, 3 F HNO$_3$	-0.80	—
	2.3 F HClO$_4$, 0.008% gelatin	-0.24	—
	12 F H$_2$SO$_4$	0.00	—
		-0.13	1.24
	0.1 F H$_2$Tart, pH 2.0	-0.22	—
		-0.52	5.07
	0.004 F EDTA, 0.1 F K$_2$SO$_4$, pH 2.5	-0.33	—
		-0.58	—
	0.05 F EDTA, pH 5.8	-0.80	1.47
	0.1 F EDTA, 0.1 F HOAc, 0.1 F NH$_4$OAc	-0.63	—
Na(I)	0.1 F N(C$_2$H$_5$)$_4$OH	-2.12	—
Nb(V)	0.1 F KCl, pH 2.6	-1.28	—
	0.3 F K$_3$Cit, pH 6.8	-1.73	—
		-2.03	—
	3 F NH$_4$F	-1.90	—
	0.1 F KNO$_3$, pH 2.6	-1.03	—
	0.1 F K$_2$C$_2$O$_4$, pH 1–5.5	-1.53	—
	1 F K$_2$Tart, pH 7.0	-2.00	—
Ni(II)	0.1 F NH$_3$, 0.1 F NH$_4$Cl	-0.92	—
	1 F NH$_3$, 1 F NH$_4$Cl	-1.10	3.56
	5 F CaCl$_2$	-0.56	—
	0.1 F KCl, 0.0003% sodium methyl red	-1.10	3.38
	0.1 F KCN, 0.1 F KCl	-1.42	—
	1 F KCN, 0.1 F KCl	-1.36	—
	1 F NaF, 0.01% gelatin	-1.12	2.29
	12 F HCl	-0.80	—
	0.1 F KH phthalate	-1.14	—
	0.5 F pyridine, 1 F KCl, 0.01% gelatin	-0.78	—
	1 F KSCN, 0.01% gelatin	-0.68	3.59
	Saturated H$_2$Tart	-1.05	—
	0.3 F triethanolamine, 0.7 F NH$_3$, 1 F NH$_4$Cl	-1.18	—
NH$_2$OH	1 F NaOH	(-0.43)	—
N$_3^-$	0.1 F KNO$_3$	$(+0.25)$	—

Table 7-4 (*Continued*)

A. Inorganic Ions

Ion	Supporting Electrolyte	$E_{1/2}$ Versus SCE	I
NO	Dilute HCl	-0.90	
HNO_2	$0.1\ F\ H_2SO_4$, $0.2\ F\ Na_2SO_4$	-0.98	—
	$0.1\ F\ KCl$, $0.01\ F\ HCl$, $0.05-0.2\ mF\ UO_2$-$(OAc)_2$	-1.00	7.45
NO_3^-	$0.04\ F\ LaCl_3$	-1.58	—
	$0.025\ F\ ZrOCl_2$, pH 1.7	-1.00	—
NH_4^+	$N(CH_3)_4Br$	-2.21	—
$N(C_2H_5)_4^+$	$N(CH_3)_4Br$	-2.67	—
$N(C_4H_9)_4^+$	$N(CH_3)_4Br$	-2.57	—
OH^-	$0.1\ F\ KNO_3$	$(+0.08)$	—
H_2O_2	Phosphate-citrate buffer, pH 7	$(+0.18)$	—
		-1.00	—
	$0.1\ F\ NaOH$	(-0.18)	—
		-1.00	—
O_2	$0.1\ F\ KNO_3$, KCl, or most other common supporting electrolytes	-0.05	6.22
		-0.90	12.30
	$7.3\ F\ H_3PO_4$	-0.23	—
		-0.64	—
	$0.1\ F\ NaOH$	-0.18	—
		-1.00	—
Os(VI)	Saturated $Ca(OH)_2$	-0.40	—
		-1.16	—
	$0.1\ F\ Na$ gluconate, $1\ F\ NaOH$	-0.50	2.09
Os(VIII)	$0.5\ F$ acetate buffer, pH 4.7	>0	—
		$+0.10$	—
	Saturated $Ca(OH)_2$	>0	—
		-0.40	—
		-1.16	—

Table 7-4 (*Continued*)

A. Inorganic Ions

Ion	Supporting Electrolyte	$E_{1/2}$ Versus SCE	I
Pb(II)	2 F HOAc, 2 F NH$_4$OAc, 0.01% gelatin	-0.50	2.70
	5 F CaCl$_2$	-0.53	—
	0.1 F KCl, 0.005% gelatin	-0.40	3.85
	1 F KCl, 0.005% gelatin	-0.44	—
	0.25 F (NH$_4$)$_3$Cit, pH 5	-0.77	—
	1 F Na$_3$Cit, 0.1 F NaOH	-0.78	—
		-1.50	—
	1 F KCN	-0.72	—
	1 F NaF, pH 1–3, 0.01% gelatin	-0.41	4.08
	1 F HCl, 0.01% gelatin	-0.44	3.86
	12 F HCl	-0.90	—
	1 F NaOH, 0.005% gelatin	-0.76	3.40
	0.1 F KNO$_3$ or NaNO$_3$	-0.38	—
	1 F HNO$_3$, KNO$_3$, or NaNO$_3$	-0.40	3.67
	1 F K$_2$C$_2$O$_4$, pH 7–10.5	-0.58	—
	1 F NaClO$_4$ or HClO$_4$	-0.38	—
	7.3 F H$_3$PO$_4$	-0.53	—
	0.1 F KH phthalate	-0.40	—
	0.1 F Na$_4$P$_2$O$_7$, pH 10	-0.69	2.57
	0.5 F Na$_2$Tart, pH 9	-0.58	2.40
	0.5 F Na$_2$Tart, 0.01 F NaOH	-0.70	2.40
	0.3 F triethanolamine, 0.1 F KOH	-0.88	—
	0.3 F triethanolamine, 0.7 F NH$_3$, 1 F NH$_4$Cl	-0.56	—
	0.05 F EDTA, 1 F HOAc, 1 F NaOAc, pH 4	-1.10	—
Pd(II)	1 F NH$_3$, 1 F NH$_4$Cl	-0.75	—
	1 F KCN	-1.77	—
	1 F monoethanolamine, 1 F KCl	-0.75	—
	0.1 F diethylamine, 1 F KCl	-0.74	—
	0.1 F ethylenediamine, 1 F KCl, 0.005% methyl red	-0.64	—
	1 F pyridine, 1 F KCl	-0.34	—
Pr(III)	0.1 F LiCl, 0.01% gelatin	-1.80	3.59
	0.1 F N(CH$_3$)$_4$I, 0.01% gelatin	-1.86	3.47
Ra(II)	Dil. KCl	-1.84	—
Rb(I)	0.1 F N(CH$_3$)$_4$OH	-2.03	—

Table 7-4   (*Continued*)

A.   Inorganic Ions

Ion	Supporting Electrolyte	$E_{1/2}$ Versus SCE	I
Re($-$I)	2.4 F HCl	(-0.17)	—
		(-0.34)	—
		(-0.47)	—
		-0.66	—
	1.2 F HClO$_4$	(-0.54)	—
		(-0.42)	—
		(-0.26)	—
		($+0.03$)	—
Re(III)	2 F HClO$_4$	-0.28	—
		-0.46	—
Rh(III)	1 F pyridine and 1 F KCl or 1 F KBr	-0.41	—
	0.9 F KSCN	-0.39	—
Rh(NH$_3$)$_5$Cl^{2+}	1 F NH$_4$Cl	-0.93	—
	1 F NH$_3$, 1 F NH$_4$Cl	-0.93	—
	1 F KCN	-1.47	—
Ru(III)	0.2 F Na gluconate, pH 14	-0.67	1.00
Ru(IV)	1 F HClO$_4$	>0	0.91
		$+0.20$	1.53
		-0.34	2.89
S^{2-}	0.1 F KOH or NaOH	(-0.76)	—
SCN$^-$	0.1 F KNO$_3$	($+0.18$)	—
S$_2$O$_3{}^{2-}$	0.1 F KNO$_3$	(-0.14)	—
S$_4$O$_6{}^{2-}$	1 F H$_3$PO$_4$ containing (NH$_4$)$_2$HPO$_4$, pH 1–8, 0.001% quinoline	-0.26	—
S$_2$O$_4{}^{2-}$	1 F NH$_3$, 0.5 F (NH$_4$)$_2$HPO$_4$, 0.01% gelatin	(-0.43)	4.09
SO$_2$	0.1 F HCl or HNO$_3$	-0.37	5.49
	Phthalate buffer, pH 3.0	-0.48	—
	Acetate buffer, pH 3.6	-0.54	—

Table 7-4 *(Continued)*

A. Inorganic Ions

Ion	Supporting Electrolyte	$E_{1/2}$ Versus SCE	I
	0.05 F phosphate buffer, pH 6.0, 0.1 F KNO$_3$	-1.23	—
SO$_3{}^{2-}$	0.1 F KNO$_3$	$(+0.01)$	—
Sb(III)	2 F HOAc, 2 F NH$_4$OAc, 0.01% gelatin	-0.40	—
		-0.59	4.20
	1 F Na$_3$Cit, 0.1 F NaOH	(-0.44)	—
		-1.24	4.40
	1 F KCN	-1.09	—
	1 F HCl, 0.01% gelatin	-0.15	5.57
	12 F HCl	-0.52	—
		-0.66	—
	0.1 F NaOH, 0.003% thymolphthalein	(-0.37)	-2.93
		-1.07	5.90
	1 F KOH	(-0.45)	—
		-1.15	6.00
	1 F HNO$_3$, 0.01% gelatin	-0.30	5.10
	7.3 F H$_3$PO$_4$	-0.29	—
	1 F H$_2$Tart, 1 F HCl	-0.14	3.66
	1 F Na$_2$Tart, 0.1 F NaOH	(-0.30)	-2.60
		-1.30	3.50
Sb(V)	6 F HCl	>0	3.00
		-0.26	7.50
	12 F HCl	>0	—
		-0.48	—
		-0.57	—
Sc(III)	0.1 F LiCl or KCl, containing trace of HCl	-1.80	—
Se($-$II)	0.05 F NH$_3$, 1 F NH$_4$Cl	(-0.84)	-4.90
		(0.00)	-6.00
	0.5 F H$_3$Cit, pH 2.5	(-0.64)	-2.70
		(0.00)	-4.40
	0.5 F Na$_2$CO$_3$, pH 10.7	(-0.89)	-4.50
	1 F HCl	(-0.49)	-3.80
		(-0.10)	-5.10
	1 F NaOH	(-1.02)	-1.95
		(-0.94)	-3.78

Table 7-4 (*Continued*)

A. Inorganic Ions

Ion	Supporting Electrolyte	$E_{1/2}$ Versus SCE	I
Se(IV)	0.1 F NH$_3$, 0.1 F NH$_4$Cl	-1.64	11.00
	1 F NH$_3$, 1 F NH$_4$Cl	-1.53	11.02
	1 F HCl	>0	—
		-0.10	—
		-0.40	—
		-0.50	—
	1 F (NH$_4$)$_2$Tart, 2 F NH$_3$	-1.53	10.20
Sm(III)	0.1 F N(CH$_3$)$_4$I, 0.0005 F H$_2$SO$_4$, 0.01% gelatin	-1.80	3.85
		-1.96	12.10
Sn(II)	2 F HOAc, 2 F NH$_4$OAc, 0.01% gelatin	(-0.16)	—
		-0.62	2.60
	0.5 F NaCl, 2 F HClO$_4$	-0.35	—
	1 F Na$_3$Cit, 0.1 F NaOH	(-0.91)	—
		-1.12	—
	1 F NaF, pH 4–6, 0.01% gelatin	(-0.20)	-4.10
		-0.73	4.10
	12 F HCl, 0.002% triton X-100	-0.83	—
	1 F NaOH, 0.01% gelatin	(-0.73)	-3.45
		-1.22	3.45
	1 F HNO$_3$, 0.01% gelatin	-0.44	4.02
	1 F HClO$_4$	$(+0.14)$	—
		-0.43	—
	7.3 F H$_3$PO$_4$	-0.58	—
	0.5 F H$_2$SO$_4$, 0.01% gelatin	-0.46	3.54
	0.4 F H$_2$Tart, 0.1 F NaHTart, pH 2.3 0.01% gelatin	-0.49	—
	0.4 F Na$_2$Tart, 0.1 F NaClO$_4$, pH 9.0, 0.01% gelatin	(-0.33)	—
		-0.92	—
	0.5 F Na$_2$Tart, 0.1 F NaOH, 0.01% gelatin	(-0.71)	—
		-1.16	2.86
Sn(IV)	4 F NH$_4$Br, 0.005% gelatin	>0	—
		-0.50	6.52
	0.5 F NaCl, 2 F HClO$_4$	-0.47	—
	1 F Na$_3$Cit, 0.1 F NaOH	-1.22	—

Table 7-4 (*Continued*)

A. Inorganic Ions

Ion	Supporting Electrolyte	$E_{1/2}$ Versus SCE	I
Sr(II)	0.1 F N(C$_2$H$_5$)$_4$I	-2.11	3.46
Ta(V)	0.9 F HCl	-1.16	—
	0.5–1 F K$_2$C$_2$O$_4$, pH 0.5–3	-1.40	—
	0.1 F K$_2$Tart, pH 3–5	-1.57	—
Te($-$II)	0.1 F NH$_3$, 1 F NH$_4$Cl, 0.003% gelatin	(-1.10)	—
	0.5 F citrate buffer, pH 3.3, 0.03% gelatin	(-0.95)	—
	1 F HCl, 0.003% gelatin	(-0.73)	
	1 F NaOH, 0.003% gelatin	(-1.20)	-3.50
		(-0.40)	—
Te(IV)	1 F NH$_3$, 1 F NH$_4$Cl, pH 9.4	-0.67	—
	0.5 F NaBO$_2$ or Na$_2$CO$_3$, pH 9.4	-0.88	—
	0.5 F H$_3$Cit, pH 1.6	-0.05	—
		-0.40	5.72
	1 F NaF, pH 6.9, 0.01% gelatin	-0.89	6.70
	1 F NaOH, 0.003% gelatin	-1.10	—
		-1.19	9.75
	1 F (NH$_4$)$_2$Tart, 2 F NH$_3$, pH 9.3	-0.70	6.40
Te(VI)	Acetate buffer, pH 5.6	-1.18	15.40
	NH$_3$ – NH$_4$Cl buffer, pH 8.0, 0.0005% gelatin	-1.21	17.50
	Carbonate buffer, pH 8.3	-1.37	16.60
	0.1 F KCl or KClO$_4$	-1.10	—
		-1.45	—
	0.5 F (NH$_4$)$_3$Cit, pH 6.2, 0.003% gelatin	-1.19	—
	1 F Na$_3$Cit, 0.1 F NaOH	-1.54	—
	1 F KCN	-1.36	—
	1 F NaF, pH 6.5–9.5, 0.01% gelatin	-1.50	2.35
	12 F HCl	>0	—
		-0.43	—
		-0.79	—
	1 F NaOH, 0.003% gelatin	-1.57	—
	Saturated (NH$_4$)$_2$C$_2$O$_4$ + NH$_3$, pH 8.0	-1.23	16.30
	1 F (NH$_4$)$_2$Tart, pH 8.4	-1.38	13.00
Ti(III)	Saturated CaCl$_2$	(-0.12)	—
	0.01 F HCl	(-0.14)	—

Table 7-4 (*Continued*)

A. Inorganic Ions

Ion	Supporting Electrolyte	$E_{1/2}$ Versus SCE	I
	0.2 F H$_2$C$_2$O$_4$, pH 1	(−0.30)	−1.60
	Saturated H$_2$Tart	(−0.44)	—
	0.1 F KSCN	(−0.46)	—
	EDTA, pH 1.0–2.5	(−0.22)	—
Ti(IV)	Saturated CaCl$_2$	−0.12	—
	0.2 F H$_3$Cit	−0.37	—
	0.4 F Na$_3$Cit, pH 5.5–6, 0.005% gelatin	−0.90	1.02
	0.1 F HCl, 0.005% gelatin	−0.81	1.56
	3.5 F lactic acid	−0.40	—
	0.2 F H$_2$C$_2$O$_4$, pH 0.5	−0.28	1.75
	Saturated phthalic acid←	−0.93	—
	Saturated salicylic acid	−0.35	—
	0.03 F H$_2$SO$_4$	−0.79	—
	Saturated H$_2$Tart	−0.44	—
	0.1 F KSCN	−0.46	—
	EDTA, pH 1.0–2.5	−0.22	—
	0.1 F EDTA, 2 F NaOAc	−0.53	—
Tl(I)	2 F HOAc, 2 F NH$_4$OAc, 0.01% gelatin	−0.47	2.30
	0.1 F NH$_3$, NH$_4$Cl, KCl, HCl, KOH, HNO$_3$, KNO$_3$, HClO$_4$, or NaClO$_4$	−0.46	2.70
	1 F NH$_3$, NH$_4$Cl, KCl, HCl, KOH, HNO$_3$, KNO$_3$, HClO$_4$, or NaClO$_4$	−0.48	—
	1 F Na$_3$Cit, 0.1 F NaOH	−0.56	—
	0.1–1 F KCN	>0	—
	1 F NaF, pH 3.5–6.5, 0.01% gelatin	−0.50	2.67
	7.3 F H$_3$PO$_4$	−0.63	—
	0.2 F Na$_4$P$_2$O$_7$, 0.2 F KOH	−0.55	—
	17 F H$_2$SO$_4$	−0.98	—
	0.01 F EDTA, 1 F HOAc, 1 F NaOAc	−0.46	—
Tl(III)	0.6 F HCl	>0	3.83
		−0.45	5.72
	1 F HClO$_4$	>0	—
		−0.48	—
U(III)	1 F HClO$_4$	(−0.87)	−1.50

Table 7-4 (*Continued*)

A. Inorganic Ions

Ion	Supporting Electrolyte	$E_{1/2}$ Versus SCE	I
U(IV)	0.1 F HClO$_4$	-0.86	1.57
	1 F HCl	-0.89	—
U(V)	0.5 F NaClO$_4$, 0.01 F HClO$_4$	(-0.18)	-1.57
U(VI)	0.4 F HOAc, pH 2.7	-0.15	—
		-0.69	—
		-1.02	—
	2 F HOAc, 2 F NH$_4$OAc, 0.01% gelatin	-0.45	1.70
	0.5 F (NH$_4$)$_2$CO$_3$	-0.83	1.50
		-1.45	—
	1 F NH$_3$, 1 F NH$_4$Cl	-0.80	—
		-1.40	—
	1 F Na$_2$CO$_3$	-0.95	1.50
	0.1 F H$_3$Cit, 0.1 F K$_3$Cit	-0.38	2.40
	1 F Na$_3$Cit, 0.1 F NaOH	-0.98	—
	0.01 F HF	-0.21	2.48
	1 F NaF	-0.94	1.40
	0.1 F HCl	-0.18	1.54
		-0.94	—
	2 F HCl	-0.21	3.08
		-0.90	—
	12 F HCl	>0	—
		-0.63	—
	2 F hydroxylamine hydrochloride	-0.26	2.05
	0.5 F H$_2$C$_2$O$_4$	-0.13	3.20
	0.5 F NaClO$_4$, 0.01 F HClO$_4$	-0.18	1.57
	7.3 F H$_3$PO$_4$	-0.12	—
		-0.58	—
	0.05 F H$_2$SO$_4$	-0.22	2.00
		-0.90	2.35
		-1.06	—
	0.1 F Na gluconate, 0.1 F NH$_4$ClO$_4$, pH 11	-1.17	1.95
	0.1 F EDTA, 2 F NaOAc	-0.41	—
V(II)	1 F acetate buffer, pH 5.4	(-0.89)	-1.09
		(-0.11)	-3.36
	1 F KBr	(-0.50)	-2.03

Table 7-4 (*Continued*)

A. Inorganic Ions

Ion	Supporting Electrolyte	$E_{1/2}$ Versus SCE	I
	0.5 F KHCO$_3$, 0.5 F Na$_2$CO$_3$, pH 9.4	(-0.75)	-1.19
		(-0.18)	-4.16
	1 F Na$_3$Cit, pH 7	(-1.17)	-0.87
	1 F KI	(-0.49)	-2.27
	1 F K$_2$C$_2$O$_4$, pH 6.5	(-1.09)	-1.43
	Saturated KH phthalate, pH 5.2	(-0.84)	-1.78
		(0.15)	-3.53
	1 F Na salicylate-salicylic acid buffer pH 4.7	(-0.76)	-1.16
		(-0.15)	-3.52
	0.5 F H$_2$SO$_4$	(-0.50)	-1.74
	1 F Na$_2$Tart, pH 6	(-0.17)	-1.07
	1 F NH$_4$SCN	(-0.47)	-2.04
	0.1 F EDTA, pH < 8.3	(-1.27)	-1.10
V(III)	1 F acetate buffer, pH 5.4	-0.98	0.57
		-1.25	1.39
	1 F KBr, pH 2.5	-0.43	1.42
		-0.87	1.94
	0.5 F KHCO$_3$, 0.5 F Na$_2$CO$_3$, pH 9.4	(-0.34)	-2.80
	1 F KCN	-1.17	0.71
		-1.77	1.33
	1 F HCl or HClO$_4$, or 0.5 F H$_2$SO$_4$	-0.51	1.41
	1 F K$_2$C$_2$O$_4$, pH 3.5–6.5	-1.14	1.95
	Saturated KH phthalate, pH 5.2	(-0.10)	-1.13
		-0.88	1.22
	1 F Na salicylate-salicylic acid buffer, pH 4.7	(-0.06)	-1.17
		-0.97	0.26
		-1.21	1.26
	1 F NH$_4$SCN	-0.46	1.78
	0.1 F EDTA, pH 5–8.5	-1.27	1.20
V(IV)	1 F NH$_3$, 1 F NH$_4$Cl, 0.08 F Na$_2$SO$_3$	(-0.32)	-0.94
		-1.28	1.82
	0.5–1 F KHCO$_3$ saturated with CO$_2$	(-0.16)	-1.41
	1 F Na$_3$Cit, 0.1 F NaOH	(-0.47)	—
		-1.76	—
	12 F HCl, 0.002% triton X-100	>0	—
		-0.62	—
		-0.75	—

Table 7-4 (*Continued*)

A. Inorganic Ions

Ion	Supporting Electrolyte	$E_{1/2}$ Versus SCE	I
	0.05 F H$_2$SO$_4$, 0.005% gelatin	-0.85	3.20
	0.1 F EDTA, pH 9.5	-1.25	2.20
	1.0 F K$_2$C$_2$O$_4$, pH 6	-1.31	4.32
V(V)	1 F NH$_3$, 1 F NH$_4$Cl, 0.005% gelatin	-0.96	1.60
		-1.26	4.72
	0.1 F HCl	>0	—
		-0.80	—
	1 F K$_2$C$_2$O$_4$, pH 5	>0	1.86
		-1.33	5.60
	7.3 F H$_3$PO$_4$	>0	—
		-0.54	—
		-0.91	—
	0.025 F H$_2$SO$_4$, 0.1 F KCl, 0.005% gelatin	>0	1.65
		-0.98	4.96
	0.1 F H$_2$Tart, pH 2.0	$+0.40$	—
		0.00	—
		-0.60	—
	0.1 F EDTA, pH 9.5	-1.22	3.29
W(III)	3 F HCl	(-0.65)	—
		(-0.45)	—
	12 F HCl	(-0.53)	—
	12 F HCl	(-0.56)	-1.10
W(V)	12 F HCl	-0.56	2.53
W(VI)	12 F HCl	>0	1.31
		-0.55	3.82
	10 F HCl	>0	1.42
		-0.60	4.33
	7.3 F H$_3$PO$_4$	-0.59	1.47
	0.1 F H$_2$Tart, 5 F HCl	-0.33	—
		-0.68	4.40
Yb(III)	0.1 F NH$_4$Cl	-1.41	1.57
		-2.00	—
Zn(II)	2 F HOAc, 2 F NH$_4$OAc, 0.01% gelatin	-1.10	1.50
	1 F NH$_3$, 1 F NH$_4$Cl, 0.005% gelatin	-1.35	3.82
	1 F KCl, 0.0003% sodium methyl red	-1.00	3.42

Table 7-4 (*Continued*)

A.   Inorganic Ions

Ion	Supporting Electrolyte	$E_{1/2}$ Versus SCE	I
	0.15 F K$_3$Cit, pH < 3	-1.06	—
	1 F Na$_3$Cit, 0.1 F NaOH	-1.43	—
	1 F NaF, pH 4–6.5, 0.01% gelatin	-1.14	3.15
	1 F N$_2$H$_4$, 1 F NaClO$_4$, pH 8–9	-1.13	—
	1 F NaOH, 0.002% gelatin	-1.49	3.04
	0.01 F KNO$_3$	-0.99	—
	Nearly saturated (NH$_4$)$_2$C$_2$O$_4$	-1.30	—
	1 F NaClO$_4$	-1.00	—
	0.1 F KH phthalate	-1.01	—
	0.1 F pyridine, 0.1 F KCl, 0.01% gelatin	-1.02	—
	1 F pyridine, 0.1 F NaOH	-1.57	—
	Saturated H$_2$Tart	-1.03	—
	0.5 F Na$_2$Tart, pH 8.8, 0.01% gelatin	-1.15	2.30
	0.5 F Na$_2$ Tart, 0.1 F NaOH, 0.01% gelatin	-1.42	2.65
	0.1 F KSCN	-1.01	—
	0.3 F triethanolamine, 0.1 F KOH	-1.57	
	0.3 F triethanolamine, 0.7 F NH$_3$, 1 F NH$_4$Cl	-1.36	—

B.   Organic Molecules

1.   Acids and Acid Derivatives

Compound	Supporting Electrolyte	$E_{1/2}$(V)	I
Acetic acid	0.1 F Et$_4$NClO$_4$/CH$_3$CN	-2.3	—
Acetylene dicarboxylic acid	HCl-KCl (pH 0.5)	-0.56	7.1
Diethyl ester	HCl-KCl (pH 1.5)	-0.48	4.3
		-0.63	4.3
Acrylamide	0.05 F Me$_4$NI/30% EtOH	-1.91	3.49
Acrylic acid, ethyl ester	0.05 F Me$_4$NI/30% EtOH	-1.82	—
Acrylonitrile	0.05 F Me$_4$NI/30% EtOH	-1.96	—
Benzhydroxamic acid	0.05 F Bu$_4$NCl/90% EtOH	-2.18	4.6
Benzoic acid	0.1 F Et$_4$NClO$_4$/CH$_3$CN	-2.1	—
Fumaric acid	NH$_3$ buffer (pH 8)/10% EtOH	-1.57	4.1
Diethyl ester	NH$_3$ buffer (pH 8)/10% EtOH	-1.01	3.7
α-Ketoglutaric acid	0.7 F KCl + HCl (pH 2)	-0.63	2.71
Maleic acid	NH$_3$ buffer (pH 8)/10% EtOH	-1.35	3.6

Table 7-4   (*Continued*)

B.   Organic Molecules

1.   Acids and Acid Derivatives

Compound	Supporting Electrolyte	$E_{1/2}$(V)	I
Diethyl ester	NH_3 buffer (pH 8)/10% EtOH	-1.02	3.6
Oxalic acid	0.1 F Et_4NClO_4/CH_3CN	-1.6	2.06
Phthalic acid			
Diethyl ester	0.1 F Me_4NCl/75% EtOH	-1.87	4.64
Diphenyl ester	0.1 F Me_4NCl/75% EtOH	-1.65	4.47

2.   Carbonyls and Carbonyl Derivatives

Acetaldehyde	0.06 F LiOH	-1.93	—
Acetone	0.1 F Bu_4NCl + 0.1 F Bu_4NOH/ 80% EtOH	-2.53	—
1,3-Diphenyl-	0.1 F Bu_4NCl + 0.1 F Bu_4NOH/ 80% EtOH	-2.10	—
Acetophenone oxime	0.1 F HCl-KCl/50% EtOH	-1.09	4.90
Benzaldehyde	0.1 F LiOH/50% EtOH	-1.51	—
Oxime	0.1 F HCl-KCl/50% EtOH	-0.77	4.60
tert-Butyl phenyl ketone	0.1 F LiOH/50% EtOH	-1.92	—
Butyrophenone	0.1 F LiOH/50% EtOH	-1.75	—
n-Caprylic aldehyde	0.05 F Bu_4NCl/90% EtOH	-2.35	—
Crotonaldehyde	0.2 F Me_4NOH/50% EtOH	-1.37	—
		-1.80	—
Dimethylglyoxime	0.1 F HCl/15% EtOH	-0.81	11.50
Formaldehyde	0.06 F LiOH	-1.75	—
Methyl vinyl ketone	0.1 F KCl	-1.42	—
Phthalaldehyde	OAc^- buffer (pH 5)/1.5% EtOH	-0.72	—
		-1.09	—
p-Quinone dioxime	0.1 F NaOH/55% EtOH	-1.12	—

3.   Halogen and α-Halocarbonyl Compounds

Acetaldehyde, chloro-	NH_3 buffer (pH 8.5)	-1.03	—
		-1.67	—

Table 7-4 *(Continued)*

B. Organic Molecules *(Continued)*

3. Halogen and α-Halocarbonyl Compounds *(Continued)*

Compound	Supporting Electrolyte	$E_{1/2}$(V)	I
Acetone			
Bromo-	OAc⁻ buffer (pH 4.5)	−0.34	3.30
Chloro-	OAc⁻ buffer (pH 4.5)	−1.15	2.90
Iodo-	OAc⁻ buffer (pH 4.5)	−0.14	3.30
Allyl bromide	Li_3Cit/50% dioxane	−1.18	—
α-Chloro-	Li_3Cit/50% dioxane	−0.88	—
Benzene			
Bromo-	0.05 F Et_4NBr/DMF	−2.24	4.20
Chloro-	0.05 F Et_4NBr/DMF	−2.55	—
Iodo-	0.05 F Et_4NBr/DMF	−1.64	—
Butyraldehyde			
α-Bromo-	0.1 F LiCl/50% dioxane	−0.48	—
		−1.91	—
α-Chloro-	0.1 F LiCl/50% dioxane	−1.33	—
		−1.90	—
Cyclohexane			
Bromo-	0.05 F Et_4NBr/DMF	−2.28	—
Cyclohexanone, 2-chloro-	0.1 F KCl	−1.45	1.86
Cyclopentanone, 2-chloro-	0.1 F KCl	−1.35	2.93
Phenacyl chloride	0.05 F Bu_4NCl/90% EtOH	−0.92	—
Propene, 1-bromo-	0.05 F Et_4NBr/DMF	−2.50	—

4. Heterocyclic Compounds

2-Acetothiophene	OAc⁻ buffer (pH 5)	−1.25	3.50
2,2′-Bipyridine	OAc⁻ buffer (pH 4.5),	−1.14	—
	0.1 F KCl	−1.25	—
2,4′-Bipyridine	OAc⁻ buffer (pH 4.5),		
	0.1 F KCl	−1.04	—
3,3′-Bipyridine	OAc⁻ buffer (pH 4.5),		
	0.1 F KCl	−1.55	—
Phthalimide	HCl-KCl/50% EtOH (pH 2)	−0.90	—
N-Propyl pyridinium bromide	0.05 F PO_4^{-3} buffer (pH 8)	−1.36	—
Pyrimidine	0.05 F PO_4^{-3} buffer (pH 7)	−1.30	3.50
Pyronin	PO_4^{-3} buffer/1% EtOH (pH 7)	−0.69	—
Quinoline	0.2 F Me_4NOH/50% EtOH	−1.50	—
Quinoxaline	0.05 F PO_4^{-3} buffer (pH 7)	−0.68	3.69

Table 7-4 (*Continued*)

B.   Organic Molecules   (*Continued*)

Compound	Supporting Electrolyte	$E_{1/2}$(V)	I

5.   Nitro and Related Compounds

Compound	Supporting Electrolyte	$E_{1/2}$(V)	I
Azobenzene	0.01 F HCl + 0.02 F KCl/		
	30% MeOH	−0.06	3.11
		−0.81	2.50
Azoxybenzene	0.01 F HCl + 0.02 F KCl/		
	30% MeOH	−0.25	6.14
		−0.83	2.54
Benzenediazonium chloride	OAc⁻ buffer (pH 4)	−0.19	—
Cyclohexyl nitrate	0.5 F LiCl, 0.01% gelatin/		
	11.25 wt. % EtOH	−0.63	3.06
Ethyl nitrate	0.5 F LiCl, 0.01% gelatin/		
	11.25 wt. % EtOH	−0.82	3.39
Nitrobenzene	OAc⁻ buffer/60% EtOH (pH 3)	−0.43	—
p-Chloro-	OAc⁻ buffer/60% EtOH (pH 3)	−0.40	—
p-Hydroxy-	OAc⁻ buffer/60% EtOH (pH 3)	−0.56	—
(tri)Nitroglycerin	0.1 F Me₄NCl/75% EtOH	−0.70	—
Nitrosocyclohexane dimer	0.1 F HCl-KCl/50% EtOH	−0.21	2.20
		−0.51	5.70
N-Nitrosodimethylamine	OAc⁻ buffer (pH 3.5)	−1.21	—

6.   Sulfur Containing Compounds

Compound	Supporting Electrolyte	$E_{1/2}$(V)	I
Cystine	NH₃ buffer (pH 9)	−0.70	—
		−1.30	—
Diethyl disulfide	0.025 F Bu₄NOH/2-PrOH-MeOH-		
	H₂O(2:2:I)	−1.78	—
Diphenyl disulfide	0.025 F Bu₄NOH/2-PrOH-MeOH-		
	H₂O(2:2:1)	−0.65	—
Diphenyl sulfone	0.1 F Me₄NBr/50% EtOH	−2.04	2.65
Diphenyl sulfoxide	0.1 F Me₄NBr/50% EtOH	−2.07	2.54
Methyl phenyl sulfone	0.1 F Me₄NBr/50% EtOH	−2.14	2.55
Sodium diethyldithio phosphate	0.1 F HClO₄	(−0.06)	−2.00

Table 7-4   (*Continued*)

B.   Organic Molecules   (*Continued*)

6.   Sulfur Containing Compounds   (*Continued*)

Compound	Supporting Electrolyte	$E_{1/2}(V)$	I
Sodium formaldehyde sulfoxylate	0.1 F NaOH	(−0.42)	−3.84
Thiourea	0.05 F H_2SO_4	(+0.04)	—

7.   Miscellaneous Compounds

Compound	Supporting Electrolyte	$E_{1/2}(V)$	I
Benzoyl peroxide	0.3 F LiCl/C_6H_6-MeOH (1:1)	0.00	3.00
tert-Butyl hydroperoxide	0.3 F H_2SO_4/5% EtOH	−0.31	—
	0.3 F LiCl/C_6H_6-MeOH (1:1)	−0.96	5.80
Catechol	PO_4^{-3} buffer (pH 8)	(+0.10)	—
Chlorotriethyllead	1 F KCl	−0.68	—
Dichlorodiethyltin	1 F KCl	−0.57	—
Succinic acid peroxide	0.3 F LiCl/C_6H_6-MeOH (1:1)	−0.19	3.10

8.   Unsaturated Hydrocarbons

Compound	Supporting Electrolyte	$E_{1/2}(V)$	I
Acetylene, phenyl-	0.175 F (*n*-Bu)$_4$NI/75% dioxane	−2.37	—
Anthracene	0.175 F (*n*-Bu)$_4$NI/75% dioxane	−1.94	—
Azulene	0.175 F (*n*-Bu)$_4$NI/75% dioxane	−1.63	—
		−2.28	—
		−2.52	—
Naphthalene	0.175 F (*n*-Bu)$_4$NI/75% dioxane	−2.49	3.02
1-(-Cyclohexenyl)	0.1 F (*n*-Bu)$_4$NI/75% dioxane	−2.42	—
		−2.48	—
1-(1-Cyclopentenyl)	0.1 F (*n*-Bu)$_4$NI/75% dioxane	−2.25	3.03
		−2.49	3.60
Phenanthrene	0.175 F (*n*-Bu)$_4$NI/75% dioxane	−2.44	—
		−2.67	—
Styrene	0.175 F (*n*-Bu)$_4$NI/75% dioxane	−2.34	—
β-Methyl-	0.175 F (*n*-Bu)$_4$NI/75% dioxane	−2.54	—

An important specialized type of voltammetric system is a self-contained cell for the determination of oxygen in both the gas and solution phases. This is the so-called Clark electrode (14, 15) which consists of a platinum or gold electrode in the end of a support rod which is covered by an oxygen permeable membrane (polyethylene or Teflon) such that a thin film of electrolyte is contained between the electrode surface and the membrane. A concentric tube provides the support for the membrane and the means for containing an electrolyte solution in contact with a silver-silver chloride reference electrode. The Clark device has found extensive application in monitoring oxygen partial pressure in blood, the atmosphere, and in sewage plants. By appropriate adjustment of the applied potential it gives a volt-ammetric current plateau which is directly proportional to the oxygen partial pressure. The membrane material prevents interference from electroactive ions as well as from surface contaminating biological materials. Figure 7-9 illustrates one configuration for this important device.

Controlled potential electrolysis provides the features of electrodeposition plus the ability to carry out analyses more rapidly while using smaller samples. Furthermore, by integrating the current-time curve the necessity for plating a weighable amount of substituent is eliminated. One of the most important applications of controlled potential electrolysis is the evaluation of the number of electrons involved in the electrode reaction, n.

In addition to the analytical applications discussed above, controlled potential methods are used increasingly for the evaluation of thermodynamic data and diffusion coefficients in both aqueous and nonaqueous solvents. Polarographic and voltammetric methods provide a convenient and straightforward means to the evaluation of the diffusion coefficients in a variety of media. The requirements are that the current be diffusion con-trolled, the number of electrons in the electrode reaction be known, and the concentration of the electro-active species and the area of electrodes be known. With these conditions satisfied, diffusion coefficients can be evaluated rapidly over a range of temperatures and solution conditions.

Voltammetric methods also provide a convenient approach for establish-ing the thermodynamic reversibility of an electrode reaction and for the evaluation of the electron stoichiometry for the electrode reaction. As outlined in the theoretical section the standard electrode potential, the dissociation constants of weak acids and bases, solubility products, and the formation constants of complex ions can be evaluated from polarographic half-wave potentials, if the electrode process is reversible. Furthermore, studies of half-wave potentials as a function of ligand concentration provide the means for determining the formula of a metal complex.

Although the use of voltammetric methods for the study of the electronic and molecular properties of organic and biological molecules was suggested

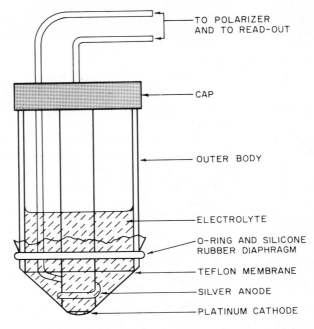

Figure 7-9. Clark electrode for the voltammetric measurement of oxygen partial pressure.

many years ago, only recently has interest in this application developed among organic chemists and biochemists. One reason for this has been that the technology of voltammetry until recently has been such that meaningful thermodynamic and kinetic data in general were only possible for aqueous systems. From the organic chemist's viewpoint water is not an ideal solvent nor does it represent a medium that is particularly relevant to organic processes. Thus the application of voltammetric methods to physical chemical studies of organic systems has required the development of electronic instruments capable of using three electrode configurations. This has permitted the use of the high resistance solutions that are characteristic of nonaqueous solvent systems. Under these conditions studies of the effects of substituents and of solvents on the half-wave potential of electroactive organic molecules are possible. Such studies provide a relative measurement of the free energies of removal (or of addition) of valence electrons, which can be correlated with molecular orbital calculations. Table 7-5 summarizes several examples of such applications, and indicates the general applicability of voltammetric methods for the generation of fundamental electronic data for structural studies (16).

Table 7-5   Half-Wave Reduction Potentials for Hydrocarbons in
75% Dioxane-H_2O and Energies of Lowest Vacant Molecular
Orbitals

Hydrocarbon	$-m^*_{m+1}$ E, Lowest Orbital	$-E_{1/2}$ versus SCE
Triphenylmethyl	0	1.05
1-Phenyl-6-biphenylenehexatriene	0.202	1.35
1-Phenyl-4-biphenylenebutadiene	0.251	1.46
Acenaphthylene	0.285	1.65
Tetracene	0.295	1.58
Perylene	0.347	1.67
1,2-Benzpyrene	0.365	1.85
Fluoranthene	0.371	1.77
1,4-Diphenylbutadiene	0.386	2.00
Azulene	0.400	1.64
Anthracene	0.414	1.96
Pyrene	0.445	2.11
1,2-Benzanthracene	0.452	2.00
1,2,5,6-Dibenzanthracene	0.474	2.03
4,5-Benzpyrene	0.497	2.00
Stilbene	0.504	2.16
Chrysene	0.520	2.30
2,2'-Binaphthyl	0.521	2.21
p-Quaterphenyl	0.536	2.20
Coronene	0.539	2.04
1,1-Diphenylethylene	0.565	2.25
p-Terphenyl	0.593	2.33
Phenanthrene	0.605	2.46
Naphthalene	0.618	2.50
Butadiene	0.618	2.63
Styrene	0.662	2.37
Triphenylene	0.684	2.49
Biphenyl	0.705	2.70

The new techniques of voltage sweep and cyclic voltammetry provide the analytical and physical chemical capabilities of classical voltammetry and in addition provide the means to perform these measurements much more rapidly for a broader range of conditions. Cyclic voltammetry is particularly useful for the rapid assessment of thermodynamic reversibility, and for the evaluation of the stoichiometry for the electrode reaction.

As outlined in the theoretical section of this chapter, controlled potential methods have extensive application in the study of the kinetics and mechanisms of the electron-transfer reaction of electrochemical processes. Furthermore, pre- and postchemical reactions that are associated with the electron transfer reaction are readily studied by controlled potential methods. For a number of systems the rate constants for these associated chemical processes can be evaluated.

For processes that are sufficiently slow, polarographic and voltammetric current-potential curves provide a measure of the heterogeneous rate constant and the transfer coefficient for the electrode reaction. Equations 7-20 and 7-21 indicate the pertinent relationships for such measurements. Although homogeneous chemical rate constants and mechanisms have been studied by these methods there are much more convenient approaches that provide a quantitative measure of such associated processes.

In particular, voltammetric sweep and cyclic voltammetry are ideally suited for the study of the kinetic phenomena of electrochemistry (4). The kinetic parameters for the heterogeneous rate constants and the transfer coefficient can be evaluated either from single sweep or cyclic voltammetry; Equations 7-29, 7-31, and 7-34 indicate the pertinent relationships. Cyclic voltammetry is especially well suited for the study of homogeneous post-electrochemical processes. Equations 7-39 and 7-40 outline the quantities that must be measured for quantitative evaluation of the rate constants for the reaction $O + ne^- \rightleftarrows R \overset{k_f}{\underset{k_r}{\rightleftarrows}} Y$.

$$E_p = E_{1/2} - \left(\frac{RT}{nF}\right)\left[0.780 + \ln K\sqrt{\frac{nFv}{RT(k_f + k_r)}} - \ln(1 + K)\right]$$

reversible chemical reaction

(7-39)

$$E_p = E_{1/2} - \left(\frac{RT}{nF}\right)\left(0.780 - \ln\sqrt{\frac{k_f RT}{nFv}}\right)$$

irreversible chemical reaction

(7-40)

Because of the high scan rates associated with cyclic voltammetry unstable intermediates can be observed and some measure of their lifetimes can be made.

Voltammetric methods provide the means of elucidating the mechanism for the rate controlling electron transfer process as well as for any associated chemical processes that are rate controlling. However, the overall electron stoichiometry frequently must be determined for the electrode reaction. This normally is accomplished by the use of controlled potential coulometry, which provides a direct measure of the number of electrons per electroactive species. The latter knowledge is essential if a truly realistic mechanism is to be developed for the overall electrochemical process.

Both the ring-disc and thin-layer electrodes provide a convenient means of observing unstable intermediate products from electrochemical reactions. Quantitative evaluations of the lifetimes of these intermediates and of the products from such intermediates are readily evaluated by each of these methods (8).

One of the most important, yet latent, applications of controlled potential electrolysis is electrochemical synthesis. Although electrolysis has been used for more than a century to synthesize various metals from their salts, application to other types of chemical synthesis has been extremely limited. Before the advent of controlled potential methods the selectivity possible by classical electrolysis precluded fine control of the products. The only control was provided by appropriate selection of electrode material, solution acidity, and supporting electrolyte. Thus by these means the effective electrode potential could be limited to that which caused the electrolysis of the supporting electrolyte or the solvent. Today potentiostats and related controlled potential electrolysis instrumentation are commercially available which provide effective control of the potential of the working electrode to ± 1 mV, and a driving force of up to 100 V for currents up to several amperes. Through such instrumentation electrochemical synthesis becomes possible using a wide range of solvent systems and supporting electrolytes. During the past decade significant effort has been expended to develop practical syntheses based on electrochemistry. Some of the more promising of these efforts are summarized in Table 7-6. Thus the high degree of control possible through appropriate selection of controlled potential, electrode material, supporting electrolytes, and pH permits highly efficient syntheses with few if any byproducts.

When electrochemical studies have established the ideal electrochemical conditions for a particular synthesis a simpler, more rapid and efficient chemical synthesis can be developed on the basis of the electrochemical knowledge. For example, if a given organic substance can be quantitatively reduced to give the desired product at a mercury pool electrode using a potential of -0.900 V versus SCE, then an electrochemical synthesis could be accomplished with adequate instrumentation in approximately 20 to 30 min. However, these conditions could be duplicated chemically through the use of a cadmium amalgam in contact with an electrolyte solution containing EDTA which is adjusted to pH 8. That this is true was first established by polarographic studies of cadmium ion in the presence of the specified

Table 7-6 Examples of Practical Electrochemical Syntheses

Product	Reactant	Process	Ref.
3-Acetoxycyclohexene-1	Cyclohexene, acetic acid	Oxidative substitution	17
Adiponitrile	Acrylonitrile	Reductive coupling	18
Benzyl alcohol	Benzoic acid	Reduction	19
Chromium hexacarbonyl	Chromium acetyl-acetonate, carbon monoxide	Reduction	20
Cyanogen Bromide	Hydrogen cyanide, ammonium bromide	Oxidation	21
Cyclohexadiene dicarboxylic acid	Phthalic acid	Reduction	22
"Dewar Benzene"	[2.2.0] Hexa-5-ene-2,3-dicarboxylic acid	Oxidation	23
Diethyl adipate	Ethyl acrylate	Reductive coupling	24
1,3-Dimethylbicyclobutane	1,3-Dibromo-1,3-dimethylcyclo-butane	Reductive cyclization	25
Dimethyl sebacate	Monomethyl adipate	Oxidation	26
Dimethyl 1,14-tetradecan-dioate	Monomethylsuberate	Oxidative coupling	27
Polyacrylonitrile	Acrylonitrile	Oxidative polymeri-zation	28
Polymethyl methacrylate	Methyl methacrylate	Reductive polymeri-zation	29
Polystyrene	Styrene	Reductive polymeri-zation	30
Propylene oxide	Propylene	Oxidation	31
Salicylaldehyde	Salicylic acid	Reduction	19,32
Sorbitol	Glucose	Reduction	33
Tetraethyllead	Ethyl chloride, lead	Oxidation	34

supporting electrolyte. Such an example illustrates that the equivalent of crude potential control can be accomplished by the appropriate selection of the amalgamated material. A finer degree of potential control is then accomplished by appropriate selection of the supporting electrolyte ligand, and fine potential control is provided by adjustment of the electrolyte pH.

V. A-C and Pulse Methods

The imposition of an A-C voltage across an electrochemical cell results in an A-C current whose magnitude is proportional to the ionic concentration of the solution and to the ionic mobility of the ions. This serves as the basis for the techniques of conductometry and conductometric titrations. Although important historically, these have limited application to chemical characterization. Furthermore, numerous monographs and reviews provide thorough discussions of the theory, practice, and application of conductometric methods. Two of these are especially useful (35,36) and are

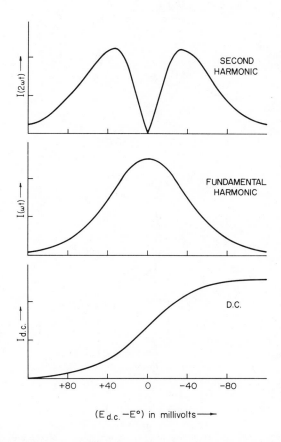

Figure 7-10. Conventional, fundamental harmonic A-C, and second haromonic A-C polarograms for a reversible system at a dropping mercury electrode. Curves are the envelope of the upper limits of the current oscillations.

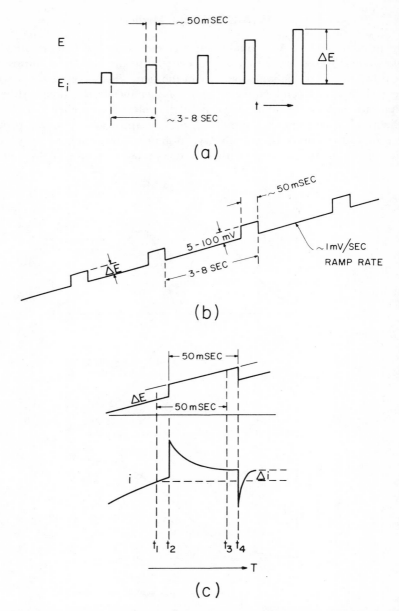

Figure 7-11. Potential-time sequence for (a) normal pulse polarogram and (b) differential pulse polarogram. The current-time response for the latter is given by (c) with t_1 and t_3 the times at which current is measured, t_2 the time at which pulse is applied, and t_4 the time at which pulse is removed.

recommended to those who have an interest in and need for conductometric measurements.

Another dimension of A-C electrochemistry is the modern technique of small amplitude A-C polarography. This consists of the imposition of a sinusoidal A-C voltage of about 10 mV (frequency, between 100 and 1000 Hz) upon the linear voltage ramp of classical polarography. The resulting A-C current is measured as the fundamental harmonic, or it can be measured as the second harmonic. Figure 7-10 illustrates the different response curves for an electrochemically reversible system. These curves are the envelopes of the dropping mercury electrode current oscillations. Effective use of A-C polarography requires a degree of electrochemical expertise beyond the scope of this monograph. However, the method is becoming more popular for special analytical problems and for the study of the fundamentals of electrode processes. For those interested two complete reviews and discussions are available, one an older monograph (37) and the other a recent review article (38).

Another derivative of A-C measurements is the method of pulse polarography. This takes two forms, one known as *normal* pulse polarography and the other *differential* pulse polarography. The first involves the imposition of square voltage pulses, each of increasing magnitude, upon a static D-C voltage (see Figure 7-11a). The resulting current response is illustrated by Figure 7-12a. For differential pulse polarography uniform square voltage pulses are imposed upon the linear voltage ramp of classical polarography (see Figure 7-11b). This approach yields the current response that is illustrated by Figure 7-12b. The purpose of the pulse technique is to minimize the amount of capacitative current (from the charging of the electrochemical double layer) in the current measurement. Figure 7-11c indicates the current-time response relative to the potential-time pulse for the differential pulse method. By sampling the current near the end of the pulse most of the capacitative charging current has decayed.

Pulse polarography has enjoyed increasing popularity during the past 5 years because of the availability of commercial instruments (Southern Analytical Limited, Melabs, and Princeton Applied Research). Furthermore, there is general agreement that the differential form of the method provides the greatest analytical sensitivity in the electrochemical field. Two review articles provide more detail of the method and its applications (39,40), and are recommended as an introduction.

For a comparison of the sensitivity and usefulness of the various A-C and pulse polarographic methods, a study (41) has been made of Cu(II) ion in 1 F NaNO$_3$. The results of this comparison are summarized in Table 7-7 and confirm that differential pulse polarography is the most sensitive technique with a detection limit of 1×10^{-7} F. Recent work (42) has shown that this

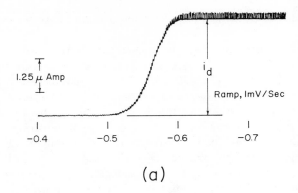

$$1.25\,\mu\,\text{Amp}$$

i_d

Ramp, ImV/Sec

−0.4	−0.5	−0.6	−0.7

(a)

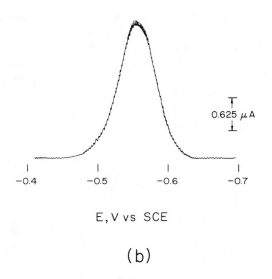

$$0.625\,\mu\text{A}$$

−0.4	−0.5	−0.6	−0.7

E, V vs SCE

(b)

Figure 7-12. Pulse polarograms for $10^{-4}F\,Cd^{2+}$ ion. (*a*) Normal and (*b*) differential modes.

concentration limit can be lowered two or three orders of magnitude by interfacing a minicomputer to the differential pulse polarograph. The general characteristics and controlling parameters for pulse polarography have been discussed in an earlier article (43).

Table 7-7   Summary of Polarographic Data for Determination of Copper

	Limit of Detection (M)	Concentration Limit Recommended for Quantitative Analysis (M)	Comments
Conventional D-C	2×10^{-6}	6×10^{-6}	—
Pulse	5×10^{-7}	1×10^{-6}	Superior to D-C methods below 10^{-5} M. Long drop times most favorable.
Derivative pulse	$\cong 5 \times 10^{-7}$	—	Poor reproducibility at low concentrations. Not recommended for trace analysis.
Differential pulse	1×10^{-7}	—	Most sensitive technique but waves broad under some conditions. Long drop times most favorable.
Phase-sensitive A-C with natural drop time	5×10^{-7}	1×10^{-6}	$\Delta E = 10$ mV and freq. $\cong 100$ Hz recommended. Far superior to nonphase-sensitive A-C polarography.

References

1. A. Fick, *Pogg. Ann.*, **94**, 59 (1855).
2. J. Heyrovsky, *Chem. Listy*, **16**, 256 (1922).
3. D. Ilkovic, *Collect. Czech. Chem. Commun.*, **6**, 498 (1934).
4. R. S. Nicholson and I. Shain, *Anal. Chem.*, **36**, 706 (1964).
5. R. S. Nicholson, *Anal Chem.*, **37**, 1351 (1965).
6. H. Eyring, S. Glasstone, and K. J. Laidler, *J. Chem. Phys.*, **7**, 1053 (1939).
7. J. M. Saveant, C. P. Andrieux, and L. Nadjo, *J. Electroanal. Chem.*, **41**, 137 (1973).
8. (a) V. G. Levich, *Physiochemical Hydrodynamics*, Prentice-Hall, Englewood Cliffs, N.J., 1963; (b) W. J. Albery, S. Bruckenstein, and D. T. Napp, *Trans. Faraday Soc.*, **62**, 1932 (1966).
9. L. Meites, *Polarographic Techniques*, 2nd ed., Interscience Publishers, Inc., New York, 1965.
10. L. B. Anderson and C. N. Reilley, *J. Electroanal. Chem.*, **10**, 295, 538 (1965).

11. J. J. Lingane, *Electroanalytical Chemistry*, 2nd ed., Interscience Publishers, Inc., New York, 1958.

12. I. M. Kolthoff and J. J. Lingane, *Polarography*, 2nd ed., Interscience Publishers, Inc., New York, 1952.

13. I. M. Kolthoff and P. Zuman, *Progress in Polarography*, Interscience Publishers, Inc., New York, 1962.

14. L. C. Clark, Jr., *Trans. Am. Soc. Artif. Intern. Organs*, **2**, 41 (1956).

15. D. T. Sawyer, R. S. George, and R. C. Rhodes, *Anal. Chem.*, **31**, 2 (1959).

16. H. B. Mark, Jr., *Rec. Chem. Progr.*, **29**, 217 (1968).

17. T. Shono and T. Kosaka, *Tetrahedron Lett.*, **1968**, 6207.

18. J. H. Prescott, *Chem. Eng.*, **1965** (November 8), 238.

19. K. Natarajan, K. S. Udupa, G. S. Subramanian, and H. V. K. Udapa, *Electrochem. Tech.*, **2** (5–6), 151 (1964); T. D. Balakrishman *et al.*, *Chem. Ind. (London)*, **1970**, 1622.

20. R. Ercoli, M. Guainazzi, and G. Silvestri, *Chem. Commun.*, **1967**, 927.

21. R. W. Foreman and J. W. Sprague, *Ind. Eng. Chem. Prod. Res. Develop.*, **2**, 303 (1963).

22. P. C. Condit, *Ind. Eng. Chem.*, **48**, 1252 (1956).

23. P. H. Radlick, R. Kelm, S. Spurlock, J. J. Sims, E. E. van Tamelen, and T. Whiteside, *Tetrahedron Lett.*, **1968**, 5117.

24. M. M. Baizer, *Tetrahedron Lett.*, **1963**, 973.

25. M. R. Rifi, *J. Am. Chem. Soc.*, **89**, 4442 (1967).

26. A. I. Kamneva, M. Ya. Fioshin, L. I. Kazakova, and Sh. M. Itenberg, *Neftekhim.*, **2**, 550 (1962).

27. J. D. Anderson, M. M. Baizer, and J. P. Petrovich, *J. Org. Chem.*, **31**, 3890 (1966).

28. S. Goldschmidt and E. Stocke, *Ber.*, **85**, 630 (1952).

29. C. L. Wilson, *Rec. Chem. Progr.*, **10**, 25 (1949).

30. G. C. Bond, *Discuss. Faraday Soc.*, No. **41**, 20 (1966).

31. J. A. M. Le Duc (to Pullman, Inc.), U.S. Patent 3, 342, 717 (September 19, 1967).

32. K. S. Udupa, G. S. Subramanian, and H. V. K. Udapa, *Ind. Chem.*, **1963** (May), 238.

33. R. L. Taylor, *Chem. Met. Eng.*, **44**, 588 (1937).

34. L. L. Bott, *Hydrocarbon Proc.*, **44** (1) 115 (1965).

35. T. Shedlovsky and L. Shedlovsky, "Conductometry," in *Techniques of Chemistry*, Vol. I, *Physical Methods of Chemistry*, Part IIA, A. Weissberger and B. W. Rossiter, eds., Wiley-Interscience, New York, 1971, pp. 163–204.

36. J. W. Loveland, "Conductometry and Oscillometry," in Part I, Vol. 4, *Treatise on Analytical Chemistry*, I. M. Kolthoff and P. J. Elving, eds., Interscience, New York, 1963, pp. 2569–2629.

37. B. Breyer and H. H. Bauer, "Alternating Curent Polarography and Tensammetry," Vol. XIII, *Chemical Analysis*, P. J. Elving and I. M. Kolthoff, eds., Interscience, New York, 1963.

38. D. E. Smith, *Crit. Rev. Anal. Chem.*, **2**(2) 247 (1971).

39. D. E. Burge, *J. Chem. Ed.*, **47**(2) A81 (1970).

40. J. G. Osteryoung and R. A. Osteryoung, *Am. Lab.*, **4**(7) 8 (1972).

41. A. M. Bond and D. R. Canterford, *Anal. Chem.*, **44**, 721 (1972).

42. H. E. Keller and R. A. Osteryoung, *Anal. Chem.*, **43**, 342 (1971).

43. E. P. Panz and R. A. Osteryoung, *Anal. Chem.*, **37**, 1634 (1965).

Eight

CONTROLLED CURRENT
METHODS

The principle of controlled current electrolysis has been known since the beginning of this century (1). However, the utilization of this form of electrochemistry remained dormant for 50 years until three groups of investigators illustrated its many advantages for analytical and physical chemical measurements (2–4). A recent review summarizes the extent of many applications of chronopotentiometry to chemical problems (5).

The basic form of controlled current electrolysis is called chronopotentiometry. As the name infers, while a constant current is passed between a pair of electrodes immersed in a quiescent solution, the potential of the measuring electrode ("working electrode") is monitored as a function of time. The basic form of instrumentation is illustrated in Figure 8-1 and indicates that the potential of the working electrode normally is measured relative to a noncurrent carrying reference electrode to minimize errors from the iR drop between the other two electrodes. Figure 8-2 illustrates the potential-time curve ("chronopotentiogram") for the reduction of ferric ion at a platinum electrode. At the beginning of the electrolysis the potential changes very little with time because of the "buffered" (or poised) condition, whereby both ferrous and ferric ions are in the vicinity of the working electrode. However, as the electrolysis continues the ferric ions are depleted, which forces the electrode potential to shift sharply in a negative direction until some other species can be reduced to maintain the constant current. As indicated in Figure 8-2, if other reducible ions are absent in the sample solution, then either hydrogen ions or water will be reduced.

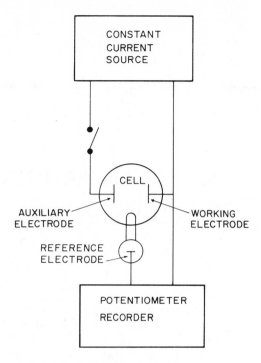

Figure 8-1.   Instrumentation for chronopotentiometry.

The point at which the sample ion is depleted in the vicinity of the working electrode is called the transition time, τ; this quantity is related to a number of variables including the sample-ion concentration. In 1905 Sand (1) derived the equation which describes the functional dependence of the transition time for a constant current electrolysis of a diffusion-controlled process

$$i\tau^{1/2} = \frac{\pi^{1/2}nFAD^{1/2}C}{2} \tag{8-1}$$

where i is the electrolysis current, n the number of electrons in the electrolysis reaction, F the faraday, A the area of the working electrode, D the diffusion coefficient of the electroactive species, and C the concentration of the electroactive species. This relation can be expanded for a two-step electrolysis involving two sample species. Because the first species continues to diffuse to the electrode even after the first transition time, the expression is

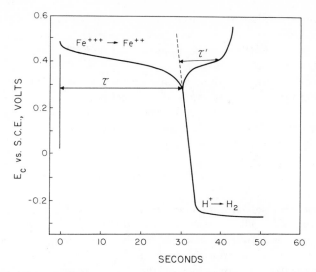

Figure 8-2. Chronopotentiogram for the reduction of Fe (III) at a platinum electrode. τ represents the transition time for the reduction process and τ' the transition time upon current reversal for the oxidation of the reduction product.

more complex than a simple additive relation

$$i(\tau_1 + \tau_2)^{1/2} = \frac{\pi^{1/2}FA}{2}\left(n_1 D_1^{1/2}C_1 + n_2 D_2^{1/2}C_2\right) \qquad (8\text{-}2)$$

The relation for the second transition time can be obtained by substracting Equation 8-1 from Equation 8-2 to give

$$i\tau_2^{1/2} = \frac{\pi^{1/2}FA}{2}\left(2n_1 n_2 D_1^{1/2}D_2^{1/2}C_1 C_2 + n_2^2 D_2 C_2^2\right)^{1/2} \qquad (8\text{-}3)$$

For the special condition where C_1 equals C_2 and D_1 equals D_2 (or for consecutive electrolysis by two steps), the ratio of transition times simplifies to

$$\frac{\tau_2}{\tau_1} = \frac{2n_2}{n_1} + \left(\frac{n_2}{n_1}\right)^2 \qquad (8\text{-}4)$$

Thus for two one-electron steps the second transition time is three times as long as the first. For a one-electron step followed by a two-electron step the ratio increases to eight. In contrast to polarography, chronopotentiometry offers enhanced analytical sensitivity for a second (or third) species. Advantage of this characteristic can be taken by adding a known amount of a species electrolyzed prior to the unknown species; in some cases this can provide a tenfold increase in sensitivity.

Reference to Figure 8-2 indicates that reversal of the current at (or before) the transition time will cause the product species to be electrolyzed until it is depleted from the vicinity of the electrode. If both the primary and reverse processes are diffusion controlled, then the transition time for the reverse process, τ', is related to the forward process by

$$\tau' = \frac{t}{3}(t \leqslant \tau) \tag{8-5}$$

Thus current reversal chronopotentiometry is useful for characterizing the electrode process as well as for ascertaining the nature of the product species.

If the chronopotentiometric reaction involves a thermodynamically reversible process without complications, Equation 8-1 may be used to relate concentrations to transition times in the Nernst expression. For a system where both reactant and product are soluble species (such as Figure 8-2), the chronopotentiogram is described by the relation

$$E = E_{\tau/4} + \frac{RT}{nF} \ln \frac{\tau^{1/2} - t^{1/2}}{t^{1/2}} \tag{8-6}$$

$$= E_{\tau/4} + \frac{0.059}{n} \log \frac{\tau^{1/2} - t^{1/2}}{t^{1/2}} \qquad \text{at } 25°C \tag{8-6a}$$

where $t^{1/2}$ is the time of electrolysis and $E_{\tau/4}$ is the quarterwave potential (the point on the curve where the logarithmic term becomes zero); equivalent to the half-wave potential in polarography. For a process where the product species is insoluble, Equation 8-6 takes the form

$$E = E_{\tau/4} + \frac{0.059}{n} \log(\tau^{1/2} - t^{1/2}) \qquad \text{at } 25°C \tag{8-7}$$

In current reversal chronopotentiometry the reverse wave of a reversible process is described by the relation

$$E = E_{\tau/4} + \frac{0.059}{n} \log \frac{\tau^{1/2} - \left[(\tau + t')^{1/2} - 2t'^{1/2}\right]}{(\tau + t')^{1/2} - 2t'^{1/2}} \qquad \text{at } 25°C \tag{8-8}$$

where t' is the reverse electrolysis time. Also, for the reverse wave

$$E_{0.222\tau'} = E_{\tau/4} \tag{8-9}$$

When the electrolysis process is irreversible (in a thermodynamic sense) but still limited by linear diffusion, the potential-time relationships take a form that includes the heterogeneous electron-transfer kinetic parameters. For a cathodic process the relation is

$$E = \frac{0.059}{\alpha n_a} \log \frac{nFACk_{f,h}^0}{i} + \frac{0.059}{\alpha n_a} \log \left(\frac{\tau^{1/2} - t^{1/2}}{\tau^{1/2}} \right) \qquad \text{at } 25°C \quad (8\text{-}10)$$

where E is the electrode potential versus the NHE, α the cathodic transfer coefficient (a kinetic symmetry parameter), $k_{f,h}^0$ the heterogeneous forward electron-transfer rate constant, and n_a the number of electrons in the rate controlling step; the other terms have their usual meaning. For an anodic process the relation is

$$E = \frac{-0.059}{(1-\alpha)n_a} \log \frac{nFACk_{b,h}^0}{i} - \frac{0.059}{(1-\alpha)n_a} \log \left(\frac{\tau^{1/2} - t^{1/2}}{\tau^{1/2}} \right) \qquad \text{at } 25°C \quad (8\text{-}11)$$

where $(1-\alpha)$ is the anodic transfer coefficient and $k_{b,h}^0$ the heterogeneous backward electron-transfer rate constant. Thus $k_{f,h}^0$ and $k_{b,h}^0$ represent the rate of electron transfer at 0.000 V versus NHE for unit concentration and unit electrode area. If the formal thermodynamic potential, E', for the process is known (or can be evaluated), then $k_{f,h}^0$ and $k_{b,h}^0$ may be converted to a simplified rate constant, $k_{s,h}$, which is a measure of the rate of electron transfer at the formal potential.

$$k_{s,h} = k_{f,h}^0 e^{-\alpha n_a FE'/RT} = k_{b,h}^0 e^{(1-\alpha)n_a FE'/RT} \qquad (8\text{-}12)$$

Consideration of Equations 8-10 and 8-11 indicates that chronopotentiometry can be useful for evaluating the kinetic parameters for electron transfer reactions. By plotting the electrode potential, E, versus $\log[(\tau^{1/2} - t^{1/2})/\tau^{1/2}]$ a linear curve is obtained whose slope permits evaluation of αn_a (or $(1-\alpha)n_a$). Extrapolation of such a plot to the point where $t=0$ causes the logarithmic terms of Equations 8-10 and 8-11 to go to zero and thereby permits $k_{f,h}^0$ and $k_{b,h}^0$ to be evaluated from the values of $E_{t=0}$. Plots of $E_{t=0}$ versus $\log C$ and $\log 1/i$ provide another means of evaluating αn_a.

The chronopotentiometric wave for current reversal of a cathodic process also may be evaluated using the relation

$$E = \frac{-0.059}{(1-\alpha)n'_a} \log \frac{2k_{b,h}^0}{\pi^{1/2} D_R^{1/2}} - \frac{0.059}{(1-\alpha)n'_a} \log \left[(\tau + t')^{1/2} - 2t'^{1/2} \right] \quad (8\text{-}13)$$

Table 8-1 Diagnostic Criteria of Chronopotentiograms for Various Kinetic Schemes

Kinetic Scheme	Linear Log Plot	Slope, Log Plot	$\dfrac{\partial E_{1/4}}{\partial \log i}$	$\dfrac{\partial E_{1/4}}{\partial \log C}$	$\dfrac{\tau_r}{\tau_f}$
$O \overset{rapid}{\underset{slow}{\rightleftarrows}} R$	$(\tau^{1/2}-t^{1/2})/t^{1/2}$	RT/nF	0	0	$1/3$
$O \overset{slow}{\rightleftarrows} R$	None	—	0 to $-RT/\alpha nF$	0 to $+RT/\alpha nF$	$1/3$
$O \rightarrow R$	$\tau^{1/2}-t^{1/2}$	$RT/\alpha nF$	$-RT/\alpha nF$	$+RT/\alpha nF$	$1/3$ or 0
$O \overset{rapid}{\rightleftarrows} R(\text{insol.})$	$\tau^{1/2}-t^{1/2}$	RT/nF	0	$+RT/nF$	1
$O \overset{slow}{\rightleftarrows} R(\text{insol.})$	None	—	$-RT/nF$ to $-RT/\alpha nF$	RT/nF to $RT/\alpha nF$	1
$O \rightarrow R(\text{insol.})$	$\tau^{1/2}-t^{1/2}$	$RT/\alpha nF$	$-RT/\alpha nF$	$+RT/\alpha nF$	1 or 0
$O \overset{rapid}{\rightleftarrows} R \overset{slow}{\rightleftarrows} Y$	None	—	0 to $-RT/nF$	0 to $+RT/nF$	0 to $1/3$
$O \overset{rapid}{\rightleftarrows} R \overset{rapid}{\rightarrow} Y$	$\tau^{1/2}-t^{1/2}$	RT/nF	$-RT/nF$	$+RT/nF$	0
$O \overset{rapid}{\rightleftarrows} R \overset{slow}{\rightarrow} Y$	None	—	0 to $-RT/nF$	0 to $+RT/nF$	0 to $1/3$
$mO \overset{rapid}{\rightleftarrows} pR$	$(\tau^{1/2}-t^{1/2})^m/t^{p/2}$	RT/nF	0	$(m-p)RT/nF$	$1/3$
$2O \overset{rapid}{\rightleftarrows} 2R \overset{slow}{\rightleftarrows} Y$	None	—	0 to $-RT/3nF$	$-RT/nF$ to $+RT/nF$	0 to $1/3$
$2O \overset{rapid}{\rightleftarrows} 2R \overset{rapid}{\rightarrow} Y$	$(\tau^{1/2}-t^{1/2})$	RT/nF	$-RT/3nF$	$+RT/nF$	0
$2O \overset{rapid}{\rightleftarrows} 2R \overset{slow}{\rightarrow} Y$	None	—	0 to $-RT/3nF$	$-RT/nF$ to $+RT/nF$	0 to $1/3$

where D_R is the diffusion coefficient of the reduced species and t' is the time of the reverse electrolysis.

Table 8-1 summarizes a number of diagnostic criteria for analysis of chronopotentiometric potential-time curves that are complicated by various kinetic schemes (6).

Chronopotentiometry is especially useful for studying prechemical steps that occur prior to the electron-transfer reaction. These can be generalized by the expression

$$Y \underset{k_b}{\overset{k_f}{\rightleftharpoons}} 0 \overset{ne^-}{\rightarrow} R; \qquad K = \frac{k_f}{k_b} \tag{8-14}$$

where Y is a nonelectroactive species that is in slow equilibrium with O, which is electroactive. As a result of this complication, Equation 8-1 is modified to

$$i\tau_k^{1/2} = \frac{\pi^{1/2} nFAD^{1/2}C}{2} - \frac{\pi^{1/2}i}{2AK(k_f + k_b)^{1/2}} \text{erf}\left[(k_f + k_b)^{1/2}\tau_k^{1/2}\right] \tag{8-15}$$

where erf represents the error function. If the argument of erf is greater than 2, erf approaches unity. Then

$$i\tau_k^{1/2} = \frac{\pi^{1/2} nFACD^{1/2}}{2} - \frac{\pi^{1/2}i}{2AK(k_f + k_b)^{1/2}} \tag{8-16}$$

or

$$\tau_k^{1/2} = \tau_d^{1/2} - \frac{\pi^{1/2}}{2K(k_f + k_b)^{1/2}} \tag{8-17}$$

This approach can be expanded to include a process of the form

$$pY \underset{k_b}{\overset{k_f}{\rightleftharpoons}} 0 \overset{ne^-}{\rightarrow} R, \qquad K = \frac{k_f}{k_b} \tag{8-18}$$

Then,

$$\tau_k^{1/2} = \tau_d^{1/2} - \frac{\pi^{1/2} nFAD^{1/2}}{2pi}\left[\frac{i}{nFAD^{1/2}K}\text{erf}(k_b\tau)^{1/2}\right]^{1/p} \tag{8-19}$$

For the condition where $(k_b\tau) > 2$

$$\tau_k^{1/2} = \tau_d^{1/2} - \frac{\pi^{1/2}}{2p}\left(\frac{nFAD^{1/2}}{i}\right)^{1-1/p}(k_b^{1/2}K)^{-1/p} \tag{8-20}$$

Table 8-2   Diagnostic Criteria of Chronopotentiograms for Various Reaction
Mechanisms

Mechanism	$(\partial i\tau^{1/2}/\partial i)$	$(\partial^2 i\tau^{1/2}/\partial i^2)$	$C_{\tau=0}{}^a$	$(\partial\tau/\partial T)^b$
$O \rightarrow R$	0	0	0	+
$Y \rightleftarrows O \rightarrow R$	−	0	+	+ +
$pY \rightleftarrows O \rightarrow R$	−	+	+	+ +
$Y \rightleftarrows pO; \; O \rightarrow R$	−	−	+	+ +
O (adsorbed) $\rightarrow R$	+	Variable	−	−

^{a}Obtained by extrapolation of plot of τ versus C to $\tau = 0$.
^{b}Temperature coefficient of τ.

and

$$i\tau_k{}^{1/2} = i\tau_d{}^{1/2} - \frac{\pi^{1/2}}{2p}(nFAD^{1/2})^{1-1/p}(i^{1/p})(k_b{}^{1/2}K)^{-1/p} \qquad (8\text{-}21)$$

A diagnostic approach for the analysis of various prechemical reactions (7) is summarized in Table 8-2. Thus by analyzing the variation of the first and second derivative of the quantity $i\tau^{1/2}$ as a function of current, the type of prechemical process can be determined. Two experimental plots will provide sufficient data to use the criteria of this table; $i\tau^{1/2}$ versus i at constant C and $\tau^{1/2}$ versus C at constant i.

Through the use of current reversal chronopotentiometry, reaction sequences of the type

$$O \overset{ne^-}{\rightleftarrows} R \overset{k_f}{\rightarrow} Y \qquad (8\text{-}22)$$

can be investigated. However, the functions are complex and the use of a graphical expression is much more convenient (8). Figure 8-3 indicates the variation of the ratio of τ_r/t_1, as a function of $k_f t_1$, where t_1 represents the time of forward electrolysis and τ_r represents the reverse transition time. Thus for any value of the ratio, the quantity $k_f t_1$ can be evaluated. By plotting $k_f t_1$ versus t_1, the slope of the linear curve is equal to k_f of Equation 8-22.

Constant current electrolysis also is applicable to thin-layer cells such that Equation 8-1 becomes

$$i\tau = nFACd \qquad (8\text{-}23)$$

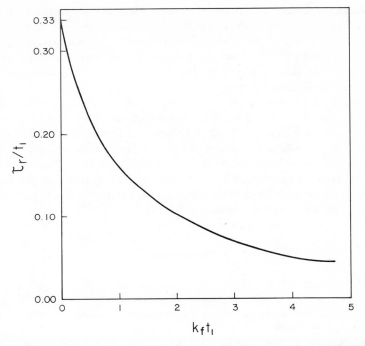

Figure 8-3. Variation of reverse transition time, τ_r, with the time of forward electrolysis, t_1. The quantity k_f represents the unimolecular homogeneous rate constant for the conversion of the product of the forward electrolysis to a nonelectroactive species (in terms of the reverse electrolysis).

where d is the thickness of the electrode spacing. Clearly, such a system provides a convenient and rapid means for evaluating the number of electrons, n, for an unknown process. The reverse transition time should be the same as that for the forward process, unless there are postelectrochemical complications.

A number of enbellishments of controlled current electrolysis have been proposed for specialized systems, but discussion of these is beyond the scope of this chapter. For those interested in pursuing this material, they are referred to discussions of programmed current techniques (9), single and double pulse galvanostatic methods (10), and chronocoulometry (11).

As indicated at the beginning of this chapter, the methodology of chronopotentiometry is straightforward. Some care is necessary to ensure that the electrode configuration and placement are such as to cause the electrolysis to be limited by semiinfinite linear diffusion. Chapter 2 indicates a number of indicator electrode designs that are appropriate for such performance. As in any diffusion-controlled process, absence of vibration of the electrolysis cell is essential to obtain meaningful quantitative data. In general, electrolysis times from 0.1 to about 30 sec represent the convenient and useful quantitative range for conventional instrumentation and laboratory conditions.

Although there has been much debáte about the proper method of measuring the transition time, τ, in chronopotentiometry, most workers now agree on the procedure outlined in Figure 8-2. For electrolysis times of less than 0.1 sec, specialized recording of the potential-time curve is necessary, usually with an oscilloscope. When the transition time is less than several milliseconds, correction must be made for the current-time integral required to charge the double layer. This problem has led to the development of the double-pulse galvanostatic method (10).

References

1. H. J. S. Sand, *Phil. Mag.*, **1**, 45 (1901).
2. L. Gierst and A. Juliard, *J. Phys. Chem.*, **57**, 701 (1953).
3. P. Delahay, *New Instrumental Methods in Electrochemistry*, Interscience Publishers, Inc., New York, 1954, Chap. 8.
4. J. J. Lingane, *Electroanalytical Chemistry*, 2nd ed., Interscience Publishers, Inc., New York, 1958, Chap. XXII.
5. P. J. Lingane, *CRC Crit. Rev. in Anal. Chem.*, **1**, 587 (1971).
6. W. H. Reinmuth, *Anal. Chem.*, **32**, 1514 (1960).
7. W. H. Reinmuth, *Anal. Chem.*, **33**, 322 (1961).
8. A. C. Testa and W. H. Reinmuth, *Anal. Chem.*, **32**, 1512 (1960).
9. R. W. Murray and C. N. Reilley, *J. Electroanal. Chem.*, **3**, 64 (1962).
10. R. L. Birke and D. K. Roe, *Anal. Chem.*, **37**, 450, 455 (1965).
11. F. C. Anson, *Anal. Chem.*, **38**, 54 (1966).

Nine

ELECTROCHEMICAL
TITRATIONS

I. Introduction

One of the most extensive applications of electrochemistry has been for endpoint detection in titrations. The latter continue to be important in analysis, in spite of their more tedious nature, because of the better precision and accuracy than is possible by direct electrochemical measurements. For example, in potentiometry a 0.25-mV error represents a 1% relative error in the concentration of the detected species.

Electrochemical endpoint detection methods provide a number of advantages over classical visual indicators and have become increasingly popular during the past two decades. In particular, such methods provide increased sensitivity and are often amenable to automation. Electrochemical methods of endpoint detection are applicable to most oxidation-reduction, acid-base, and precipitation titrations, and to many complexation titrations. The only necessary condition is that either the titrant or the species being titrated must give some type of electrochemical response that is indicative of the concentration of the species.

II. Endpoint Detection Methods

By far the most common endpoint system for titrations is a potentiometric indicating electrode because of its simplicity and its universal applicability. It only requires a redox couple that gives some response which is indicative of either the titrant or of the species being titrated. In its simplest form the endpoint detection system consists of a reference electrode plus an inert indicating electrode. In the situation where both the titrant and the species being titrated give reversible thermodynamic responses, the potential of the indicator electrode is governed by either couple. An example of such a system is the titration of ferrous ion by ceric ion.

$$Ce^{4+} + Fe^{2+} \rightarrow Fe^{3+} + Ce^{3+} \tag{9-1}$$

For this system the indicating potential is represented by either half reaction

$$E_{ind} = E^0_{Fe^{3+},Fe^{2+}} + \frac{0.059}{1} \log \frac{[Fe^{3+}]}{[Fe^{2+}]} \qquad \text{at } 25°C \tag{9-2}$$

$$E_{ind} = E^0_{Ce^{4+},Ce^{3+}} + \frac{0.059}{1} \log \frac{[Ce^{4+}]}{[Ce^{3+}]} \tag{9-3}$$

Prior to the endpoint the Nernst expression indicated by Equation 9-2 is most convenient for computing the indicator potential. In contrast, past the equivalence point Equation 9-3 is more convenient for computing this potential. To calculate the indicator potential at the equivalence point a combination of Equations 9-2 and 9-3 gives

$$2E_{ind} = E^0_{Fe^{3+},Fe^{2+}} + E^0_{Ce^{4+},Ce^{3+}}$$

$$+ \frac{0.059}{1} \log \frac{[Fe^{3+}][Ce^{4+}]}{[Fe^{2+}][Ce^{3+}]} \qquad \text{at} \quad 25°C \tag{9-4}$$

Consideration of Equation 9-1 indicates that the conditions represented by

$$[Fe^{3+}]_{ep} = [Ce^{3+}]_{ep}; \qquad [Fe^{2+}]_{ep} = [Ce^{4+}]_{ep} \tag{9-5}$$

prevail at the equivalence point. Substitution of these quantities into Equation 9-4 establishes that the indicator potential at the equivalence point is independent of the concentrations of the titrant and the reactant.

$$(E_{ind})_{ep} = \frac{E^0_{Fe^{3+},Fe^{2+}} + E^0_{Ce^{4+},Ce^{3+}}}{2} \tag{9-6}$$

This specific result can be generalized by the expression

$$(E_{ind})_{ep} = \frac{mE_1^0 + nE_2^0}{m + n} \tag{9-7}$$

where m and n represent the number of electrons for the two half-reactions that make up the titration reaction.

With any endpoint detection system several practical considerations are important for reliable results. For example, the indicator electrode should be placed in close proximity to the flow pattern from the buret, so that a degree of anticipation is provided to avoid overrunning the endpoint. Another important factor is that the indicator electrode be as inert and nonreactive as possible to avoid contamination and erratic response from attack by the titration solution. A third and frequently overlooked consideration is the makeup of the reference electrode and, in particular, its salt-bridge. For example, a salt-bridge system containing potassium chloride can cause extremely erratic behavior of any electrochemical system if the titrant solution contains perchlorate ion because of the precipitation of potassium perchlorate at the salt-bridge titrant-solution interface. Likewise, a potassium chloride salt-bridge in a potentiometric titration involving either silver ion or mercurous ion can cause serious titration errors due to leakage of salt solution into the titration system.

Potentiometric titration curves normally are represented by a plot of the indicator potential as a function of volume of titrant, as indicated in Figure 9-1. However, there are some advantages to plotting such data as the first derivative of the indicator potential with respect to volume of titrant or even as the second derivative. Such titration curves also are indicated in Figure 9-1, and illustrate that a more definite endpoint indication is provided by both differential curves than by the integrated form of the titration curve. Furthermore, titration by repetitive constant-volume increments allows the endpoint to be determined without plotting of the titration curve; the endpoint coincides with the condition when the differential potentiometric response per volume increment is a maximum. Likewise, the endpoint can be determined by using the second derivative; the latter has distinct advantages in that there is some indication of the approach of the endpoint as the second derivative approaches a positive maximum just prior to the equivalence point before passing through zero. Such a second derivative response is particularly attractive for automated titration systems that stop at the equivalence point.

A specialized version of the potentiometric endpoint detection system is the use of two dissimilar metals as the electrode pair. For example, if a platinum electrode is used in combination with a tungsten electrode a sharp potential response frequently is observed that coincides with the equivalence point. This occurs not only with oxidation reduction titrations but also with acid-base titrations. It comes about because of the generally more reversible behavior of platinum electrodes relative to tungsten electrodes. Although this is an empirical system it offers the advantage of an extremely inert reference electrode system. A related approach is to place the buret beneath the surface of the titrant solution and place a platinum wire inside the buret tip, and another platinum electrode in the titrant solution. This provides differential response with the electrode inside the buret maintaining a reasonably constant potential and serving as the reference electrode.

The subject of potentiometric titrations has been exhaustively treated by the excellent monograph by Kolthoff and Furman (1). Although the second

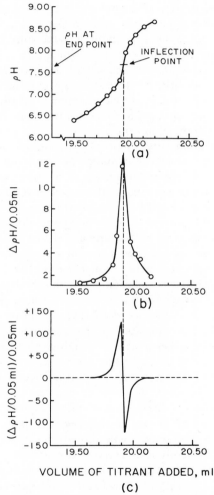

Figure 9-1. Potentiometric titration curves. (*a*) Experimental titration data; (*b*) first derivative of curve *a*; and (*c*) second derivative of curve *a*.

edition was published in 1931, it is still the definitive work. Unfortunately, it is no longer in print, but it is available in many chemical libraries. A complete and thorough discussion of the principles and theory of potentiometric titrations is provided, together with an extremely extensive summary of the many applications of potentiometric measurements.

Another specialized form of potentiometric endpoint detection is the use of dual polarized electrodes, which consist of two metal pieces of electrode material, usually platinum, through which is imposed a small constant current, usually 2 to 10 μA. The differential potential created by the imposition of the current is a function of the redox couples present in the titration solution. Examples of the resultant titration curve for two different systems are illustrated in Figure 9-2. In the case of two reversible couples, such as the titration of ferrous ion with ceric ion, curve *a* results in which

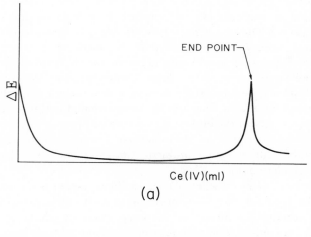

(a)

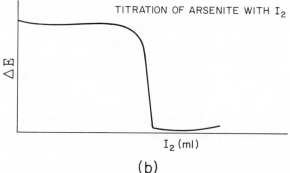

(b)

Figure 9-2.   Dual polarized-electrode potentiometric titration curves. (*a*) Titration of Fe(II) by Ce(IV); (*b*) titration of As(III) by I_2.

there is little potential difference after initiation of the titration up to the equivalence point. In contrast, the titration of arsenic(III) ion with iodine is representative of an irreversible couple being titrated with a reversible system. Hence prior to the equivalence point a large potential difference exists because the passage of current requires decomposition of the solvent for the cathode reaction (Figure 9-2b). Past the equivalence point the potential difference drops to zero because of the presence of both iodine and iodide ion.

The use of dual polarized electrodes was first suggested more than 45 years ago (2); the subject has been reviewed thoroughly by two more recent publications (3,4). Almost all modern commercial pH meters have provision for imposing a polarizing current of either 5 or 10 μA to make possible measurements by dual polarized electrode potentiometry. Such a provision is included because dual polarized potentiometry is by far the most popular endpoint detection method for the Karl-Fischer determination of water. For this titration a combination of reagents is used, including iodine; the response curve is similar to that of Figure 9-2b. In practice the response is many times more sensitive than that obtained by a conventional potentiometric indicating system and accounts for the extreme popularity of the dual polarized system for the Karl-Fischer titration. Without question a number of other potentiometric titrations would benefit from the use of the dual polarized approach. This is to be recommended not only because of the increased sensitivity of the endpoint response that frequently occurs, but also because of the simplicity of the electrode system (which avoids solution contamination and the other problems that frequently are associated with the reference electrode).

Another form of electrochemical endpoint detection is the amperometric method. In its most common form this consists of a polarizable microelectrode, usually the dropping mercury electrode characteristic of polarography, in combination with a large nonpolarizable reference electrode. Thus polarographic measurements are made of the titration system as a function of volume of titrant. Potential is applied across the indicating system such that it is on the diffusion current plateau for either the titrant, the reactant, or both. If a dropping mercury electrode is used, then all the technology and background data from polarography may be used in setting the proper solution and electrode conditions for an effective response that is characteristic of either the titrant or reactant concentration.

Figure 9-3 illustrates the kind of titration curve that is obtained with an amperometric indicating system for the titration of lead ion with dichromate ion. If the applied potential is set on the plateau for the reduction of lead ion, approximately -0.5 V versus SCE, then curve a in Figure 9-3 will result. In contrast, if the applied potential is set at 0 V versus SCE, then no

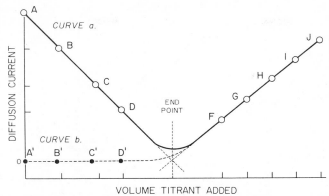

Figure 9-3. Amperometric titration curves for the titration of Pb(II) ions by $Cr_2O_7^{2-}$ ions at $E_{DME} = -0.8\,V$ versus SCE (curve a) and at $E_{DME} = 0.0$ V versus SCE (curve b).

current will flow until the point when excess chromate ion exists in the solution; curve b is indicative of the titration curve that would be obtained.

The amperometric approach to endpoint detection provides considerable latitude in the selection of the best conditions for the most specific and sensitive endpoint response. Furthermore, the response signal is directly proportional to the concentration of the observed species, whereas potentiometric responses are a logarithmic function of the concentration. Another attractive feature of amperometric endpoint detection is that the most important data are obtained prior to and after the equivalence point whereas in potentiometric titrations the most important data occur at the equivalence point, which is the most unstable condition of the titration. With amperometric titrations an extrapolation of the straight-line portion of the curve, either prior to or after the equivalence point, to an intercept will provide an accurate measure of the equivalence point. To have as straight a line as possible, it is necessary to apply a dilution correction to the observed current. This can be simplified by using an extremely high concentration of titrant relative to the concentration of the species being titrated.

The entire subject of amperometric titrations has been reviewed in a number of monographs on electrochemistry (4–6); a definitive work on this subject also appeared recently (7). Because the amperometric titration method is not dependent on having one or more reversible couples associated with the titration reaction, it permits electrochemical detection of the endpoint for a number of systems that are not amenable to potentiometric detection. All that is required is that electrode conditions be adjusted such that either a titrant, a reactant, or a product from the reaction gives a polarographic diffusion current.

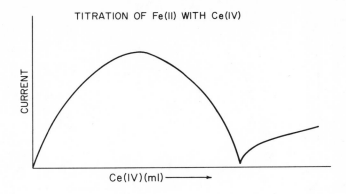

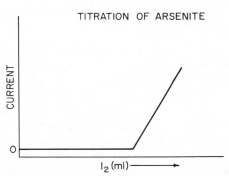

Figure 9-4. Dual polarized-electrode amperometric titration curves. Both curves result from the application of a 0.25-V potential across two identical platinum electrodes that are immersed in the titration solution.

A specialized version of the amperometric endpoint system is one using dual polarized electrodes, often a pair of small foils. In contrast to the dual polarized potentiometric system, dual polarized amperometric electrodes have a constant potential applied to them, such that one or the other of the electrodes gives a diffusion controlled current response during some portion of the titration. The applied potential, which normally ranges from zero to several tenths of a volt, is ascertained either by consideration of po-larographic half-wave potential data, or from experimental measurements. What is sought is an applied potential that is on the diffusion-current plateau for one or more couples involved in the titration. Figure 9-4 illustrates the type of current response that is obtained for a system involving a pair of reversible couples (titration of ferrous ion with ceric ion, curve *a*), as well as for the titration of an irreversible couple with a reversible couple

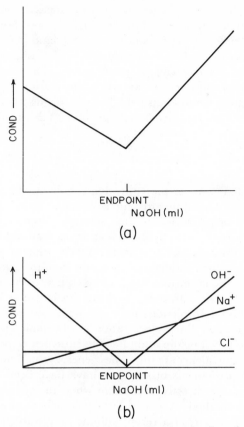

Figure 9-5. (*a*) Conductometric titration curve for the titration of HCl by NaOH. (*b*) Conductances of individual ions during the course of the titration; summation of these at any volume of titrant corresponds to the point on curve *a*.

[arsenic(III) ion with iodine, curve *b*]. The latter curve also is characteristic of the response one obtains if a dual polarized amperometric endpoint detection system is used for the Karl-Fischer titration. The dual polarized amperometric system is especially attractive because of the simplicity of the instrumentation. Only a microammeter plus a potentiometric voltage divider and a dry cell are necessary to have an extremely sensitive and versatile endpoint detection system. The method has been known for many years but has not enjoyed the popularity that it deserves. Some of the older literature refers to the method as the "dead-stop" method.

One of the oldest electrochemical detection methods is the conductometric monitoring of the ionic concentrations of species involved in a titration. The general approach is to make conductometric measurements during the course of the titration with a plot of the conductance as a function of volume of titrant. Figure 9-5a is representative of the kind of response that is obtained for the titration of hydrochloric acid with sodium hydroxide. The shape of the titration curve can be predicted by summing the ionic concentrations of the various species at any point during the course of the titration; the resulting summation will give the titration curve. Figure 9-5b indicates how these vary for the example titration. The titration itself can be represented by

$$(H^+ + Cl^-) + (Na^+ + OH^-) \rightarrow (Na^+ + Cl^-) + H_2O \qquad (9\text{-}8)$$

The net effect is that prior to the equivalence point hydrogen ion is being replaced with sodium ion while after the equivalence point sodium ion plus hydroxide ion are being added to this solution. Because the equivalent conductance of hydrogen ion is many times greater than sodium ion, there is a net decrease of the total solution conductance prior to the equivalence point; after this point the conductance increases from the addition of sodium and hydroxide ion (the latter also has a large equivalent conductance).

As with amperometric titrations, to have straight-line portions of the titration curve dilution corrections must be made because the response is directly dependent upon the concentration of the ionic species. Also, the important data are taken before and after the equivalence point rather than precisely at the equivalence point. The general conditions for effective conductometric measurements of solutions have have been discussed in an earlier chapter and are directly applicable when the system is used as the endpoint detection method. A particularly complete review of the subject has been presented as part of a recent volume on electrochemical methods (8).

One of the factors that has caused conductometric titrations to have limited popularity has been the problem of electrode contamination and the resultant decrease in sensitivity and and electrode stability. To overcome this problem, the general conductometric approach was extended some 15 years ago from the conventional 1-kHz frequency range to a frequency of several megahertz. With the latter frequency the indicating electrodes can be placed on the outside of the titration vessel rather than having them immersed in the solution. This avoids electrode contamination as well as the need for having a highly activated platinized surface. A number of forms of the electrode configuration have been developed; both parallel plates and concentric plates give satisfactory responses. In general high frequency conductometric titrations have had limited application, in part because of the attendent difficulties associated with electronic circuits at the megahertz frequency range. A second limitation has been that the high frequency conductometric response is not a simple linear function of ionic equivalent conductance, as is the case with conventional conductometric measurements. A particularly complete and authoritative discussion of the entire subject of high frequency titrations has been presented by Reilley (9).

III. Autotitrators

One of the advantages of electrochemical endpoint detection is that it lends itself to automation through electronic circuitry. Not only is automation of a titration a labor saving advantage but it eliminates human error and prejudice in selecting the endpoint of a titration. The principles and some of the approaches to automating electrochemical titrations have been discussed in two recent monographs (10, 11).

Although all three electrochemical endpoint detection methods (potentiometric, amperometric, and conductometric) have the potential for automation, to date the potentiometric method is the only one that has enjoyed much attention. Two approaches have been used in automating potentiometric titrations; one involves presetting the endpoint potential such that the titration is caused to proceed until the indicating system attains its present point. Such an instrument is marketed by Beckman Instruments and consists of a potentiometric pH meter which allows presetting of the endpoint potential together with a solenoid activated stop-cock assembly for delivery of titrant. An anticipatory circuit is provided such that as the endpoint potential is approached the rate of delivery is slowed to avoid "overrunning" the end of the titration. For reliable performance the buret tip must be placed such that the indicator electrode anticipates the endpoint concentration prior to its equilibrium attainment. This approach is convenient and simple in terms of instrumentation but it demands an accurate prior knowledge of the proper potential to set for the equivalence point. To gain maximum reliability the entire titration curve for a sample system must be determined so that the proper endpoint potential can be set. This can be accomplished by incrementally adjusting the setting for the endpoint potential and waiting for the titrator to attain that potential. A titration curve can be obtained by plotting potential settings *versus* volume of titrant delivered.

The second form of autotitrator for potentiometric titrations involves instrumentation that simultaneously delivers titrant and records the indicator potential. To obtain reliable and meaningful data the rate of titrant delivery must be sufficiently slow to allow attainment of equilibrium. The placement of the electrodes must be such that the equilibrium concentration within the titration vessel is representatively sensed. Because potentiometric strip-chart recorders are the most convenient form of recording, titrant delivery normally must be at a constant rate (e.g., a motor-driven syringe). A convenient approach is to connect the drive assembly of the syringe to the chart-drive assembly for the recorder. If the latter is done the rate of delivery of titrant can be adjusted such that the rate of potential change is made constant. Such provision will minimize the titration error and give as rapid a recording as is possible for a given system. Commercial versions of such instrumentation are available from Radiometer, Metrohm, Sargent-Welch, Mettler, Beckman, and the International Instrument Company in Canyon, Calif.; the latter has been described in the literature (12).

Autotitrations that provide the entire titration curve give the maximum information concerning the sample system and its characteristics at the equivalence point. They assure that the proper point on the titration curve is selected as the endpoint. They also provide the maximum amount of reliable data for the evaluation of the dissociation constants of weak acids, weak bases, and complexed species.

In addition to these two versions of autotitrators, titrators have been developed that take advantage of the change in potential per increment of titrant added. Such instruments terminate the addition of titrant when the first derivative of potential relative to volume reaches a maximum, or when the second derivative of potential with respect to volume goes through zero (see Figure 9-1). The Sargent Company markets instrumentation based on this approach which has the advantage over a preset endpoint potential in that the specific sample system being titrated actually determines when the endpoint will be selected based on the second derivative of the titration curve. This system also must have a controlled rate of delivery of titrant which is best accomplished through the use of a motor-driven syringe rather than from a conventional buret.

Whether a constant speed or a variable speed motor drive is used depends in part upon the sophistication of the instrumentation and the nature of the anticipatory and detection section of the instrumentation. Because automated titration instrumentation frequently is needed for a specific problem a great number of custom fabricated assemblies have been developed. This frequently is the best solution for an individual and is not a difficult task once the critical factors are appreciated.

IV. pH-Stats

A related form of an automatic potentiometric titrator is instrumentation that permits the maintenance of the acidity or basicity of a solution over a period of time. Such devices are known as pH-stats, and find application in kinetic studies of hydrolysis reactions. The general approach is (by either manual or automatic means) to add either acid or base such that the pH of the solution is maintained constant over a period of time. Normally the amount of acid or base added as a function of time is sought in order that kinetic measurements may be made for the system. In its simplest form the acidity of the solution is monitored with a pH meter and controlled at a preselected value by the addition of acid or base from a buret; the quantity delivered as a function of time is recorded in a notebook. Obviously for fast reactions this becomes difficult and dependent on the dexterity of the individual.

In general autotitrators that work with a preset endpoint lend themselves to application as pH-stats. All that is necessary is to record the volume of titrant added as a function of time. If a motor-driven syringe is used then this can be combined with a helical potentiometer and a strip-chart recorder to provide a volume-time curve; a manual version would include a digital register ganged to the motor-driven screw of the syringe. The digitizer can be read as a function of time and provides the necessary data for analysis of the kinetics of the hydrolysis reaction.

pH-Stats have found their primary application in the study of hydrolysis rates such as

$$\text{RCOOR}' + \text{OH}^- \rightarrow \text{RCOO}^- + \text{R}'\text{OH} \tag{9-9}$$

In the case of such an ester hydrolysis the rate law is expressed by the relation

$$\frac{-d(\text{RCOOR}')}{dt} = k(\text{RCOOR}')(\text{OH}^-) = \frac{-dm_{\text{OH}^-}}{V_t dt} \tag{9-10}$$

$$= \frac{C^0_{\text{OH}^-}}{V_t}\left(\frac{dV_{\text{OH}^-}}{dt}\right)_{\text{pH}} = \frac{i}{V_t F} \tag{9-10a}$$

with the right-hand term representing the rate of disappearance of hydroxide ion from the sample solution (V_t) as a function of time (m_{OH^-}, moles of hydroxide ion; $C^0_{\text{OH}^-}$, concentration of hydroxide ion in buret). Equation 9-10a indicates this rate of disappearance must be matched by an equivalent rate of addition of hydroxide if the pH of the solution is to be maintained. Furthermore, if the hydroxide ion were to be produced electrochemically in the solution, then the right-hand term in Equation 9-10a would be indicative of the rate of hydroxide addition (in terms of current used to electrolyze the solution to provide hydroxide ion). Rearrangement of these equations gives a relation for the evaluation of hydrolysis rate constants

$$k = \left(\frac{dV_{OH^-}}{dt}\right)_{pH} \frac{C^0_{OH^-}}{(OH^-)m_{ester}} = \frac{i}{(OH^-)m_{ester}F} \tag{9-11}$$

where m_{ester} is the moles of ester and F the faraday. As Equation 9-11 indicates, pH-stats, whether using titrant delivery of hydroxide ion or electrochemical generation of hydroxide ion, provide an extremely convenient means of evaluating hydrolysis rate constants (13).

Another application of pH-stats is to control solution pH without the use of a buffer system. Again, this can be accomplished by either a buret delivery system, a motor-driven syringe, or through electrochemical generation of hydroxide ion or hydrogen ion. This can be extremely useful for systems where all available buffer solutions interfere with the reaction of interest.

V. Coulometric Titrations

The coulometric method of titration is based on the electrochemical generation of the titrant and on Faraday's law of electrolysis which equates equivalents of material to coulombs of electricity. The method was first discussed some 30 years ago (14) and has been investigated in detail in the intervening years. A particularly detailed discussion is presented by Lingane (15), and a recent review and discussion are given in the treatise by Kolthoff and Elving (16). The method is of sufficient interest and importance that it is reviewed biannually in *Analytical Chemistry* (17).

The basic approach in coulometric titrations is to generate electrochemically (at constant current) a titrant in solution which subsequently reacts by a secondary chemical reaction with the species being determined. For example, a large excess of cerous ion is placed in the solution together with a ferrous sample. When a constant current is applied the cerous ion is oxidized at the anode to produce ceric ion which subsequently reacts with the ferrous ion.

$$Ce^{3+} \rightarrow Ce^{4+} + e^- \text{ (electrode reaction)} \tag{9-12}$$

$$Ce^{4+} + Fe^{2+} \rightarrow Fe^{3+} + Ce^{3+} \text{ (secondary reaction)} \tag{9-13}$$

Should any ferrous ion reach the anode it also would be oxidized and thereby not require the chemical reaction of Equation 9-13 to bring about oxidation, but this would not in any way cause an error in the titration. This method is equivalent to the constant rate addition of titrants from a buret. However, in place of a buret the titrant is electrochemically generated in the solution at a constant rate which is directly proportional to the constant current. For accurate results to be obtained the electrode reaction must occur with 100% current efficiency (i.e., without any side reactions involving solvent or other materials that would not be effective in the secondary reaction). In the method of coulometric titrations the material that chemically reacts with the sample system is referred to as an electrochemical intermediate (the cerous-ceric couple is the electrochemical intermediate for the titration of ferrous ion). Because 1 F of electrolysis current is equivalent to 1-g-equivalent of titrant, the coulometric titration method is extremely sensitive relative to conventional titration procedures. This becomes obvious when it is recognized that there are 96,500 C/F. Thus 1 mA of current flowing for 1 sec represents approximately 10^{-8} g-equivalent of titrant.

In general, coulometric titrations use currents ranging from 1 to 100 mA in magnitude. An essential part of a coulometric titration assembly is a reliable constant current source with a range of accurately preset currents known with an error of less than $\pm 0.1\%$. Another crucial element in the instrumentation is a sensitive and rapid endpoint detection system. By its very nature the coulometric titration method is a constant-rate-of-delivery system, which demands sensitive and rapid response if the endpoint is not to be overrun. Reference to the literature confirms that potentiometric, amperometric, and visual indicating systems have been applied to coulometric titrations. However, the potentiometric and the dual polarized amperometric detection systems have enjoyed the widest application because of their selective and sensitive response. The final element of a coulometric titration system is an accurate timing device. This can be provided either by a precision stop-clock or by an electronic counting system based on time. To summarize, a coulometric titration consists of passing a constant current through a solution until the endpoint is detected at which time the period of electrolysis is recorded. The product of the current times the electrolysis time gives the number of coulombs, which is directly proportional to the number of equivalents of the substituent being analyzed. For titration periods of 100 sec or less the timing device must be accurate to ± 0.1 sec, if 0.1% accuracy is to be realized.

During the past 30 years coulometric titration procedures have been developed for a great number of oxidation-reduction, acid-base, precipitation, and complexation reactions. The sample systems as well as the electrochemical intermediates used for them are summarized in Table 9-1 and indicate the diversity and range of application for the method. An additional specialized form of coulometric titration involves the use of a spent Karl-Fischer solution as the electrochemical intermediate for the

Table 9-1 Systems for Coulometric Titrations

	Electrogenerated Reagent	Precursor	Determined Species
A. Oxidizing agents			
	Ce(IV)	Ce(III), 3 F H$_2$SO$_4$	Fe(II), Fe(CN)$_6^{4-}$, Ti(III), As(III), U(IV), I$^-$, hydroquinone, phenols
	Mn(III)	Mn(II), 6 F H$_2$SO$_4$	Fe(II), C$_2$O$_4^{2-}$, As(III), H$_2$O$_2$
	Cl$_2$	Cl$^-$, HCl or H$_2$SO$_4$	As(III), I$^-$, unsaturated fatty acids
	Br$_2$	Br$^-$, H$_2$SO$_4$	As(III), Sb(III), I$^-$, Tl(I), U(IV), NH$_3$, N$_2$H$_4$, thioglycol, 8-quinolinol, phenols, unsaturated hydrocarbons
	I$_3^-$	I$^-$	As(III), S$_2$O$_3^{2-}$, H$_2$SeO$_3$, Sb(III), S^{2-}
	Fe(CN)$_6^{3-}$	Fe(CN)$_6^{4-}$	Tl(I)
B. Reducing agents			
	Ti(III)	TiOSO$_4$, 7 F H$_2$SO$_4$	Fe(III), Ce(IV), V(V), U(VI)
	Sn(II)	SnCl$_4$, 3 F NaBr,	

Fe(II)EDTA	Fe(III)EDTA		Au(III), Ce(IV), Fe(III), I_2
Fe(II)	Fe(III), 1 F H_2SO_4, H_3PO_4	HCl	Fe(III)
			Ce(IV), $Cr_2O_7^{2-}$, V(V), MnO_4^-
Cu(I)	Cu(II), 2 F HCl		$Cr_2O_7^{2-}$, IO_3^-, Br_2
U(IV)	UO_2SO_4, 0.3 F H_2SO_4		$Cr_2O_7^{2-}$, Ce(IV), Fe(III)
V(IV)	$NaVO_3$, 2 F H_2SO_4		
C. Precipitating agents			
Ag(I)	Ag, 0.5 F $HClO_4$		Cl^-, Br^-, I^-, $CH_3CS(NH_2)$, SCN^-, RSH
Hg(I)	Hg, 0.5 F $NaClO_4$, $HClO_4$		Cl^-, Br^-, I^-
$Fe(CN)_6^{4-}$	$Fe(CN)_6^{4-}$		Zn(II)
D. Complexing agents			
$H(EDTA)^{3-}$	$HgNH_3(EDTA)^{2-}$, pH 8.5(NH_4^+, NH_3)		Ca(II), Cu(II), Zn(II), Pb(II)
E. Acid-base agents			
H^+	H_2O, 1 F Na_2SO_4		Bases
OH^-	H_2O, 1 F Na_2SO_4		Acids
F. Karl-Fischer reagent			
I_2	"Spent" Karl-Fischer reagent		H_2O

determination of water at extremely low levels. For such a system the anode reaction regenerates iodine, which is the crucial component of the Karl-Fischer titrant. This then reacts with the water in the sample system according to the reaction

$$C_6H_5N \cdot I_2 + C_6H_5 \cdot SO_2 + C_6H_5N + CH_3OH + H_2O$$

$$\rightarrow 2C_6H_5N \cdot HI + C_6H_5NH \cdot SO_4CH_3 \qquad (9\text{-}14)$$

For this titration the dual polarized amperometric endpoint detection system provides good sensitivity and rapid response.

The instrumentation for coulometric titrations has been discussed in detail in Chapter 5. In general the electrochemical cell is replaced by a resistor when it is not operating. This resistance value is roughly comparable to that of the electrochemical cell and prevents large transients when the circuit is switched on and off. Figure 9-6 illustrates the simplest form of current source; the system includes a high voltage battery connected to a high resistance in series with the electrochemical cell such that small changes in the resistance of the electrochemical cell and its cell voltage have little effect on the electrolysis current passing through the cell. An assembly such as that in Figure 9-6 will give reliable results for currents up to approximately 10 mA. For higher currents and greater precision electronic current sources are preferable.

The coulometric titration method lends itself to microscale analysis. However, accurate results require a good cell design. Efficient stirring is essential, and the placement of both the generator electrodes and the detection electrode is extremely important. Figure 9-7 illustrates a cell assembly that is quite satisfactory when a potentiometric endpoint detection system is used. Likewise, the assembly illustrated in Figure 9-8 is convenient for a dual polarized amperometric endpoint detection system. In either cell a convenient means of introducing small volume samples is desirable as well as provision for eliminating interfering gases from the solution (such as oxygen).

To have an electrolysis current it is axiomatic that there must be an anode as well as a cathode reaction. If the anode reaction is the one used to generate the titrant then it is imperative that the products of the cathode not interfere with the titration. This can be accomplished in a number of ways and the cells in Figures 9-7 and 9-8 indicate one approach. That is, the electrode reaction which is not involved in generating titrant is isolated from the sample solution by a fritted disc closing the end of a glass tube. Another approach is the use of ion selective membranes or ion exchange resins in the immediate proximity of the electrode.

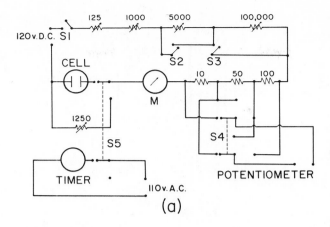

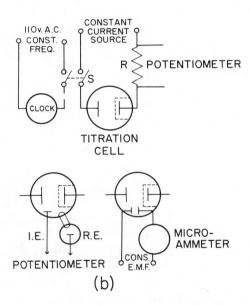

Figure 9-6. (*a*) Circuit for continuously variable constant-current generator. *M*, millimeter (0–150 mA); resistors, 10,50, 100 (±0.1%), 1250 variable (dummy resistor to substitute for cell resistance); switches, S_1 (on-off spst), S_2 (bypass spst for currents above 100 mA), S_3 (bypass spst for currents above 15 mA), S_4 (3-position double-pole for potential measurements across precision resistors), S_5 (on-off dpdt for timer); Timer, electric clock, Type S-10 (The Standard Electric Time Co.); 120 VD-C, Model SR-100 Nobatron D-C power supply. (*b*) Instrumentation for coulometric titrations. Upper part represents generation circuit, lower left represents a potentiometric end-point detection syestem, and lower right represents dual polarized amperometric end-point detection system.

423

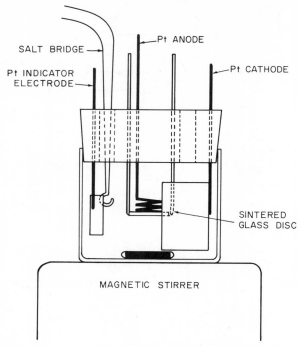

Figure 9-7. Cell system for coulometric titration by a platinum generator electrode and an isolated auxiliary electrode; system includes provision for potentiometric end-point detection.

In addition to the use of ion exchange membranes to isolate the counter electrode from the working electrode, perm-selective solid-state membranes can be used to generate coulometric titrants. For example, hydrogen ion can be generated in water or nonaqueous solvents by oxidation of hydrogen which has diffused through a palladium membrane anode. Silver ion can be generated through a silver/silver-sulfide membrane anode for the coulometric titration of chloride ion in strong oxidizing media (where a bare silver anode could not be used). In a similar fashion, fluoride ion can be generated at a europium doped lanthanum fluoride membrane.

Because the generator electrodes must have a significant voltage applied across them to produce a constant current, the placement of the indicator electrodes (especially if a potentiometric detection system is to be used) is critical to avoid induced responses from the generator electrodes. Their placement should be adjusted such that both the indicator electrode and the reference electrode occupy positions on an equal potential contour. When dual polarized amperometric electrodes are used, similar care is desirable in

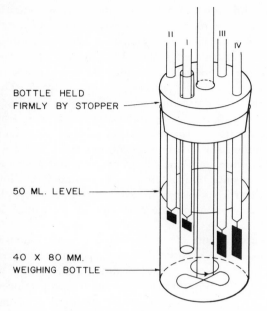

BOTTLE HELD
FIRMLY BY STOPPER

50 ML. LEVEL

40 X 80 MM.
WEIGHING BOTTLE

I, GENERATOR CA-
THODE 0.7 X 0.7 CM.

II, GENERATOR AN-
ODE 0.7 X 0.7 CM.

III, INDICATOR AN-
ODE 1.4 X 1.8 CM.

IV, INDICATOR CA-
THODE 2.5 X 1.8
CM.

Figure 9-8. Coulometric titration cell with generator (II) and isolated auxiliary (I) electrodes on the left side and a pair of identical platinum electrodes (III, IV) on the right for dual polarized-electrode amperometric end-point detection.

their placement to avoid interference from the electrolysis electrodes. These two considerations have prompted the use of visual or spectrophotometric endpoint detection in some applications of coulometric titrations.

Coulometric titrations have found their widest application where microanalysis titrations of high precision and accuracy are desired. Because small incremental additions of titrant are more convenient and precise through use of an electrolysis current than is possible through the use of a buret, coulometric titrations of small samples have the potential for enhanced accuracy. Furthermore, because the entire system (both titrant addition and detection) is electrochemical, coulometric titrations are especially amenable to automation. The technique is particularly useful for repetitive sample analysis, and has been extended to a number of industrial and commercial analytical devices. In particular, for the analysis of olefins and of sulfur compounds in petroleum streams various forms of coulometric titrators are used.

Reference to Table 9-1 indicates that olefins can be determined by the electrochemical generation *in situ* of halogens. Bromine is effective for both olefins and sulfur compounds and is the basis for an automatic coulometric titrator for continuous analysis of petroleum streams (18). For some years this instrumentation was marketed by the Consolidated Electrodynamics Corporation in the form of an instrument called the Titrilog. The basic principle of this instrument is a potentiometric sensing system which monitors bromine concentration in a continuously introduced sample stream. The bromine in the solution reacts with the sample components and causes a decrease in the concentration of bromine. When this decrease is sensed by the potentiometric detection electrodes, the electrolysis current producing bromine adjusts itself to maintain the bromine concentration. Because sample is introduced at a constant rate the electrolysis current becomes directly proportional to the concentration of the sample component. Thus the instrument records the electrolysis current as concentration of sample component and provides a continuous monitor for olefins or sulfur in petroleum streams.

Another example of the application of the principles of coulometric titrations to a continuous on-stream analyzer is the moisture analyzer developed by Keidel (19). It illustrates one of the outstanding advantages of the coulometric generation of a titrant; namely, an intermediate is produced as a titrant which would not conveniently be available in standard solutions. The principle of the moisture analyzer is to place between two closely spaced platinum electrodes (helically wound in a glass tube) a solution of phosphoric acid. When current is passed between the two electrodes the water in the phosphoric acid is electrolyzed

$$H_3PO_4 \xrightarrow{\text{electrolysis}} H_2 + \tfrac{1}{2}O_2 + HPO_3 \qquad (9\text{-}15)$$

to yield hydrogen, oxygen, and HPO_3. The latter species, even at high applied potential, is a sufficiently poor conductor of electricity that virtually no current passes between the electrodes. If a moisture-containing gas stream is passed through this cell at a constant flow rate any water in the stream will react with HPO_3

$$HPO_3 + H_2O \rightarrow H_3PO_4 \qquad (9\text{-}16)$$

to produce phosphoric acid which is a good conductor of electricity and will allow the reaction represented by Equation 9-15 to occur. Hence the electrolysis current that is observed is directly proportional to the moisture content of the sample stream. For a flow rate of 100 ml/min, at atmospheric pressure, a current of 13.2 μA results for each part per million of water in the gas phase. The general features of this form of automated coulometric titration have been discussed (20) as well as the extension of this general approach to the determination of moisture in organic liquids (21).

Another example of the use of the coulometric titration method as a continuous monitor is its application as a gas chromatographic detector for the analysis of pesticides and sulfur-halogen compounds. The approach is to burn the effluent from a gas chromatographic column in a hydrogen flame which will convert any halogen compound to hydrogen halides and any sulfur compound to H_2S. These products are carried in the gas flow path into a coulometric titration cell with a silver anode as the generator electrode and a silver indicator electrode for a potentiometric detection system. The indicating circuit is set such that the silver ion concentration in the solution is controlled to a preset level. Any halide ion or H_2S introduced into the cell solution will react with the silver ion and cause its concentration to decrease. When this is detected the electrolysis current is activated electronically and produces silver ion to maintain its preset level. The resulting electrolysis current is monitored on a strip-chart recorder as a function of time and gives elution peaks that appear identical to the elution peaks observed with a conventional gas chromatographic detection system. This system has proven extremely useful in determining minute amounts of halogen- and sulfur-containing compounds in the presence of large excesses of other extraneous materials.

Both the Titrilog instrument and the continous monitoring of pesticides and sulfur-halogen compounds indicate that the coulometric titration method is amenable to the automatic maintenance of the concentration of a component in a solution system. A manual version of this approach has been used for studying the kinetics of hydrogenation of olefins as well as for determining the rate of hydrolysis of esters (13). The latter system is in point of fact a pH-stat based on the principles of coulometric titrations. Equations 9-9 through 9-11 indicate how this approach is applied to evaluation of the rate constants for ester hydrolysis. A similar approach could be used for developing procedures for kinetic studies involving most of the electrochemical intermediates summarized in Table 9-1. The coulometric titration method provides a convenient means for extending the range of systems that can be subjected to kinetic study in solution.

References

1. I. M. Kolthoff and N. H. Furman, *Potentiometric Titrations*, 2nd ed., John Wiley and Sons, Inc., New York, 1931.

2. H. H. Willard and F. Fenwick, *J. Am. Chem. Soc.*, **44**, 2516 (1922).

3. C. N. Reilley, W. D. Cooke, and N. H. Furman, *Anal. Chem.*, **23**, 1223, 1226 (1951).

4. C. N. Reilley and R. W. Murray, *Treatise on Analytical Chemistry*, Vol. 4, I. M. Kolthoff

and P. J. Elving, eds., Interscience Publishers, Inc., New York, 1963, Part I, pp. 2181–2187.

5. J. J. Lingane, *Electroanalytical Chemistry*, 2nd ed., Interscience Publishers, Inc., New York, 1958, pp. 267–295.

6. P. Delahay, *New Instrumental Methods in Electrochemistry*, Interscience Publishers, Inc., New York, 1954, Chap. 10.

7. J. T. Stock, *Amperometric Titrations*, Interscience Publishers, Inc., New York, 1965.

8. J. W. Loveland, *Treatise on Analytical Chemistry*, Vol. 4, I. M. Kolthoff and P. J. Elving, eds., Interscience Publishers, Inc., New York, 1963, Part I, pp. 2604–2618, 2624–2625.

9. C. N. Reilley, "High-Frequency Methods," in *New Instrumental Methods in Electrochemistry*, P. Delahay, ed., Interscience Publishers, Inc., New York, 1954, Chap. 15.

10. J. J. Lingane, *Electroanalytical Chemistry*, 2nd ed., Interscience Publishers, Inc., New York, 1958, Chap. VIII.

11. N. H. Furman, *Treatise on Analytical Chemistry*, Vol. 4, I. M. Kolthoff and P. J. Elving, eds., Interscience Publishers, Inc., New York, 1963, Part I, pp. 2283–2286.

12. J. B. Neiland and M. D. Cannon, *Anal. Chem.*, **27**, 29 (1955).

13. P. S. Farrington and D. T. Sawyer, *J. Am. Chem. Soc.*, **78**, 5536 (1956).

14. L. Szebelledy and Z. Somoggi, *Z. Anal. Chem.*, **112**, 313, 323, 332, 385, 391, 395, 400 (1938).

15. J. J. Lingane, *Electroanalytical Chemistry*, 2nd ed., Interscience Publishers, Inc., New York, 1958, pp. 484–616.

16. D. D. DeFord and J. W. Miller, "Coulometric Analysis," in *Treatise on Analytical Chemistry*, Vol. 4, I. M. Kolthoff and P. J. Elving, eds., Interscience Publishers, Inc., New York, 1963, Part I, Chap. 49.

17. A. J. Bard, *Anal. Chem.*, **40**, 64R (1968); *ibid.*, **38**, 88R (1966); *ibid.*, **36**, 70R (1964); *ibid.*, **34**, 57R (1962).

18. P. A. Shaffer, Jr., A. Briglio, Jr., and J. A. Brockman, Jr., *Anal. Chem.*, **20**, 1008 (1948).

19. F. A. Keidel, *Anal. Chem.*, **31**, 2043 (1959).

20. M. Czuha, Jr., K. W. Gardiner, and D. T. Sawyer, *J. Electroanal. Chem.*, **4**, 51 (1962).

21. L. G. Cole, M. Czuha, R. W. Mosley, and D. T. Sawyer, *Anal. Chem.*, **31**, 2048 (1959).

INDEX